Atlas of the Galilean Satellites

Complete color global maps and high-resolution mosaics of Jupiter's four large moons – Io, Europa, Ganymede and Callisto – are compiled for the first time in this important atlas.

The satellites are revealed as four visually striking and geologically diverse planetary bodies: Io's volcanic lavas and plumes and towering mountains; Europa's fissured ice surface; the craters, fractures and polar caps of Ganymede; and the giant impact basins, desiccated plains and icy pinnacles of Callisto.

Featuring images taken from the pathfinding Voyager and the recent Galileo orbiter missions, this atlas is a comprehensive mapping reference guide for researchers. It contains 65 global and regional maps, nearly 250 high-resolution mosaics, and images taken at resolutions as high as 6 meters.

Paul Schenk is a Staff Scientist at the Lunar and Planetary Institute in Houston, Texas, where he specializes in impact craters and other features on icy satellites, and in 3-D imaging. He is currently analyzing released *Cassini* data of the icy satellites of Saturn, and assisting the *New Horizons* team plan their encounter with Pluto in 2015.

Atlas of the

Galilean Satellites

Paul Schenk
Lunar and Planetary Institute, Houston

CAMBRIDGE UNIVERSITY PRESS
Cambridge, New York, Melbourne, Madrid, Cape Town, Singapore,
São Paulo, Delhi, Dubai, Tokyo

Cambridge University Press
The Edinburgh Building, Cambridge CB2 8RU, UK

Published in the United States of America by Cambridge University Press, New York

www.cambridge.org
Information on this title: www.cambridge.org/9780521868358

© P. Schenk 2010

This publication is in copyright. Subject to statutory exception
and to the provisions of relevant collective licensing agreements,
no reproduction of any part may take place without
the written permission of Cambridge University Press.

First published 2010

Printed in the United Kingdom at the University Press, Cambridge

A catalog record for this publication is available from the British Library

Library of Congress Cataloging-in-Publication Data
Schenk, Paul M.
Atlas of the Galilean satellites / Paul Schenk.
  p. cm.
ISBN 978-0-521-86835-8 (hardback)
1. Galilean satellites–Atlases. I. Title.
QB404.S43 2010
523.9'850223–dc22   2009042291

ISBN 978-0-521-86835-8 Hardback

Cambridge University Press has no responsibility for the persistence or
accuracy of URLs for external or third-party internet websites referred to
in this publication, and does not guarantee that any content on such
websites is, or will remain, accurate or appropriate.

```
QB
404
.S43
2010
```

I dedicate this *Atlas* to Alice and Bernard, Carl Seyfert, William McKinnon and David Bonett, Pup, and lastly Robby the Robot (*Forbidden Planet*, 1956), for that immortal refrain, "Sorry, miss, I was giving myself an oil job."

Contents

Preface		*page* ix
Acknowledgments		xii
1	**Introduction**	1
	1.1 The revolutionary importance of the Galilean satellites	1
	1.2 Post-discovery	2
	1.3 *Voyager* and *Galileo*: Global mapping begins	3
2	**Format of the *Atlas***	7
	2.1 Nomenclature	8
3	**Making the maps**	9
	3.1 Image calibration and quality	9
	3.2 Cartographic control and geometric registration	12
	3.3 Putting it all together	13
	3.4 True colors	15
4	**Geology of the Galilean satellites: An introduction to the images**	18
	4.1 The importance of being ice	18
	4.2 Volcanism (and against cryo-ism!)	18
	4.3 Tectonism (and tides)	20
	4.4 Viscous relaxation	24
	4.5 Other global effects	24
	4.6 Polar processes and ice segregation	26
	4.7 Impact cratering: Planetary chronometer and window on the interior	26

5	**The satellites**	29
	5.1 Callisto	29
	5.2 Ganymede	32
	5.3 Europa	36
	5.4 Io	41
6	**One big happy ...**	47
	6.1 Why explore Jupiter?	47
	6.2 The future	50
Plates	**Atlas of the Galilean Satellites**	55
	Callisto	57
	Ganymede	107
	Europa	179
	Io	263
	Appendix 1: Glossary	353
	Appendix 2: Supplemental readings	359
	Appendix 3: Index maps of high-resolution images	364
	Appendix 4: Data tables	368
	Appendix 5: Nomenclature gazetteer	371
	Index	393

Preface

This *Atlas* is not what it should be. If fate had been kinder, each of the four planetary bodies represented here would have had its own *Atlas*, each larger than this volume. Don't blame the author, though; the culprit is an elegant yet critical device called the HGA, explained in Chapter 1.3. Should you pass over this book on your way to the used "pilates-at-home" bookshelf or toss it in the recycle paper bin? I hope not. Despite its shortcomings, this *Atlas* is the most complete representation we will have of the surfaces of Jupiter's large Galilean satellites for the next decade, objects that should be called planets, regardless of anyone's peculiar definition of that term.

Complex in detail and beautiful in a universe of wonders, the Galilean satellites fill the eye and mind in equal measure. They are also of considerable historical importance. My place in their history begins in 1972, the year I entered high school. A notice in the *Buffalo Evening News* announced the hiring of a manager to lead a new *Mariner* mission to the outer planets and their moons. At the time, these worlds were little more than dusky points of light. The *Voyager* mission, as it came to be called, was in reality a poor-cousin replacement for the Grand Tour, an ambitious plan to tour the entire Outer Solar System with a fleet of spacecraft.

Younger than NASA by only 31 days, I followed the USA into space along with Walter Cronkite and Jules Bergman on live TV, collecting newspaper and magazine clippings (the Internet was two decades away, information flowed a little more slowly). As awesome as the Apollo landings were to watch (I was but 10 years old), and the first Mars pictures of huge volcanoes and canyons that followed, it was the cold distant giant planets and especially their unfamiliar moons that were the great frontier of my imagination. The two *Voyager* spacecraft, launching in 1977, were the first true exploration of this frontier.

In 1979, I joined the *Voyager* mission as one of three NASA summer interns (Figure *i*). I arrived at the Jet Propulsion Lab in Pasadena two weeks before the *Voyager* 2 encounter with Jupiter and entered the beehive

Figure *i* The author, beardless, standing behind Dr. Ed Stone, *Voyager* Project Scientist, looks on dispassionately during a daily situation briefings during the heady days of Jupiter encounter, July 1979. Why a lowly summer intern was allowed into such important meetings I'll never know! I can no longer recall the subject that captured Dr. Stone's attention that day. That's a fuzzy Dr. Lonnie Lane in the far right foreground.

known as Science Investigation Support Team on the third floor of Bldg. 264. There I met Ellis Miner, Jude Montalbano, Linda (Horn) Spilker, and a bunch of crazy wonderful people supremely dedicated to the success of the project. Each day Jupiter appeared a little bit bigger in our TV monitors until the crescendo on July 9th. A highlight would be the first high-resolution views of Europa, which appeared on our monitors at about noon as I recall. It was a unique experience never repeated. JPL employees and scientists alike witnessed exploration live on TV as *Voyager* images were displayed in real time. For me there was no looking back from that rapturous summer.

This *Atlas* represents three decades of personal effort invested in these planetary bodies since 1979. It came into being because of the work I have been doing mapping the topography of these worlds. In the course of that work I accumulated knowledge of the geography of these worlds and a library of images representing their surfaces that are unavailable anywhere else. It was time to assemble that knowledge in one place and "tell the world." The digital images used in the *Atlas* were produced using software mostly developed at the US Geologic Survey in Flagstaff, AZ, and maintained by the staff of the Lunar and Planetary Institute, to all of whom I am indebted. However, image selection, geometric control and registration, mosaic and map formatting, and all other aspects of map production are the sole responsibility of the author.

The purpose of the *Atlas* is to present the collective imaging data set for these satellites as currently possessed by the human race in the year 2009 in a compact complete format. (*New Horizons* data from the Jupiter system are being processed as of this writing but nowhere exceed *Voyager* or *Galileo* coverage in resolution.) Brief descriptions are included to explain the nature of the images, introduce key topics, and provide context for the maps and images and some of the important features shown. But the basic goal here is to show the pictures, not to present an extended discourse on planetary geology or geophysics.

I have experienced my fair share of scientific insights, those unique exhilarating moments when seemingly disparate ideas or data merge into a unifying concept previously unknown. Many of those are described here, including plate tectonics and polar wander on Europa, mountain formation on Io, crater chains on Callisto formed by disrupted comets, among others. As a result, the text tends to be biased toward my own perspective, for which I make no apologies. Although I endeavor to reflect our current best understanding of the evolution of these bodies, the text simply cannot be regarded as complete, fair, or perfect, for the pen had to be put down at some point. (Please report errors to galsat400@gmail.com) Indeed it may not matter much as some, or perhaps most, of the details or even the basic outline of their planetary histories are likely different in reality than described here. Paraphrasing Dr. Morbius, "My evil self is at the keyboard, and I have no power to stop it!"

The second goal of the *Atlas* is to provide a complete and accurate reference resource of the *Galileo* and *Voyager* image library, with all high-resolution image mosaics properly located on the surface for the first time. This *Atlas* is the first compilation to show all the highest resolution image data (all those better than 750 meters per pixel) complete and in their regional context. It is hoped that these words and pictures will be only a starting point for the reader on their own voyage of discovery!

All image products in this *Atlas*, unless noted, are the work of the author and should be credited to Paul Schenk, Lunar and Planetary Institute.

Paul Michael Schenk
April 2009

Acknowledgments

I thank many for their wisdoms, including Rosaly Lopes, Ashley Davies, and David Williams for discussions and conversations on NIMS data and Io geology, Simon Kattenhorn, Wes Paterson and Louise Prockter for the same on Europa, and Brad Dalton and Carl Hibbitts on satellite composition. Elizabeth "Zibi" Turtle graciously provided descrambled I24 images of Io. Robert Morris assisted with early processing of *Voyager* control points, and Alfred McEwen and Brian Fessler provided insights into and support for ISIS image processing. Finally I thank Jeffrey Moore for comradeship, and Carl Seyfert and William McKinnon for leadership, guidance and companionship on John Mack, Jim Orgren the journey.

1
Introduction

1.1 The revolutionary importance of the Galilean satellites

Watershed moments, upon which the fates of nations, continents, or peoples hinge, are rare in human history. The Battle of Salamis in 480 BC, the Battle of Zama in 202 BC (my sympathies are with Carthage), the defeat of the Moors by Charles Martel in AD 732, the Sack of Constantinople in 1204, the death of Ogedei Khan as his armies approached Wien (Vienna) in 1241, the coming of the Black Plague in the fourteenth century, and the Czar's and Kaiser's decision to mobilize in August 1914 come to mind.

In this year 2009, we approach the 400th anniversary of another of these watershed events: the discovery of Jupiter's four large Galilean satellites, Io, Europa, Ganymede and Callisto, in January 1610 by Italian scientist Galileo Galilei. Galileo, sometimes falsely credited with invention of the telescope, perfected the basic instrument and was the first to point one at the heavens in earnest. Importantly for Galileo, he was quick to understand the revolutionary import of what he saw. Every object he observed, starting with the Moon, followed by Venus and Jupiter, revealed fundamental truths hidden to the naked eye that profoundly altered our perception of how the Universe worked, and in turn our worldview of ourselves and our place as a species in the Universe. Other revolutions were to follow in astronomy, chief among them Edwin Hubble's discovery that spiral nebulae are in fact millions of island galaxies like our own in a vast Universe, but Galileo's revolution permanently broke our myopic anthropocentric view of our place in the Universe, although it would take a few centuries for this new view to finally permeate the collective mass consciousness (and in some minds it never has).

The four moons Galileo saw are binocular objects, and would be visible to the naked eye outside the glare of Jupiter itself. Although there are claims (for Chinese astronomer Gan De in 362 BC, for example) that those with exceptional sight can actually detect the brighter moons with

Figure 1.1 A page from Galileo's notes of his discovery of the Galilean satellites January 7, 1610. Jupiter is the large star and the four moons are the small shifting stars to either side. These are among the most valuable scientific documents in history. Credit: NASA.

the eye, their existence was inconceivable as December 1609 rolled to a close. Looking at Jupiter, Galileo saw three new "stars" in a line very close to the planet (Figure 1.1). After several days of observation, it was clear that there were in fact four new objects, and that they were all in orbit around Jupiter, not the Earth. Galileo's observations, and those of the Jovian moons in particular, thus gave a critical boost to the emerging Copernican Sun-centered worldview.

For more than a thousand years, it had been generally assumed that everything revolved around the Earth, which a casual observation of the heavens would imply. Copernicus helped relaunch the ancient Greek theory (by Aristarchus) of the Sun-centered (or heliocentric) Solar System in the early 1500s, but by the early 1600s the theory had received a decidedly ambivalent response. There were a few believers to be sure, but most had never heard of it, or remained unimpressed or uninterested. What Galileo saw on the Moon, Venus and Jupiter demonstrated that the celestial bodies are not immutable and that they do not all revolve around the Earth. (We now know that nothing is the center of anything, but the major point had been achieved.) Although it would be decades before the debate was won, mainly against yet another of the many reactionary responses from the conservative wing of the church to independent human thought, the first great astronomical revolution was now inevitable (and required only Kepler's "invention" of the elliptical orbit to be complete).

1.2 Post-discovery

A few years after Galileo's announcement in the *Sidereus Nuncius* (*Stellar Message*), German astronomer Simon Marius claimed to have discovered the four moons at about the same time. Today, Galileo is given credit, but it is Marius who is credited with the names by which we know these four moons, all named after Jove's indiscretionary loves in Greek mythology. These names did not enter common use till the twentieth century. Today, Marius and Galileo are both honored with names of large provinces on Ganymede.

The Galilean satellites then lay dormant in human thought for several centuries. True, they were useful for terrestrial longitude determination and for estimating the speed of light (based on eclipse timings). In the seventeenth century, Laplace explained the curious mathematical timing, or resonance, between the three inner moons in which their orbital periods are related by simple integers (this Laplace Resonance is named for him). The profound consequences of this orbital dance were not understood for another 200 years, however.

With the advent of "modern" telescopic instruments and techniques, the Jovian moons began to emerge as real planetary bodies. Still, by the dawn of the Space Age in the late 1950s, little was known about these worlds (Figure 1.2). *Pioneers* 10 and 11 were the first visitors to Jupiter a few years before (I listened to the hourly radio news summaries for word of *Pioneer* 10's successful launch). Although the imaging systems were "primitive," they did show a few fuzzy global features that can now be identified on our maps (Figure 1.3). Earth-bound observers saw dark "polar caps" on Io, bright caps on Ganymede and a dark equatorial band (or patches) on Europa. These features proved real, but most other apparent markings did not.

Spectroscopic observations found water ice on all the moons except Io, which also looked oddly yellowish. Instead, sodium clouds were found in Io's orbit. These scant facts lead to perhaps the best-known pre-*Voyager* speculation, which suggested that Io was covered by the salty deposits of a dried-up ocean. There was also the curious coordinated timing between Jupiter's radio emissions and Io's rotation period. By the mid 1970s, it was apparent something odd was going on in the Jupiter system.

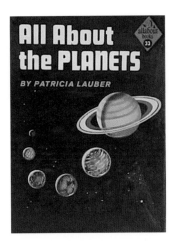

Figure 1.2 The Solar System was a relatively simple, sedate place as the Space Age dawned. Only one paragraph is granted the four Galilean satellites in this volume that once occupied my bookshelf, space enough to assert that Ganymede has canal-like lines, an obliquely prophetic statement as it turns out. (*All About the Planets*, P. Lauber, Random House, 1960.)

1.3 *Voyager* and *Galileo*: Global mapping begins

The Galilean satellites have launched another revolution in our own time, the importance of which is not yet fully manifest. This revolution began in 1979. Prior to spring that year, it was commonly assumed that the satellites orbiting the four giant outer planets were essentially relics of planetary formation, perhaps even cold dead worlds. *Voyagers* 1 and 2 were the first to explore the Jovian system with what we would call modern scientific instruments, including high-definition television cameras. What they revealed fundamentally altered our perception of the Outer Solar System. All four moons proved to be unique planetary bodies, as these pages document. The monopoly of Mars on our imagination was broken.

Voyager acquired high-resolution images of all four satellites, but the politics of celestial dynamics, competing mission requirements, and a date with Saturn demanded that the *Voyagers* give Europa less attention than the other Galilean satellites. It required the focused and detailed

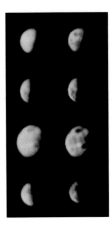

Figure 1.3 *Pioneer* 10 and 11 images of the Galilean satellites in 1973 and 1974 (from top to bottom: Io, Europa, Ganymede, and Callisto). These images have effective resolution of 200 to 400 kilometers. The right side versions have been heavily processed to enhance what little detail is present. Nowadays, the Hubble Space Telescope routinely acquires images with 10-fold improvement in resolution. Credit: NASA/Ames Research Center.

observations of another Galileo, in this case a robotic explorer launched from the human home world, to reveal the fundamental nature of this ice-covered moon, demonstrating that Europa most likely possesses an ocean of liquid water beneath its surface. This marks Europa as one of several objects in the Outer Solar System possessing liquid water, hydrocarbons (perhaps), and internal heat sources. Each is a required element of any potential habitat for life, as we understand such things. What really lies or grows (?) inside Europa is not yet known, but Europa leads an impressive group of active icy worlds, including Triton, Titan, Enceladus, and perhaps even Ganymede. Where this fundamental shift in thinking will take us in the next decades no one can say, at least until we return to Europa.

A total of eight spacecraft have visited the Jupiter system since 1972, the most recent in 2007. Of those carrying dedicated cameras, only three have passed within the confines of the Galilean satellites, and only one has lingered for more than a day (Appendix 4). These three spacecraft, the two *Voyagers* and *Galileo*, have changed our perceptions of these moons, yet no truly global mapping data sets exist for the Galilean satellites. The global mosaics presented here are cobbled together from hundreds of images taken by the *Voyager* and *Galileo* spacecraft during their flybys of these satellites, beginning in 1979 and resuming in 1996.

Voyagers' discoveries at Jupiter are spread throughout this *Atlas*, but the story began in 1966 as a simple concept to use Jupiter to accelerate a spacecraft towards the other outer planets. Although the concept of gravity assist was known, Ph.D. student Gary Flandro discovered the opportunity that became the germ of the *Voyager* project. *Voyager* started life as the Grand Tour, a fleet of four spacecraft to visit all five outer planets, including Pluto, during a grand alignment of planets that occurs only every 173 years or so. The budget was not awarded to fit this profile, so in 1972 four spacecraft became two, and five planets became two (plus two: Uranus and Neptune were optioned for *Voyager 2* only if *Voyager 1* succeeded at Jupiter and Saturn). Pluto was not physically within reach of either *Voyager* and only now is a spacecraft on its way to that remote orb.

The two *Voyagers* were targeted to observe opposite hemispheres of each satellite, but effective resolution seldom exceeded 1 km, and significant mapping gaps remained, especially on Io and Europa. Even before *Voyager* arrival, a follow-on mission, the Jupiter Orbiter Probe, was designed in the mid 1970s for a 1982 mission to capitalize on and complete the *Voyager* discoveries. Renamed *Galileo*, it would remain in Jupiter orbit for at least 2 years of extended studies of the planet and its moons.

It was *Galileo*'s mission during its repeatedly delayed grand orbital tour of Jupiter (later extended by four more years) to pass within a few hundred kilometers of Europa, Ganymede, and Callisto with a battery of remote sensing instruments. (Io was targeted for a close pass during the first orbit in 1995 but due to the extreme radiation environment, additional passes were awarded only after the primary mission had succeeded. A tape recorder anomaly caused the cancellation of these first high-resolution Io observations.) Among other investigations, *Galileo* was expected to essentially replace the partial *Voyager* maps with nearly global mapping at resolutions of a few hundred meters and acquire very high resolution images of high-priority targets at 10 to 100 meter resolutions. Information on interior structure and magnetic fields were acquired but compositional mapping was severely restricted. *Galileo* was never able to achieve more than a tiny fraction of its global mapping mission.

The principal devil in this is the High-Gain Antenna, or HGA (Figure 1.4). The HGA onboard *Galileo* was designed to furl like an umbrella inside the Space Shuttle and be opened in space to provide the primary data link to Earth at 140 000 bits per second. The additional delays incurred due to the *Challenger* accident weeks before the scheduled launch in 1986 had unforeseen consequences. After three more years on the ground and two years in space (furled to protect the gold-plated mesh from the Sun), the antenna refused to open properly for reasons that today remain obscure. The secondary antenna on *Galileo* could only transmit at roughly 10 to 20 bits per second, no better than during the first Mars mission back in 1965, when it took more than a month to transmit 22 small images back to Earth from Mars. After years of frustrated effort, the antenna remained unusable and *Galileo* would return only a tiny fraction of its intended data.

Once it was realized that the antenna would never work, JPL engineers did a superb job in teasing as much information from the probe as possible. Onboard and ground-based upgrades increased data transmission to ~150 bits per second by the time *Galileo* arrived in late 1995, an improvement but still crippling (compare this to your current cable or wireless capacity). The onboard tape recorder was also very limited, with a total capacity of only 115 megabytes, less than a CD-ROM. Using onboard data compression similar to JPEG, together with upgrades

Figure 1.4 The *Galileo* engineering test model, on display at the Jet Propulsion Laboratory, Pasadena, CA. The wire mesh antenna at top is shown in its real-life jammed configuration.

to receiving antennae, a valuable data set was obtained, including amazing high-resolution images of each satellite. Still, the imaging instruments had to share this downlink capacity with ten other instruments. The intense radiation environment at Jupiter also inflicted a toll on the spacecraft, requiring further engineering efforts to keep the machine operational. Towards the end of the mission, *Galileo* succeeded in obtaining its programmed objectives about as often as it failed. Despite the great success of all these efforts, the loss of potential data was staggering.

The science teams responsible for guiding *Galileo*'s tour of Jupiter faced a cruel choice: how best to use the sparse resources provided by the tiny backup antenna and recorder to achieve some of the original mission goals. Typical imaging results for any given Jupiter orbit during the original two-year prime mission (not including NIMS data) were only 150 to 180 images for Jupiter, its rings, and satellites, and quite a few of those were only partly returned. The allocation ratio for imaging increased slightly during the extended missions, which focused heavily on the new Europa and Io discoveries. With the exception of very limited success at Europa and Io, however, global mapping was sacrificed in favor of (reduced) high-resolution imaging (see Appendix 3). In fact, the global maps of Ganymede and Callisto are still heavily dominated by *Voyager* images, and the hemisphere of Io observed by *Voyager* 1 was never seen well by *Galileo* at all. As a result, the best resolution that can be sustained at global scales on any of these satellites is about 1 km, the resolution of all global and quadrangle maps in this *Atlas*.

2
Format of the *Atlas*

Naturally enough, the *Atlas* proper is divided into four major parts, one for each satellite. Global maps at 1-kilometer resolution introduce each satellite. These are in cylindrical projection, in which latitude increases at a constant rate from pole to pole, and are reduced in scale to fit the page. These are followed by a set of five orthographic global maps from various perspectives (including leading and trailing hemispheres), simulating the views a passing astronaut might have. Following planetary mapping convention, each satellite is then divided into 15 quasirectangular maps of roughly equal size called *quadrangles* (Figure 2.1). These show the full 1-kilometer resolution detail of the global maps. These maps are named and numbered according to convention. For example, quadrangle "Je9" refers to "J"upiter satellite "E"uropa quadrangle "9."

No two-dimensional map can fully represent a three-dimensional surface without distorting either feature sizes or shapes. The quadrangles come in three map projections. Polar maps are in polar stereographic projection. Equatorial maps are in mercator, while mid-latitude maps are in lambert conformal conic projection. All three projections are conformal in that they preserve shapes fairly well. None of these

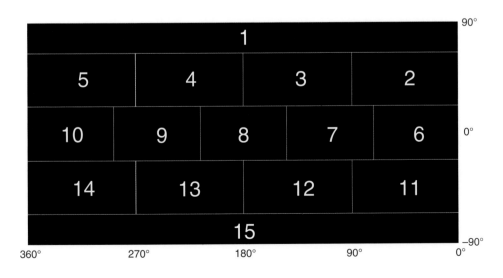

Figure 2.1 Index map showing global locations and dimensions of numbered quadrangle maps for each satellite. Each quadrangle as shown in the *Atlas* includes extra borders that overlap with neighboring maps by 2 degrees. This map is in simple cylindrical projection.

projections preserves areas, but each map size is sufficiently small that this distortion is minimized.

Following each quadrangle map, all *Galileo* and *Voyager* targeted high-resolution mosaics that occur within that quadrangle are presented. With a few exceptions, these are shown at their original resolution, which can vary between ~6 meters to roughly 500 meters (plus a few at up to 850 meters). The mosaics are in orthographic projection, but most are very small in area and map distortions are minimal. The mosaics are shown overlain on lower-resolution images to provide context. Collectively, these images cover less than 10% of the surfaces of these bodies. The locations of these targeted mosaics are shown on index maps for each satellite in Appendix 3.

The *Atlas* proper is followed by a discussion of the satellites as a planetary system and the relevance of future exploration. Appendices containing a glossary, a list of interesting related reading references, charts and data tables, and a gazetteer of feature names are in the final section.

2.1 Nomenclature

All names for features on these satellites are taken from the official International Astronomical Union sanctioned listing of names (see Appendix 5), which resides at the US Geological Survey website in Flagstaff, AZ. They are complete as of 2008. Aside from craters and active plume sites, each name includes a Latin-derived descriptor term (e.g., Maasaw *Patera* for caldera, Cadmus *Linea* for linear markings or bands, Tiamat *Sulcus* for parallel grooves and ridges, Memphis *Facula* and Castalia *Macula* for bright and dark spots, etc.) related to the type of geologic feature involved (see the USGS Planetary Nomenclature website for definitions). Proper names of Ionian features relate to the Io myth, but mostly to volcanic, fire or thunder gods, as appropriate. Those on Ganymede and Callisto are related to Mesopotamia and Egypt, and on Callisto with Nordic legend. Those on Europa are related to the myths of Europa and also to Celtic mythology and place names. On all maps, feature names are labeled to the right of the center of that feature, except when such placement would obscure other interesting features or run off the edge of the map. If the feature is ambiguous, a small dot is used to highlight its identification.

3
Making the maps

3.1 Image calibration and quality

Assembling the global maps shown here required the production of and combining of image mosaics from numerous observations by *Voyager* and *Galileo* obtained over periods of years. Before making a true global map, the images must be calibrated to remove background features inherent to the cameras. Unusual characteristics of both *Voyager* and *Galileo* cameras and the nature of each mission plan also affect the location and quality of these images, characteristics that must be taken into account during calibration and global map construction.

3.1.1 *Voyager*

The *Voyager* vidicon tube (old-style television) imagers were designed in the late 1960s and used an electron beam to read out the image data. Although robust and relatively stable, the *Voyager* cameras suffered from problems typical of vidicon cameras. On occasion, the upper left corner of the image was anomalously bright. This unpredictable phenomenon is not accounted for in standard calibration tools, and the corner must be either deleted or smoothed. Near Io, the radiation environment created a temporary background surge in the detector, occasionally saturating bright parts of the image. This can be corrected by knowing the timing and intensity of the exposure anomaly.

The most difficult problem relating to *Voyager* vidicon images is a general distortion due to bending of the electron-scanning beam. This distortion can be corrected using 220 or so 3- to 4-pixel-wide fiducial marks (reseaux) etched across the vidicon chip (and whose true positions are relatively well known). The correction is most severe and sometimes unstable near the edges, where the image sometimes remains slightly distorted. Fortunately, there is sufficient overlap in most *Voyager* mapping mosaics that we can delete the outer 40 or so pixels from the images, except in rare cases where no other imaging exists.

Also, the *Voyager* reseaux themselves represent permanent "dead" areas, amounting to a few percent of the total image. These black spots are usually filled by blurring data from neighboring pixels. For this *Atlas*, these areas were nulled out wherever possible. Valid data from adjacent or lower-resolution images were allowed to fill these holes.

Both *Voyager* encounters with Jupiter were highly successful, except for a timing offset on *Voyager* 1 between camera exposure and slewing of the camera to the next target. Triggered by high radiation before the first satellite encounter, this caused the smearing of some of the best images of Io, Ganymede and Callisto. Overlap of adjacent clean frames covers much of this loss, but for some areas we must rely on lower-resolution approach images.

3.1.2 *Galileo* SSI

Galileo's high-resolution imaging camera (solid-state imager or SSI) used the first CCD imaging system selected to fly in deep space. Built in the late 1970s, the *Galileo* CCD was then relatively new technology. Although much more stable against distortion and intensity flare than the *Voyager* vidicon cameras, the *Galileo* cameras were not without adversity. Late in the mission, images occasionally exhibited strange random square artifacts 6 or 8 pixels wide, which must be removed. Also, if bright surface features saturated or overexposed parts of the CCD, the photon charges on those pixels leaked downward into neighboring pixels, causing bright streaks across part of the scene, a phenomenon known as "bleeding" or "icicles" (an effect not seen in modern CCDs). This is usually most evident in images with highly contrasting bright and dark terrains, such as on Callisto and parts of Ganymede.

Jupiter's intense radiation resulted in a number of anomalies, the most important of which was sensitivity to radiation-induced noise in images of Io and occasionally Europa. This came in the form of random brightened pixels towards the bottom of some images, resembling shaken salt. Oddly, the highest-resolution images acquired near Io itself tended to be less affected than context images acquired a few thousand kilometers further away.

Project engineers were kept very busy keeping the spacecraft operating properly in the unforgiving high-radiation environment of Jupiter's magnetic field. Radiation occasionally triggered computer upsets on the spacecraft causing it to enter involuntary hibernation (or "safing"). This tended to occur within the high-intensity radiation zone and before or during planned encounters with Europa and Io, causing the complete loss of all planned mapping and high-resolution observations until the computer was reset. Two and a half planned encounters each were lost for Europa and Io, including all planned high-resolution imaging of the parts of Io first seen by *Voyager* 1. Two passes were saved by fast action by engineers after safing events. Software patches helped mitigate these effects, but toward the end of the mission the cumulative damage started to overcome the various spacecraft systems.

Radiation caused progressive damage to the camera itself. The first close pass of Io on orbit I24 seemed to run smoothly. During data playback it became obvious that something had gone wrong. Most of these images were acquired in a form of compression known as summation mode, whereby every 2-by-2 patch of pixels is averaged together to form one data point, reducing the amount of information by a factor of four for each image. Unfortunately, this mode malfunctioned, and the images were scrambled during readout, whereby the left and right halves were overlain. Engineers found a clever way to unscramble most of the images, but not completely. The original intensity values were lost, and only relative values regained; these images cannot be properly calibrated. Nonetheless, they provide invaluable information and are presented here. The case of orbit I31 was less fortunate. There, radiation caused the camera to fail to expose properly and all high-resolution images of Io were white. By the end of the mission, more than half of *Galileo*'s planned observations of Io were lost or corrupted in some way due to radiation effects.

The loss of *Galileo*'s antenna compromises global mapping in several ways. Some images were returned to Earth with missing lines or sections, forming significant gaps. The data compression algorithms used to increase the total number of images were equivalent to JPEG compression, which can be adjusted in severity. Areas that are relatively smooth morphologically are sometimes smoothed over even more. On a few occasions, overcompression significantly degraded the quality of *Galileo* images (for example, most images returned from *Galileo*'s second [G2] orbit), but this rarely compromises our mapping here. The chief exception is the *Galileo* data for Ganymede between longitudes 260° and

300°, which were unaccountably overcompressed (see Plates Jg5 and Jg9). These are the only mapping data we have for these areas, and should have been better treated.

3.1.3 *Galileo* NIMS

The low-resolution Near-Infrared (compositional) Mapping Spectrometer, NIMS, on *Galileo* mapped parts of the satellites in up to 408 wavelengths from 0.7 to 5.2 microns, far beyond our human vision range. NIMS provided important compositional maps, but at resolutions 50 times poorer than the SSI camera. While safing events cancelled NIMS and SSI imaging indiscriminately, NIMS did not experience the I24 and I31 anomalies of the imaging system. However, by the time *Galileo* arrived at Io for its first close pass since orbiting Jupiter, the NIMS grating had stuck and only a handful of wavelengths could be measured. NIMS resolution exceeds that of SSI or *Voyager* in only one location, near the south pole of Callisto (see Plate Jc15). NIMS imaging in the infrared shows the locations of thermal hotspots on Io that cannot be detected otherwise, and these data are shown to aid interpretation. While some NIMS sites for the icy satellites are presented to highlight compositional differences, space is not available to show this data set fully (index maps of high-resolution NIMS observations are shown in Appendix 3).

3.2 Cartographic control and geometric registration

In order for this *Atlas* to be accurate and useful for mapping, we need to know where features are on the surface of each moon. Since the images returned from space do not have surface coordinates labeled on them, we must register and link the images together in a global network. Each planetary image is defined by three vectors: the rotation rate and direction of the body, the position of the spacecraft and camera with respect to the center of the body, and the angle at which the camera is pointing. The first two are easily determined from the project records (assuming they are reasonably accurate).

The pointing vector of the camera is also available but only as an estimate based on the original commands sent to the spacecraft. These estimates are good enough to acquire the images, but are not accurate enough to make a proper map of the surface. These vectors must therefore be updated. This process involves selecting a global network of match points in overlapping images and then running a program to adjust the vectors until the differences in the estimated surface locations of these control points approach zero. This is a laborious iterative process involving adjustments to or rejection of control-point locations, and is complicated by the highly non-uniform nature of the imaging library. Once a global control network is established, the scattered high-resolution images are then integrated into this network to establish their locations.

The control network and resulting maps shown here were produced by the author. The US Geological Survey in Flagstaff, AZ, has also produced an independent control network for these satellites. It is not yet possible to say whether either of these two solutions is any better than the other: both efforts are handicapped by the highly variable resolutions and perspectives of the images and by the geometric distortions of the *Voyager* images which make up a large part of that mapping coverage. Neither solution is accurate to tens of meters, as is now the case for Mars, but are probably good to within a kilometer or two (except for a few very high resolution mosaics lacking context imaging). The key advantage of this *Atlas* is that all the images acquired by *Galileo* and *Voyager* at better than 1-kilometer resolution (Appendix 3) have, with few exceptions, been accurately located on the surfaces of these moons for the first time.

3.3 Putting it all together

Unlike global mapping of Venus completed in the 1990s and of Mars currently underway via the mapping cameras of a fleet of spacecraft, the *Voyager* and *Galileo* images we must use were acquired under widely differing illumination and viewing conditions during rapid but differently configured flybys. For most of Ganymede and Callisto and roughly half of Io, we must rely on *Voyager* imaging. Although high quality, these images sometimes suffer from the problems described above and seldom exceed 1-kilometer resolution.

For many areas on these bodies, we only have one good view from either *Voyager* or *Galileo*. The appearance of many features changes as the Sun rises in the sky: *low-Sun* viewing (near sunrise or sunset) results in shadowing and excellent relief discrimination, whereas noontime or *high-Sun* viewing washes out topography and reveals the colors and contrasts inherent in the composition of the surface (see for example Plates Jg4.2, Jg8.3, Ji9.2.5, Ji10.1.2 and others). Similarly, a telescopic view of the Moon looks very different at full and crescent phase, as do mountains on Earth from a passing aircraft. Ideally, we desire both views of planetary surfaces, providing both topographic and compositional information, though this was rarely achieved. Thus there can occasionally be abrupt transitions from high-Sun to shadowy low-Sun views where images end.

Many of the geologic units that make up these surfaces have different phase functions and colors (especially on Io). This phase function describes how the brightness of any terrain changes as our viewing geometry changes. By adjusting the image brightness to correct for this phase function, we can restore inherent brightness patterns and get a much better match when producing large image mosaics from different observations.

Not all terrains have the same phase function, however. For example, the polar frost caps on Ganymede are brighter in some filters than others. This can sometimes cause mismatches in brightness across image seams in mosaics. All these effects, including changes in resolution, can be jarring where different image sets overlap but digital photo-processing techniques have been used to minimize these transitions. In general, every effort was made to preserve true relative brightness across each satellite, except where necessary to provide a more visually interpretable map.

3.3.1 *Galileo* at Io

The challenges in mapping the Galilean satellites described here are most dramatically manifest for Io, and several choices had to be made. The images used to compile the Io map were obtained over a period spanning more than two decades, during which ongoing volcanism repeatedly changed parts of the surface. Mismatches can result if adjacent images were obtained at different times. The highest-resolution images or those with the best feature definition were used in these maps. In some cases, topographic definition was sacrificed for spatial resolution. Because of this compromise, a second global mosaic at 2-kilometer resolution was produced for Io. This map focuses on low-Sun terminator images, and is presented alongside each Io quadrangle map to highlight topographic features.

Compounding our difficulties is Io's variable and complex surface, including vibrant and variable colors. Io has much greater color variability in visible and near-infrared wavelengths than the icy Galilean moons, and some features that are dark in violet wavelengths are bright at infrared wavelengths (and vice versa). Not all images were acquired with the same color filters, and *Voyager* used different filters and had no infrared sensitivity. To produce a spectrally uniform global color map to merge with the high-resolution map, only *Galileo* color was used (these color images are usually variable and lower in resolution than the primary mapping images). Thus the colors used may not match in time the patterns in the high-resolution mapping mosaic, particularly in the sub-Jovian hemisphere centered on 0° longitude imaged by *Voyager*. Although a compromise, any mismatches appear to be relatively subtle.

Wait, there's more! The variability of phase function among the different terrains is particularly pronounced on Io. This phase function cannot yet be determined for all terrains on Io (or the other satellites), so a single phase function must be applied across the entire surface. Hence, some terrains can be over or under corrected for phase effects, leading to occasional mismatches in brightness across image boundaries.

The global map of Io presented here emphasizes resolution rather than attempting to represent one particular day or year in Io's history. Some features (e.g., Pele) are shown as they were in 1979, whereas others as they were during the years 1996–2001 (e.g., Prometheus). Some features (e.g., Pillan and Tvashtar) changed dramatically during the *Galileo* mission and are shown at one particular time in their history (that is the date when the best-resolution images were acquired). In several cases, time sequences are shown illustrating these changes.

3.4 True colors

The colors of the Galilean satellites portrayed in this *Atlas* are exaggerated. Neither *Voyager* nor *Galileo* acquired images in red, green and blue channels identical to our human vision (0.4 to 0.8 micron) and *Galileo* seldom acquired images in the red at all. The sensitivity of *Galileo*'s SSI camera (0.4 to 1 micron) extended into the infrared, outside the range of human vision (NIMS extended even further into the infrared). Surface geologic contrasts at these wavelengths are stronger than in the visible color range and strongest on Io; hence the images show "reddish" and

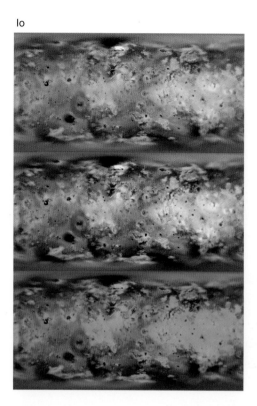 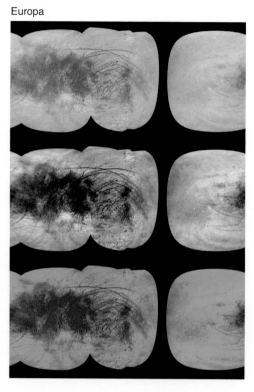

Figure 3.1 Global color maps of the Galilean satellites. Maps are pole-to-pole simple cylindrical map projections similar to Figure 2.1, extending from 0° to 360°, right to left. The right half is the leading hemisphere. Three versions are shown.

The top views are crude approximations of natural color, similar to what we might see if we were to pass by. The colors of Io and Europa are yellowish. Those of Ganymede and Callisto are brownish, similar to a milk chocolate, due in part to asteroidal and cometary dust embedded in the surface.

The middle views are what an alien species sensitive to violet and infrared light, beyond the range of human vision, might see. The images are also contrast-enhanced to maximize color differences. Additional details are visible in these views, due to contrasts in the ice phases in the infrared. This version is used in the Io and Europa global and quadrangle maps; color variations at Ganymede and Callisto are very subtle and hence not used in quadrangle mapping.

The bottom views are special color-ratio enhancements that map geologic units as different colors. Fresh icy materials such as younger impact craters or polar frosts are bluish or cyan, while non-ice units are reddish. The false color is created from infrared/violet ratio images and their inverse (violet/infrared), which are then combined so the infrared/violet, green, and violet/infrared are assigned to red, green, and blue in a composite product.

Io. Unlike the other three ice-rich satellites, the only icy phase observed on Io is frozen sulfur dioxide, which has a bright white color in the standard views and a cyan cast in the color-ratio product. Volcanic deposits, including very dark basaltic flows, dominate the surface. Magnesium-rich silicates may dominate dark diffuse deposits. The yellow and orange colors that coat nearly everything are plume and other deposits of sulfur dioxide and sulfur polymers. Sulfur-rich polar deposits and plume deposits from Pele are orange to red. The reddish polar deposits are spectrally different, and may be radiation-damaged sulfur polymers. The map includes images acquired prior to the Tvashtar eruption (see Plate Ji3.2).

Europa. Europa's dark equatorial band (or bright polar "caps") is evident in each color version. Europa's surface is yellowish to the naked eye due in part to sulfur ions from Io implanted by the magnetosphere and to dark non-ice material brought up from below. The dark red splotchy terrains in equatorial regions tend to be associated with occurrences of chaos material originating from within. The strength of these hues changes significantly at different longitudes, however, a pattern consistent with bombardment of charged particles by the Jovian magnetosphere.

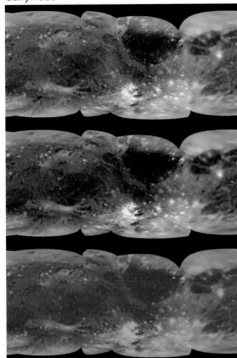
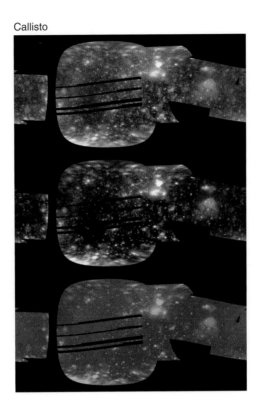

Caption for Figure 3.1 (*cont.*)

Ganymede. Especially noticeable are Ganymede's less-red polar frost caps (bluish in the enhanced versions). Bright and dark terrains on Ganymede also show a significant color contrast (most evident in the leading hemisphere of the color-ratio mosaic), due to the higher ice content of bright terrains. Fresh bright craters always show as light bluish or cyan due to higher water-ice content. The prominent dark-ray crater Kittu stands out in the color-ratio image (dark spot at center left; see Plate Jg10.1).

Callisto. Most of Callisto is similar in color to dark terrain on Ganymede. Even heavily cratered Callisto has global variations in color, however, particularly the areas near the south pole and right of center (south of Asgard), which show as darker and less red in the color views and bluish in the color-ratio version. The origin of these variations is unknown but they do not appear to follow the pattern expected for charged-particle bombardment. The central bright spots of Valhalla and Asgard also stand out in both versions. It may reflect an ancient compositional variation. Young bright craters appear bluish or cyan on Callisto, as they do on Ganymede.

"greenish" colors on Io where our human eyes would see dominantly yellows. Near-global color mapping coverage in these wavelengths was achieved for Io, although there are gaps in the polar region and significant differences in resolution (from 2 to ~35 kilometers) across the surface (Figure 3.1). *Galileo* color mapping of the three icy satellites was severely crippled, ranging from only ~50 to 75% of the surface, with very little polar or high-resolution coverage (Figure 3.1). Phase angles and resolution are very inconsistent across the globe, making mosaicing difficult.

4
Geology of the Galilean satellites: An introduction to the images

4.1 The importance of being ice

The inner planets and moons with which we are familiar (Earth, Mars, the Moon, etc.) consist of rock-forming minerals that are dominantly silicates. Silicates are composed of silicon and oxygen bonded together, usually in combination with aluminum, magnesium, iron, calcium, sodium, and other metals. On the Galilean satellites (Io excepted), the crust-forming rock is water ice, where hydrogen substitutes for silicon in its bonds with oxygen. Other ices are present in small quantities, such as methane, ammonia, solid nitrogen, carbon monoxide, carbon dioxide or sulfur dioxide, and other more complex molecules. Silicates and carbonaceous material are also present in large quantities but, except for dirtying the surface ices, they tend to reside deep in the interior, the less dense ices in the outer layers. These ices are volatile under average Earth conditions, but at surface temperatures of 100 K (–285 °F) or so, ice is much harder and more durable on Europa than the crunchable ice cubes in our drinks. Thus, the outer layers of these bodies behave much like the terrestrial planets, and are subject to many of the same processes, as well as some peculiar to the icy satellites. The dominance of ice as a crust-forming rock has several important consequences, however, for the geologic processes that occur within these planetary bodies.

4.2 Volcanism (and against cryo-ism!)

The term "cryovolcanism" was invented as a label for lava flows and eruptions on icy satellites. This is the last time you will see that term used here (after all, we do not use terms such as "cryotectonism," "cryoerosion," or "cryoimpact"). "Cryo" was meant to imply volcanism on cold surfaces, but cold is relative, and the term subtly implies instead that the volcanic process is very different on icy worlds. Volcanism is the melting of crustal and mantle minerals and the mobilization of this melt toward the surfaces of planetary bodies. For Europa and Ganymede, the crustal layers of these satellites are composed of water and a few other ices instead of silicates. Magmas and lavas there are composed of molten water ices, but they are still volcanic. Flows of molten water ice simply represent volcanism on an ice-rich body. Ganymedeans would flee from an advancing water lava flow just as vigorously as Earth creatures from a silicate lava flow on Sicily or Hawai'i.

Whether in ice or silicate, volcanism can be manifest in many ways. Lateral lava flows tend to fill depressions and can erupt from vents or fissures. Also common are more violent ash eruptions. Both types can be accompanied by caldera formation, in which subterranean magma chambers collapse, forming walled depressions. Lava flows can form conical volcanic piles called shield volcanoes. Stiff low-temperature lavas can form thick piles.

There are some unique characteristics to water-rich volcanism. Ice melts at much lower temperatures than silicates, so that less heat is required to generate melt in the crust of Europa or Ganymede. Water is also one of the few minerals that is denser in the molten (liquid) state and will tend to sink into the interior. This makes it more difficult to erupt water lavas onto an icy surface and more likely that lava will then drain back into the interior. Indeed, evidence for the eruption of water lavas on the surface of Europa is scant. On Ganymede, however, brighter younger materials have replaced two-thirds of the surface. Although highly tectonized and original flow morphologies are lost, evidence indicates most of these terrains were emplaced as water lavas flooding wide shallow fault-bounded depressions (i.e., graben). Caldera-like features are also rather common on Ganymede (Figure 4.1; Plate Jg13.1), but beyond this no evidence of shield volcanoes or explosive eruptions has been identified.

How Ganymede's watery lavas got to the surface is as yet uncertain, but local pressure differences may have encouraged upward migration. It is well to remember that dense basalts sit today on top of the low-density crust of the Columbia plateau in Washington State (and many other places), so it can clearly happen. Conversely, the density of Ganymede's original crust may actually be slightly higher than liquid water, due to the inclusion of small amounts of denser rocky materials. Until we know the interior structure and composition better, these issues will remain unresolved.

In contrast, the pervasive volcanism on Io is dominated by very high temperature silicate eruptions similar in composition to the lunar mare or Earth's ocean floor. Shield volcanoes, calderas, lava lakes and lava flows

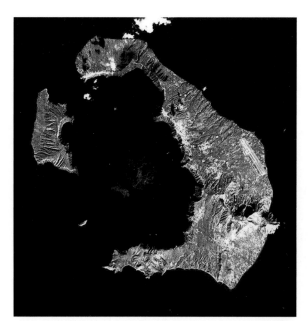 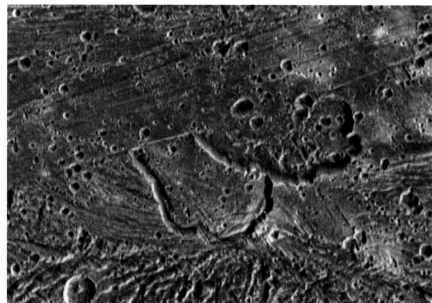

Figure 4.1 Volcanic calderas on Earth (left) and Ganymede (right). The water-filled caldera on Santorini Island, Greece, formed explosively some 3600 years ago. The similarly shaped crosscutting walled depressions on Ganymede are part of a set of eight similar features in this area (see Plate Jg13.1). Whether these features formed explosively or by simple ground collapse is not known, but they are certainly not impact related. The surfaces on the floors of these calderas are raised and ridged, an indication of viscous flow of (volcanic?) material. The band of smooth bright terrain truncating both calderas appears to be a second expression of volcanism involving shallow filling of structural depressions by thin water lavas. The Santorini caldera measures 12 × 8 kilometers, those on Ganymede 25 × 35 kilometers. Santorini Island, Greece, EOS image, NASA (left).

are common. A big difference on Io is the presence of significant quantities of sulfur and simple sulfur compounds. These result in global whitish, yellowish and reddish deposits, and umbrella-shaped plumes of gas and dust jetting into space, especially when the hot silicate lavas flow over sulfur-rich deposits, as is the case for Prometheus. The source of heat and energy to melt so much rock and ice is of central importance and linked to the tectonic disruption seen on the icy satellites.

4.3 Tectonism (and tides)

In addition to volcanism, the release of internal heat can also deform planetary surfaces. As these pages show, pervasive faulting and deformation have scarred both Europa and Ganymede. Deformation can be dilational, extensional, compressional, or transverse (Figure 4.2). Extensional faulting has been found on Europa and Ganymede in abundance and indicates local stretching of the crust, but appears to

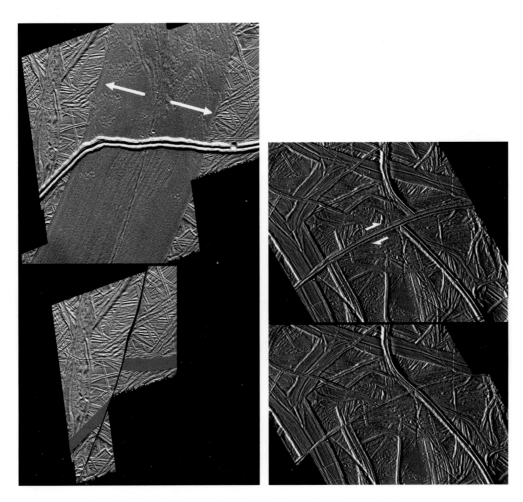

Figure 4.2a Dilational band formation on Europa. High-resolution *Galileo* mosaic shows current configuration at top. At bottom is a digital reconstruction after ~30 kilometers of crustal extension have been removed realigning older ridges and bands. The smooth band is new crustal material (see Plate Je15.3).

Figure 4.2b Strike-slip (transverse) faulting on Europa. High-resolution *Galileo* image near the south pole shows the current configuration at top. At bottom is a digital reconstruction of the area after 3 kilometers of right-lateral slip along the curved ridge has been removed, realigning previously formed ridges and bands. Original mosaic (Plate Je15.1) is at 40 meters resolution.

take different forms (Figure 4.3). Usually this takes the form of closely spaced sets of normal faults, such as on grooved terrain. The whole of Ganymede may have expanded by a few percent due to cooling of the interior. The extension implied by the bands and fractures across Europa, however, is too vast to be due to global expansion alone, but unambiguous compressional folding or faulting to take up this much extension has been found in only a handful of locations on Europa. Some features have

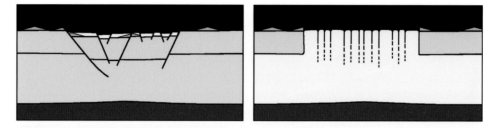

Figure 4.3 Simplified hypothetical cross-sectional diagrams of extensional tectonics on icy satellites. Bright terrain formation on Ganymede (left) apparently involves stretching and fracturing of older icy crust (dark terrain: orange). Flooding by water lavas buries these depressed blocks of older terrains. Dilational (wedge-shaped) bands on Europa (right) form by pulling apart of the ice shell and intrusion of new mobile ice (or water) from the deeper ice shell. Note the fate of the small preexisting hypothetical cones on the surface. These processes help explain resurfacing on Ganymede (Plates Jg8.5, for example) and on Europa (Plates Je8.4 and Je15.3).

yet to be explained, but this apparent discrepancy raises one of the most puzzling questions remaining from *Voyager* and *Galileo*: where is the compressional deformation on Europa?

Many of the observed tectonics on Ganymede and especially Europa are organized in global patterns. Fracturing and faulting of planetary surfaces require considerable geologic forces, and key to understanding this deformation is finding plausible stress mechanisms. Usually these forces come from within, such as the convection that overturns Earth's mantle, moves the lithospheric plates around, and folds and faults the overlying crust. Direct evidence for mantle-driven tectonics on these satellites has as yet not been discovered, but cannot be ruled out.

On the Galilean satellites, there are additional forces from without driving most of the volcanism and tectonics. Jupiter's immense gravity imposes distortions on the shapes of the satellites, forming topographic bulges at the equator. In addition, these satellites appear to rotate synchronously: that is they rotate on their axis at the same rate that they orbit Jupiter. Earth's Moon does the same thing and so we always see the same face of it from Earth. In isolation, the distortions would have no significant effects. The orbits of the three inner satellites are timed in such a way as to be integer multiples of each other (Figure 4.4), however. This orbital or Laplace "resonance" forces their orbits out of round, resulting in predictable daily tides. The tides take two forms, a daily increase and decrease in magnitude and a shift in the location and shape of the bulge due to oscillation about the Jupiter–satellite axis. Daily tides (distortions of the global shape) may be as high as 30 meters on Europa

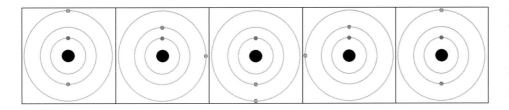

Figure 4.4 This time sequence from left to right shows the resonance orbits (clockwise) of the three inner Galilean satellites. Each panel represents one orbit of Io. The repeated conjunctions of these satellites distort their orbits slightly and lead to distortions of their physical shape and heating of the interior.

and 100 meters on Io, but are negligible on today's Ganymede. These global distortions have two important effects, especially on Europa. First this repeated daily cycle can and does fracture the surface in predictable patterns. Global stress patterns are symmetric and can be compared to mapped fracture patterns, and the fracturing history reconstructed. Second, tides input tremendous amounts of frictional heating into the interiors, potentially melting large parts of the interior, a subject we will return to.

Two additional global stress fields have been proposed. The apparent synchronous rotation of these satellites may not always be precisely true. The outer shells of the satellites can rotate slightly faster than the deep interior (at rates of once around every few thousand years). The tidal bulge on each satellite remains oriented towards Jupiter even if the shell rotates faster, however, generating stress and fractures. Nonsynchronous rotation has been demonstrated only for Europa, but may have occurred (or is still occurring) on Io or Ganymede.

A different form of nonsynchronous rotation, called true polar wander, can cause the outer shell to "flip" over, so that polar terrains are rotated over to the equator. This mechanism can be triggered by a mass anomaly near the surface, due to a large impact basin, volcanic pile, or variations in ice-shell thickness with latitude (due to the coldness of the poles). Polar wander can impose considerable stresses. Evidence for polar wander has been found so far only for Europa, and may indeed explain many of the large-scale tectonic features (see Chapter 5.3), but is possible for other satellites as well. Both forms of nonsynchronous rotation require that the outer icy shell be decoupled; that is, it rests on a layer of very soft ice or liquid water.

Surface deformation can also involve slow movement or transfer of material vertically within the icy crust. Overturn of the outer layers, called diapirism, usually occurs as fields of ovoid or mushroom-shaped blobs of rock or ice rising from or sinking back into the deeper parts of the crust (Figure 4.5). Cantaloupe terrain on Triton and pits and lenticulae on

Figure 4.5 Cross section of hypothetical diapirs within an icy crust or shell. Dark blue is water, while light blues are different layers of ice. Layering or zonation within the ice shell has been greatly simplified. Schematic diagram is based on observed seismic profiles and morphologies of terrestrial salt domes and diapirs (from the work of Martin P.A. Jackson and others) and also represents many of the observed surface morphologies of lenticulae and ovoid depressions seen on Europa (see, for example, Plate Je4.1).

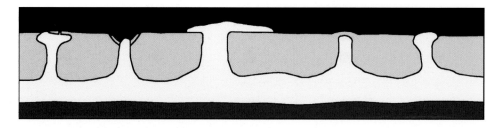

Europa (Plate Je4.1) were likely formed by this mechanism. This process is similar to convection, such as in porridge boiling on a stove, a bowl of miso soup, or in lava lamps, and usually requires a vertical instability in the layer due to heat or a density inversion. Here the line between viscous molten lavas and warm mobilized solid material can become blurred, as both can behave in similar manners.

4.4 Viscous relaxation

Glaciers on Earth flow down-slope under the force of gravity, albeit very slowly. Silicate rocks seldom do this, but water ice can flow under its own weight more readily and the warmer the ice, the more rapidly the ice will deform. This flow has the effect of softening and reducing the original topography of the surface, a process known as viscous relaxation. On chilly Triton, a scant 35K above absolute zero, water ices on the surface are very rigid and do not flow, but other ices such as nitrogen or methane ice are softer and probably do flow, leading to diapirism and other processes.

Craters provide a useful tool to map out viscous relaxation because we have a very good idea of what their original dimensions should be. The icy crusts of Ganymede and Callisto are currently too cold to flow very much, but on Ganymede, as can be seen on these pages, many older craters in dark terrain are flattened from their original shapes (Figure 4.6), whereas those formed after bright terrains are not. This indicates that heat flow and internal temperatures were considerably higher on Ganymede than they are today. The case on Callisto is not so clear due to pervasive erosion that masks potential evidence of relaxation.

Figure 4.6 Impact craters in dark terrain of Marius Regio, Ganymede. Only three craters in this scene have retained their original deep shape; all others have been flattened or relaxed by viscous creep. The two large deep craters are ~20 kilometers across and ~1 kilometer deep.

4.5 Other global effects

None of the Galilean satellites currently possess an atmosphere of significance capable of shielding their surfaces. This leaves them vulnerable to bombardment from meteorites and charged particles, or plasma. Very small rock and dust-sized particles are predicted to vastly outnumber the crater-forming comets we can see from space. This constant rain of debris, though not strong enough to bother a team of prospecting astronauts, will over time grind up the surface and form a thin loose soil or regolith. Io and Europa may be too young to have a significant regolith, but Ganymede and Callisto probably do. The second type of bombardment, by high-energy plasma, is high-energy protons and electrons swept up in the fast-rotating Jovian magnetosphere, including particles ejected from the surface of Io. As these particles impact satellite surfaces, they can be implanted directly or alter existing molecules, changing the surface composition. They can also control polar processes (see Chapter 4.6).

The rotation rate of Jupiter's magnetic field (about 10 hours per day), and the charged particles embedded within it, is much faster than the revolution of the satellites (2 to 16 days). Hence, the plasma of sulfur, oxygen, potassium, and other ions within the belts will always overtake and strike the "slow-moving" satellites on their trailing hemispheres. This process has altered global color, brightness, and compositional patterns on Europa and Ganymede. Only Ganymede has a magnetic field capable of partially shielding the equatorial zone from these particles, funneling them towards the poles and controlling formation of the polar caps.

If rotating synchronously, the side of each satellite that faces in the direction of orbital motion (the apex, or leading hemisphere) is also always the same. As a result, the bombardment of meteorites and charged particles will be rather asymmetric. Due to the high orbital velocities (from 8 to 17 kilometers per second), these satellites will tend to sweep meteorites and dust preferentially onto their leading, or front-facing, hemispheres. Throw a non-spinning baseball into a swarm of irritated bees to visualize this effect. By prediction, there should be 30 to 40 times the number of craters on the leading hemisphere than on the trailing hemisphere.

Nonsynchronous rotation and polar wander both can mitigate and even erase the patterns described above. The fact that a cratering asymmetry is not observed on Callisto is probably due to its great age and the shear number of craters formed. The muted asymmetry signal measured on Ganymede is more troubling. The observed asymmetry is only a factor of four or so. Either the theory of impact crater distribution is wrong (unlikely) or Ganymede has not always rotated synchronously, smearing out the expected pattern. Europa is not imaged sufficiently to map the global crater distribution, although indications so far are that craters are uniformly distributed. Similarly, if the outer shell migrates over time, the tectonic patterns will also shift from their original location and then be overprinted by new fractures, making for a spaghetti-like web of overlapping tectonic patterns. Such appears to be the case on Europa, and possibly Ganymede as well. Detailed planet-wide mapping will be required to fully untangle these overlapping patterns, patterns that can only be glimpsed with the data we currently have in hand.

4.6 Polar processes and ice segregation

The Jupiter system differs from most other planets in the very low inclination of the rotational axis. Earth's seasons are not a result of changing distance from the Sun, but rather the tilt of our axis, shading each pole from the Sun every 6 months. Hence, unlike Mars, Earth, or Titan, the Galilean satellites experience no seasons. Each pole is continually locked in a permanent equinox (true polar wander is a permanent, not a seasonal shift). Polar regions and any preferentially pole-facing slopes are thus cold traps. Despite the cold average surface temperatures of 100 to 130 K (–280 to –225 °F), water-ice molecules can be knocked off the surfaces of the icy Galilean satellites by charged plasma bombardment and by solar radiation, a process called sublimation. Molecule by molecule, these frosts can travel for short distances until they find areas that are colder and more stable, such as shady and cooler slopes (whether inside craters or on ridges or grooves) facing toward the poles. This can lead to segregation of rocky and icy materials on the surface. The warmer a surface, the more susceptible it is to this erosive process. Examples of such deposits can be seen on Callisto (Plate Jc3) and in high-resolution images of Ganymede (Plates Jg3.2 and Jg14.1).

4.7 Impact cratering: Planetary chronometer and window on the interior

Impact cratering can't be avoided. The rain of comets (and a few stray asteroids) into the Jupiter system has been continuous, if not constant, for 4.56 billion years. The breakup of comet Shoemaker-Levy 9 in 1993 is an example of such a comet that recently ventured too close to Jupiter and paid for it. The result of this violent bombardment is the slow, steady accumulation of large and small craters across the surfaces of all four satellites, formed by comets striking at velocities of several kilometers per second. A general rule of thumb is that a comet creates a crater roughly 10 times its original size, although this varies depending on angle and relative velocity.

Craters can obliterate the geologic history of a terrain, given enough time, but they also tell us things we might not know otherwise. The chief of these is the ages of these surfaces (currently our only tool for doing so). By surface age, we usually mean the time since the last resurfacing or surface modification event. If we know or can guess how intense this rain of comets has been, all we need to do is count the craters to determine how long that surface has been accumulating them. Thus, terrains with few craters are younger than those with lots of craters. The key problem is estimating the flux rate at which comets hit the surface. There are three competing models. The first is that the asteroids that are striking the Moon also strike the moons of the Outer Solar System. This is demonstrably wrong, based on observations of today's asteroids and comets, but asteroids may have been more important in the most ancient times. The second model, "our current best guess," is based on observations of modern-day comets near Jupiter, work begun by Eugene Shoemaker and continued by Kevin Zahnle. The ages reported in this *Atlas* are based on that model, but we can hope that better estimates will be forthcoming in the decades to come. The third model incorporates what is called the Late Heavy Bombardment, a violent pulse of comets and asteroids that ransacked the Solar System some 500 to 700 million years after formation. Regardless of the true flux history of impacts, it is clear that Europa's surface is very young, probably less than 100 million years old. It may have formed when dinosaurs roamed Earth. The younger bright terrains of Ganymede may be 1 to 2 billion years old, but the ancient surfaces of both Ganymede and Callisto probably date back at least 4 billion years. Determining the true ages of terrains across the Outer Solar System may require landing and analyzing the surfaces of one or more of these bodies directly.

Craters themselves age in predictable ways. Fresh craters, such as Pwyll on Europa, Osiris on Ganymede, and Buri on Callisto, excavate and

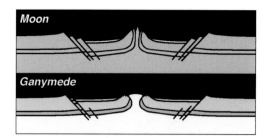

Figure 4.7 Cross-sectional diagrams of final shapes of large complex craters. The top diagram is a typical crater on the Moon, such as Copernicus; the bottom is of a similar-sized crater on Ganymede, such as Osiris. The different shades are intended to show how layers at different depths are deformed during impact. On the Moon, the lower units can be rich in olivine, which has been detected on central peaks (and rims). On Ganymede, the deeper units are warm soft ice that creeps when exposed on the surface, forming rounded domes. Domical central uplifts, lack of rim terraces, and shallow depths characterize craters on Ganymede and Callisto. This model will likely change as computer models improve and when we return to Jupiter.

expose very bright water-ice frosts as ray patterns. These frosts darken with time at a steady, if unknown but likely latitude- and longitude-dependent, rate. Thus the general age of a crater can be guessed by the brightness of its ejecta. On the Moon, bright-rayed craters are generally less than 500 million years old. Whether that is true on the Galilean satellites has not yet been determined, but can be assumed so for the time being.

Impact craters can also tell us about the structure or composition of the deeper interior. As described earlier, the degree of topographic relaxation in older craters, especially palimpsests and penepalimpsests, can be used to estimate heat flows in the past, much more cheaply than installing a thermal probe on the surface and building a time machine. Similarly, we would like to known the interior structure of these satellites and we can use craters to do the drilling for us. Impact craters here on Earth tell us that the central portion undergoes several kilometers of steep uplift, forming a narrow plug of material and bringing deeply buried rock to the surface (Figure 4.7). On the Moon and planets, this central uplift forms a mountain or central peak. On Ganymede and Callisto, we see this uplift as central peaks and bright central domes.

Comparison of impact crater morphologies on the three icy satellites is very revealing. Contrary to expectation, there are significant differences in the shape and morphology of larger craters on Europa and Ganymede (compare, for example, Pwyll on Europa and Achelous on Ganymede; Plates Je14.1 and Jg2.2). These differences are clearly related to the liquid-water ocean not far beneath Europa's surface, which must be a lot deeper on Ganymede. Impact craters larger than roughly 20 km across are not as stable in the thinner floating ice shell on Europa compared to what we observe on colder Ganymede, creating more chaotic landforms (see Chapter 5.3).

5
The satellites

5.1 Callisto

From a distance, Callisto has a mesmerizing unearthly beauty, like Wuthering Heights from the moors on a foggy night (see the global views in Plates JcO1, O2, and O3). The surface appears to be illuminated from within and resembles nothing so much as a sphere of stars, as if our Galaxy had collapsed inward on itself and then turned inside out. The "stars" are in fact thousands of impact craters, which overwhelmingly dominate the surface. Like the stars of our Galaxy, Callisto's craters come in all varieties of size and brightness, each a function of the size of the comet or asteroid that created it and the age since each crater was formed. Yet even from afar, this global pattern of innumerable star-like spots is deceptively simple. Strange diffuse nebula-like dark and bright features create a patchy pattern not unlike our Milky Way Galaxy on a cold crisp night.

The ethereal beauty of Callisto is appropriate. Callisto, rooted in the Greek Kalliste, means "most beautiful." In various versions of the myth, she is a nymph or attendant of Artemis, and daughter of Lycaon, King of Arcadia, who took a vow of chastity. Zeus/Jupiter lured her and took her, but wrathful Juno/Hera turned her into a bear. To avoid being killed by a hunter (her own son), Zeus transformed both into Ursa Major and Ursa Minor, another gift to mortals from the gods.

Only when we get close does her gossamer beauty give way to a more haggard appearance. Callisto's cratered surface is very ancient, at least 4 billion years old, but in many places the surface is highly degraded, almost desiccated in appearance (Figure 5.1.1). Crater rims have been eaten away, leaving only rings of knobs. Some of the diffuse bright patches appear to be large, ancient, eroded impact features (Plate Jc7, for example) but aside from several large ringed impact basins and radial crater fractures, almost no coherent tectonic patterns are left, if any ever existed. Despite *Galileo*'s observations, the origin of many of these strange brightness patterns, so apparent in many of the quadrangles, remains a mystery. Callisto in reality is far removed from Lin Carter's jungle moon of his fictional "Jandar of Callisto" series (Figure 5.1.2).

Figure 5.1.1 Perspective view of cratered plains near the Asgard multi-ring impact structure. The trough crossing the scene is one of a dozen or so eroded graben encircling Asgard. The large crater at top is 16 kilometers across. View is based on topographic analysis of *Galileo* data by the author.

Figure 5.1.2 Cover from Part 5 of the "Jandar of Callisto" series. Has Callisto been fooling us all? Courtesy Orbit Publishing.

Callisto suffers in comparison with her brother and sisters. The relative lack of geologic complexity makes Callisto a target of neglect and occasional derision, a perception that dates from *Voyager* days. Pre-*Voyager* expectations had Callisto (and Ganymede) as one of the more dynamic and interesting bodies, while Io was expected to be Moon-like. *Voyager* simply reversed the order. Callisto has few fans. (I once infamously co-authored a presentation entitled "Callisto is not boring!" There weren't many converts.) But the apparent blank slate of Callisto's surface is itself a profound puzzle and gives us important clues to events in the Jupiter system long ago. One must look closely to unveil her secrets.

Geologically, Callisto's cratered plains resemble dark terrain on Ganymede, though there are differences. Impact craters are numerous and similar in morphology on the two satellites, but the intense fracturing of dark terrain that parallels bands of bright terrain on Ganymede is absent on Callisto. Large ringed basins dominate the geography of Callisto, forming concentric sets of graben similar to Ganymede's furrow sets (Figure 5.1.1). The largest of these, Valhalla, extends up to 3000 kilometers across and consists of several dozen concentric sets of furrows (Plates Jc2 and Jc6).

On a local scale, however, it is the often-severe erosion of craters and topography that dominates (Figure 5.1.3). Callisto's conversion from icy cratered surface to dark eroded plains is not globally uniform, nor is it replicated on Ganymede. Some regions appear more heavily eroded than others, although the global pattern is not understood due to the limited imaging coverage. The erosion is probably driven by sublimation, whereby solar heating slowly drives off more volatile water-ice molecules toward local crests and summits, leaving the non-ice components behind to form a residual dark surface. The thickness of this residue in not known but is at least tens if not hundreds of meters thick. The fact that this dark surface is itself so heavily cratered suggests that this process occurred a long time ago. This segregation of bright and dark materials is much more pronounced on Callisto than on Ganymede.

The composition of the non-water-ice material that covers much of Callisto is partially known. Magnesium- and iron-bearing hydrated silicates, as well as the condensed gases carbon dioxide, sulfur dioxide, and oxygen, have been identified on the surface from telescopic and NIMS observations (Figure 5.1.4). Ammonia is suspected, as are unidentified organic compounds. Volatiles like carbon dioxide are often associated with younger impact craters. Some of these components may derive from Callisto's formation or be related to the composition of certain primitive asteroids brought to Callisto by impact.

One Callisto mystery from *Voyager* days, the formation of unusually long chains of craters, or catenae (Figure 5.1.5), has been solved. Neither ordinary impact nor tectonic/volcanic processes could explain these strange chains. In 1993, the equally strange comet D/Shoemaker-Levy 9 (SL9) was discovered orbiting Jupiter. A close passage to Jupiter had disrupted this comet into a very linear chain of smaller fragments. A few weeks after this discovery, both Jay Melosh and myself, sitting 2000 kilometers apart, remembered these chains on Callisto and leapt to the same conclusion: if the breakup of comets happened once, it had happened before. The chains are the impact scars of ancient and recent tidally disrupted comets at Jupiter, comets that Callisto and the other satellites had gotten in the way of. Ganymede also records these chains but Europa and Io are too young to retain any record of them.

Chief among Callisto's unresolved mysteries is the stark contrast between itself and its slightly larger twin, Ganymede. *Galileo*'s measurement of Callisto's gravity provides a clue. Callisto does not have a dense core and light ice-rich mantle like Ganymede, but is it totally undifferentiated, a completely homogenous mix of ice and rock? Callisto does appear to have

Figure 5.1.3 Oblique image of Callisto's highly eroded cratered plains. The bright knobs that peak above the dark plains are remnants of an older icy cratered terrain. The dark plains are heavily cratered and hence old. They consist of dark non-volatile materials such as hydrated silicates and organic compounds, although the exact mixture remains unknown. At 9 meters resolution, this is one of our highest-resolution views of Callisto.

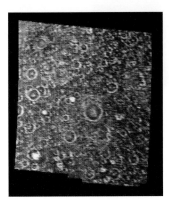

Figure 5.1.4 NIMS spectral map of cratered plains on Callisto. These maps illustrate the compositional heterogeneity of Callisto's surface. Red tones are non-ice materials; blue highlights areas of enhanced ice concentration. Bright blue spots are relatively young bright-rayed craters. The yellowish area at upper left has a slightly enhanced water-ice signature and is an ancient impact scar. Map is centered in quadrangle Jc8 at 10°N, 210°W. The NIMS data have a resolution of 12 kilometers and have been combined with the global image mosaic.

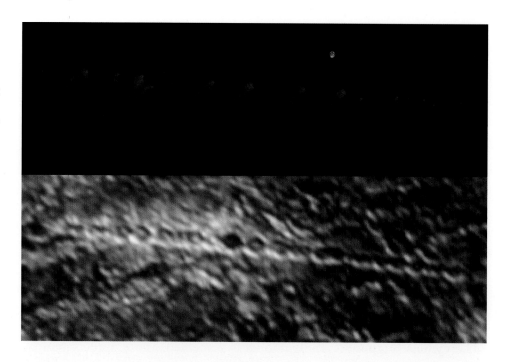

Figure 5.1.5 Comet Shoemaker-Levy 9 (top) and Gomul Catena (bottom), a crater chain on Callisto. Crater chains like this are formed by comets like that. The small bright spot in the comet image is Earth, shown to scale. The comet and Callisto images are not to scale, however! Comets disrupted by Jovian tides hit the satellites when they are much closer to Jupiter and hence much shorter in length. Comet image courtesy Space Telescope Science Institute.

Figure 5.1.6 Cutaway view of probable interior structure of Callisto. Callisto may be partially differentiated, that is, only partly segregated into ice and rock layers. A thin water layer may separate the thick icy outer layers (white) from the mixed rock and ice interior. Whether Callisto has a small core is unknown. Original NASA/JPL diagram modified by author.

an ice-rich outer layer (despite the dark surface coating), and possibly a liquid-water ocean layer buried roughly 100 to 150 kilometers beneath the ice. Otherwise the deep interior is probably a well-mixed blend of ice and rock (Figure 5.1.6). If so, Callisto (and the other satellites) did not get very warm during formation, or melting and differentiation would have occurred here. Otherwise, Callisto has been geologically quiet for most of its history. Callisto has never been part of the orbital resonance responsible for melting parts of Io, Europa, and briefly, Ganymede. This may be why Callisto was "left out," banished perhaps by an angry Artemis.

5.2 Ganymede

Although it lacks the youthful vigor of Europa or Io, Ganymede may be the most complex and planet-like of the four "sisters." Ganymede, the largest of the natural satellites, is a larger and flashier twin to outermost Callisto. Unlike Io or even Europa, these twins are water worlds, composed of roughly 50 to 60% water (and a few other) ices, the remainder being various rocky components ranging from carbonaceous to silicate materials.

Ganymede is named for the Cupbearer of the Gods (Figure 5.2.1), a Trojan prince kidnapped from Mount Ida by Jove to serve in that role. Ganymede also caught Jove's notorious wandering eye, and Hera's wrath, to become one of his many conquests (in a subtle variation on a recurring theme). Ganymede, handsomest of mortals, can thus be considered the unofficial gay moon. Ganymede was, by some accounts, later placed in the sky as Aquarius and has been associated in mythology as a source of the Nile, all curious in light of Ganymede's high water-ice content and evidence for a water ocean deep inside the moon.

Although it lacks Titan's organic-rich atmosphere, Ganymede is the only satellite known to have its own magnetic field, which triggers and controls aurora above the surface (Figure 5.2.2), and controls formation of "polar caps" of water frost (Plates Jg3.2 and Jg14.1.2). *Galileo* also revealed that the interior is very clearly differentiated into a crust and mantle of water (and perhaps other) ices, a rocky outer core and an inner core mostly of iron (Figure 5.2.3). Ganymede's internal water ocean is at least 150 km deep, however, considerably deeper than Europa's. Ganymede also records a complex geologic history spanning billions of years, a far longer period than does Europa.

Ganymede's complex geologic history is transitional between youthful Europa and ancient Callisto. The surface can be divided into two basic terrains: older *dark terrain*, and younger *bright terrain*. Dark terrain is heavily cratered and resembles the ancient surface of Callisto, at least superficially. Younger less-cratered bright terrain covers 66% of Ganymede's surface and breaks the dark terrain into polygonal blocks of various size and shape, the largest being Galileo Regio (Plate Jg3).

Dark terrain may or may not have experienced some cryptic form of ancient icy volcanism, but its dominant record is one of extensive impact cratering, long arcuate graben formation (related to basin impacts), and widespread landform degradation through down-slope creep and viscous relaxation of topography (Figure 5.2.4). Ancient sets of arcuate graben called furrows were formed in a matter of hours as large impact basins collapsed. Remnants of the largest of these systems can be found in Galileo, Marius and Perrine Regios (Plate Jc3, for example). Dense sets of younger, narrow, parallel fractures occur in many areas of dark terrain, but most are in close proximity to bands of bright terrain and are probably relics of its formation. In a few cases, this fracturing is intense enough to obliterate preexisting morphology. The intense segregation of water ice and rock and pervasive erosion so prevalent on Callisto has not occurred on Ganymede, but down-slope creep of non-ice material has. Relaxation of Ganymede's icy surface has softened much of dark terrain topography. Relaxation is strongly temperature dependent, implying that

Figure 5.2.1a Roman mosaic at Kato Paphos, Cyprus, of a booted Ganymede being abducted by Zeus, in disguise.

Figure 5.2.1b Ganymede greeting Zeus disguised as an eagle. Statue by Hubacher graces Zurich harbor. Photo by D. Bonett.

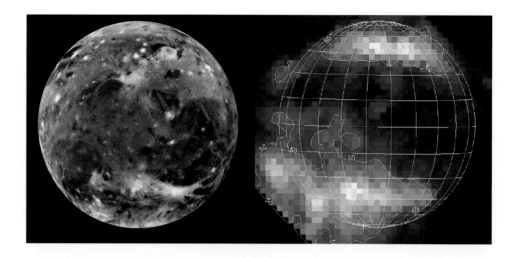

Figure 5.2.2 Ganymede (left) and its aurora (right). The bright ultraviolet emissions seen in the Hubble Space Telescope image are likely due to excitation of oxygen molecules. The brighter aurora coincide roughly with the boundaries of Ganymede's polar caps. Aurora image courtesy of HST/STIS, copyright AAS 2000.

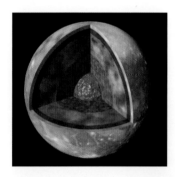

Figure 5.2.3 Cutaway view of probable interior structure of Ganymede. The dense iron-rich core is surrounded by a rocky outer core, a water-ice mantle and a thick icy crust. A liquid-water layer is strongly suspected beneath the icy shell. The thicknesses and depths of each layer, including that of the liquid-water layer (deep blue) depend on composition and are not known precisely (as is true for all four satellites). Original NASA/JPL diagram modified by author.

heat flow and interior temperatures were significantly higher in the past than at present.

Both bright terrain and dark terrain contain water ice, but dark terrain also contains a significant amount of non-ice material (Figure 5.2.5). Hydrates salts, as well as magnesium and possibly sodium and hydrogen sulfates, have been identified. Carbon and sulfur dioxides are also present. Organic materials are suspected but have not been specifically identified. Whether these materials relate to the original composition of Ganymede or have been altered in some way remains unknown.

The major event that defines Ganymede as we see it today is the extensive resurfacing associated with bright terrain. Younger bright terrain (Figure 5.2.6) can be smooth or grooved, but a characteristic feature of bright terrain is the polygonal shapes and sharp fault-bound edges that define each section of this terrain. Strike-slip and compressional deformation is conspicuously inconsequential, absent, or not yet detected on Ganymede.

A major focus of *Galileo* was to determine how these bright terrains came to be. Much remains unclear, but a story has evolved that appears to explain what we see. There were several consequences of this event: viscous relaxation, volcanism, and tectonism. Unlike plate tectonics on Earth, bright terrain on Ganymede has not destroyed, replaced, or displaced dark terrain, but rather has buried and resurfaced it in place through faulting and water and ice volcanism. Bright-terrain formation apparently starts with fracturing of dark terrain into long narrow bands that formed walled depressions called graben. Relaxation affected all preexisting dark terrains at about the same time. Low-viscosity water lavas subsequently flooded many of these down-dropped sections of dark terrain. It is possible that these topographic depressions may have induced pressure gradients that pulled the lavas to the surface. Vestiges of this volcanism may be found in the form of dozens of open-ended irregular-shaped caldera-like features (see Plate Jg13.1) scattered across bright terrain and associated with narrow bands of smooth undeformed terrain. Finally, nearly continuous and pervasive stretching and faulting at different scales during bright-terrain emplacement formed crosscutting sets of narrow parallel troughs and ridges called grooved terrain (Figure 5.2.7).

Why and when did all this take place so late in Ganymede's history? Our best guess is that most of bright-terrain activity took several hundred million years and occurred roughly 2 billion years ago (this age is uncertain by 500 million years or more, however). Bright-terrain formation involved the release of a lot of heat energy, but the current orbital resonance responsible for melting in Io's and Europa's ice has very little thermal impact on today's Ganymede. One scenario suggests that these events were triggered by migration of Ganymede into the three-way orbital resonance with Io and Europa described in Chapter 4. This event will have induced unusual orbital digressions called chaotic motion, causing an intense but temporary increase in tidal deformation and heating. Perhaps bright terrain records this event in Ganymede's history.

A different scenario suggests that the resonances were locked in very early in the history of the Jovian system, within the first several hundred million years or so. It is possible that our favored age for this resurfacing ($\sim$2 Gyr) is not correct and that bright terrain is more ancient. This seems highly unlikely given the low density of craters in many areas and what we know about the flux of comets in the Jupiter system. If bright terrain formed when we think it did, then the early resonance formation would have occurred far too early to trigger the observed resurfacing. If Ganymede had simply cooled since early days, the tendency would be for it to shrink, but the large amounts of crustal extension seen in bright terrain suggests otherwise. Delaying differentiation from an early Ganymede to the current rocky core and icy mantle may well have produced the upheaval we see recorded in Ganymede's tectonics, but again we are left with the question of why it was delayed or triggered at all.

Figure 5.2.4 Oblique images of dark terrain on Ganymede. The surface in this region has a rounded characteristic suggestive of degradation by creep or relaxation. The bright-rimmed ridges crossing scene center are the eroded parallel rims of an ancient impact furrow. The large crater on the right edge is 17 kilometers across. Image resolution is 77 meters/pixel. See also Plate Jg8.1.

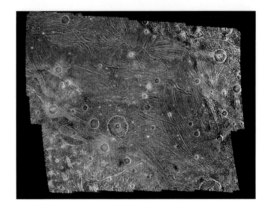

Figure 5.2.5 NIMS spectral map of Uruk Sulcus. NIMS map (shown in color) has been merged with the global Ganymede mosaic. In this false-color rendition, blue tones represent water-ice concentration. These data confirm that bright terrains have higher concentrations of water ice than do dark terrains. The highest concentrations of water ice occur in bright-rayed craters. A few smaller craters, however, have very high concentrations of non-ice material. These form small purple splotches. Map is centered at 8°S, 146°W. See southeast corner of Plate Jg8 for location.

Figure 5.2.7 Oblique image of highly grooved bright terrain in Nippur Sulcus. Closely spaced faults have created a landscape of long parallel ridges. The faulted terrain has partly cut into a block of smoother older terrain at upper right. The largest crater is ~15 kilometers across. See also Plate Jg4.3. Image resolution is 180 meters.

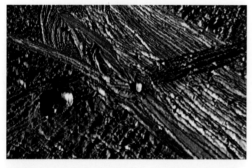

Figure 5.2.6 This perspective view of Ganymede shows narrow bands of both smooth and highly grooved bright terrain across Tiamat Sulcus. The "grooves" are closely spaced sets of extensional faults forming a landscape of long parallel ridges. Cratered dark terrain lies to either side of this band. A narrow band of faulted terrain crosses Tiamat Sulcus. The largest crater, Maa, is ~28 kilometers across. Image resolution is 500 meters. Relief is based on digital topography produced by the author. See also Plate Jg8.8.

The origins of Ganymede's resurfacing remain vexingly nebulous, especially as Callisto did not experience a similar upheaval. How Ganymede came to be so complex and Callisto so boring should give us a key to Jupiter system evolution. The thermal and composition evolution of large icy bodies is bound to be complex, due in no small part to the unusual properties of the various phases of water ice. Several possible pathways have been identified but, ultimately, precise dating of the ages of these events on and a comprehensive assessment of the composition of Ganymede, perhaps requiring surface samples, will be critical.

5.3 Europa

The first impressions of Europa to an approaching explorer might bring to mind Jackson Pollack during his drip phase. Arcuate and interwoven dark reddish-brown strands densely cross the surface, interspersed with scattered dark spots. These long features resemble bubbly strands of dark paint on a white canvas. The impression is particularly striking where the midday Sun enhances the dark contrasting overlapping linear swaths (see especially Plate Je6). On close examination, the surface of Europa is very intense. The canvas of dark spots and streaks on white devolves into very dense and complex crosscutting patterns. Virtually no acreage has been left undamaged by deformation.

Simon Marius (or Kepler?) apparently proposed Europa, a Phoenician and distant relation to Io and possibly a daughter of Agenor, Queen of Tyre, as the name for this frosty moon. By some means she came to Crete, where Jove found her and, turning into a white bull, ran off with her on his back (Figure 5.3.1). Ultimately Europa may have given her name to the continent of Europe, and her face to coins.

One thing Europa and Io share is a fixation with youth. Europa has few large craters, and current paradigm suggests that the present surface is on the order of only 60 to 100 million years old. This is extremely young by geologic standards, given that these moons formed roughly 4.56 billion years ago. *Galileo* observed no direct evidence of geologic activity, such as new fractures forming or water-vapor plumes as seen at Saturn's moon Enceladus, but *Galileo*'s communications difficulties made it impossible to acquire the types of observations necessary to make such a discovery. Current activity on Europa thus remains a very open question. It seems likely, however, that Europa is geologically active today in some form.

Europa's interior is dominated by a large rocky (and perhaps Io-like) core while most of the water and ice is concentrated in an outer layer up to 150 kilometers thick or so (Figure 5.3.2). Is Europa a glaciated Io? Maybe, but it can also be thought of as transitional between rocky Io and ice-rich Ganymede and Callisto, both of which are roughly 50% by volume water ice. Europa is only slightly smaller than sister Io, and roughly the same density, yet the two have very different personalities. The intense heat of Io gives way to an icy surface tormented by pervasive fracturing and disruption (Figure 5.3.3). Although the ancient cratered terrains of Ganymede and Callisto are nowhere to be seen on Europa, Ganymede and Europa do share a history of tectonic activity. The big difference between Europa and the other icy Galilean satellites is that the salty liquid-water oceans believed to reside inside these bodies lies unusually close to Europa's cold icy surface hidden beneath a global ice shell. This remains one of *Galileo*'s most important discoveries.

Figure 5.3.1 Europa rides Jove on the Greek 2-euro coin.

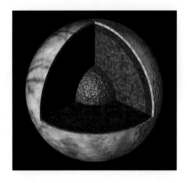

Figure 5.3.2 Cutaway view of probable interior structure of Europa. The outer ice and ocean layers (white and blue) are exaggerated slightly for clarity, but have a total thickness of roughly 150 kilometers. An iron core probably lies beneath the thick rocky mantle. Modified by the author from an original from NASA/JPL.

Figure 5.3.3 Oblique view of Europa during *Galileo* orbit E12. Chaos is in the foreground, ridged plains in the background. The dark scarp across the lower half is a geologic fault that cuts across numerous features. Dark deposits have formed in valley floors between the numerous ridges, as is sometimes observed on Ganymede. This is one of our highest-resolution views (~20 meters) of Europa's surface.

Despite *Voyager*'s poor view of Europa, with a best resolution of only 1.7 kilometers, the low crater density immediately suggested the possibility that Europa could be tidally heated like Io. Europa is further from Jupiter and the tidal heating is less severe, but water ice is much easier to melt than rock. This encouraged speculation that tidal heating might be sufficient to keep *some* part of Europa's deeper ice layer liquid.

This line of thinking was encouraged by *Voyager*'s evidence of mobility within the icy crust. Paul Helfenstein and Alfred McEwen both found that the orientations of *Voyagers*' global-scale dark lineations were consistent with nonsynchronous rotation (see discussion in Chapter 4.3). The second bit of evidence germinated during my summer as a NASA undergraduate intern in 1979 during the *Voyager 2* Jupiter encounter. I recall the first high-resolution images of Europa (recorded the day before) as they came up on monitors during our daily meeting on July 10, 1979. A few weeks after returning home, I was relating my adventures with my advisor and mentor, Dr. Carl Seyfert, whom I miss greatly. Carl's enthusiasm was like gasoline on the fires of a young imagination. Examining the delicate dark lineations on the images, I realized that a seemingly broken cluster of lineations could be precisely realigned if we slid a large block of Europa's icy crust 25 kilometers to the west, closing an adjacent dark band (see quadrangle Je8 and Plate JeO1). It was unmistakable: Europa's icy shell has broken into large coherent blocks that were moving and pulling apart. This was plate tectonics, of a peculiar sort to be sure, yet on an alien world.

Galileo's high-resolution views of Europa were among the most eagerly anticipated of the mission. The complexities of the surface will take years to unravel, but the variety of landforms observed include icy diapirs, dilational bands, strike-slip and normal faults, oddly shaped impact craters, and curved cycloidal ridges. All support the idea that the ice we see on the surface is just the top of a global shell floating on an ocean of liquid water. The geology evidence is strong but circumstantial; the most direct evidence comes from the magnetic field signature of a conducting layer near the surface: a salty ocean.

The outer icy shell is riven by scarps, ridges, and curvilinear bands. Most of these are extensional, dilational, or strike-slip faults (very few are compressional or can be recognized as such). Many of these features relate to Europa's daily tides or to rotational forces, all of which have mobilized and disrupted the icy shell. Many of these bands and lineaments make a good match to stress patterns from nonsynchronous rotation.

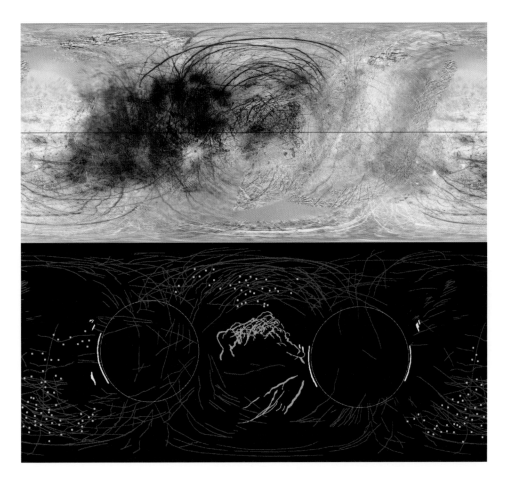

Figure 5.3.4 This "altered" global mosaic shows how Europa's geography might have looked before polar wander of the outer ice shell shifted its rotational orientation. Below the mosaic is a simplified map showing the major features of Europa. Compare this mosaic to Plate Je, the global map of Europa as we see it today. From this new perspective, a strong symmetry is now apparent either side of the paleo-equator (the dark line across center) despite major gaps in our mapping coverage. Very long arcuate lineations, mostly dark bands (blue in the map), form symmetric patterns that converge toward two opposite regions (large circles on the map) located on the paleo-equator. Small dark freckles (cyan) resolved as lenticulae are also common through this zone on both sides. The long bands also enclose a concentration of both shorter dark dilational bands (green) and the unusual bright bands Agenor and Corrick Linea (red) along the paleo-equator. This symmetric pattern fits very well with the stress patterns predicted if Europa's icy shell rotated almost 90°, shifting polar terrains to the equator. The author and colleagues Isamu Matsuyama and Francis Nimmo proposed this scenario in 2008, almost 30 years after *Voyager* first mapped Europa. Europa continues to surprise.

Polar wander, however, looks to be an even better match to the global pattern (Figure 5.3.4) of dilational bands, bright bands, arcuate dark bands, and arcuate troughs (the "crop circles" in Plate Je10 and Je12). Polar wander, in which the spin orientation of a planet or its outer layers changes, moving old polar terrains to the equator, has been postulated for

Figure 5.3.5 Oblique view of Conamara Chaos, Europa, during *Galileo* orbit E12. The plateaus are remnant blocks of older ridged plains, some of which have split apart and separated. The large block at center is 2.5 kilometers across. The matrix material in between blocks is composed of ridged plains material that has been pulverized into small aggregate fragments of various sizes. Some of the large blocks are high standing, others tilted and half buried, aspects that are quite common within exposed diapirs of salt on Earth. Loose debris (talus) has accumulated at the base of most of these cliffs.

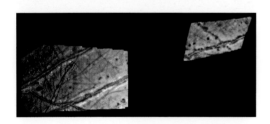

Figure 5.3.6 Two NIMS spectral maps of large global-scale dark-flanked bands on Europa. NIMS spectra (shown here merged with the global image mosaic) confirm that bright materials are usually ice-rich, but also show that dark regions include high concentrations of hydrated salts or sulfate minerals (shown in orange here), possibly related to the composition of the deep crust or even the ocean. These include long dark bands and some smaller spots. Area shown corresponds to northwest corner of Plate Je8.

diverse bodies ranging from Earth to Mars, Enceladus, and even Miranda. Polar wander may be a common event throughout the Solar System. Like plate motions, this mechanism is possible only if the icy shell is decoupled from the rocky interior, mostly likely by a liquid-water layer.

Other geologic processes have marred the surface of Europa. Fields of dark "freckles" appear to be diapirs breaking through the surface. At high resolution, these freckles come in a range of shapes: dark ovoid-shaped features 10 to 20 kilometers across and larger amoeboid patches called "chaos" up to 150 kilometers across (Figure 5.3.5). These features are often domed and usually have crinkly, severely disrupted surfaces. Chaos often include large intact blocks of ridged plains embedded within them. Despite appearances, chaos are probably not refrozen holes in the ice shell, but have the classic expression of salt dome diapirs on Earth, formed by rising mushroom-shaped blobs of deep crustal ice. What makes these features interesting is that they involve slow overturn of Europa's icy shell and may have uplifted and exposed icy or watery material from deep within the ice shell, which we can measure and perhaps some day sample.

Galileo also got a close look at some of Europa's impact craters. What is most interesting about Europa's craters in how they differ from those on Ganymede and Callisto. Craters should be very similar on all three satellites, given their icy compositions. As shown here, large craters like Pwyll (Plate Je14.1) are shallow and irregularly shaped, very different from similar-sized craters on the other two icy satellites. Measurements and numerical models of these shapes (by the author, Zibi Turtle, Boris Ivanov and others) indicate that the icy shell is much weaker and thinner on Europa than on Ganymede or Callisto, probably on the order of 10 to 20 kilometers thick. The largest-known crater, Tyre (Plate Je3.1), is only 38 kilometers across but it is possible that it nearly punched through Europa's icy shell to the ocean beneath.

Key to understanding this ocean and its biological potential is its composition. We have no samples of Europa, but fractures, diapirs and large impact craters can all potentially bring bits of ocean material to the surface. While mostly water ice, dark reddish material is also widely distributed on Europa. Our best estimate is that this material is hydrated sulfuric acid plus hydrated magnesium or sodium sulfate salts (Figure 5.3.6). Small amounts of carbon dioxide, sulfur dioxide, hydrogen peroxide and molecular oxygen and chain sulfur have also been identified, but some (or all) of these are alteration products or have been imported from Io. Which of these compounds relate to possible ocean composition remains to be seen, but the question is central to understanding how the ocean is evolving and whether it can support organic carbon-rich chemistry. Going back to ice-shrouded Europa remains a highest priority (Figure 5.3.7).

5.4 Io

Io was once called a pizza planet, but that does an injustice to the sweeping watercolor pastels of yellow, orange, and red that can give Io the intensity of an abstract volcanic sunrise. Io stands apart from its three sibling satellites. Far from being locked in a perpetual ice age, Io is a Hadean world of hot volcanic lava flows, fiery lava fountains, and noxious gases. Io's intense global-scale volcanic activity is rivaled only by the Earth and perhaps Venus in the amount of heat released through volcanism. *Voyager*, *Galileo*, and most recently *New Horizons* observed Io at more than two dozen distinct opportunities over a period of 28 years, and each time the surface of Io had changed in some way. Earth-based observations have filled some of the time gaps but at low resolution. Some of the more dramatic of these changes are documented in this *Atlas*.

In mythology, Io was another of Jupiter's (Zeus's) conquests, whom he turned into a white heifer to trick his jealous wife Hera (Juno). In her wanderings she found Prometheus, tied to a tree as punishment for rashly giving humans knowledge of fire (Figure 5.4.1). This is an interesting poetic turn considering that Simon Marius (or was it Kepler?) knew nothing of Io's fiery surface when he suggested the name Io. She was eventually restored and married King Telegonis (Plate Ji12.1).

For more than 300 hundred years, Io remained a small bright orb in the sky. The orbital resonances with Europa and Ganymede were discovered in the late 1700s but their importance went unrealized for another 200 years. Io's odd yellowish color was noticed at the turn of the twentieth century, and keen telescopic observers had by then noticed reddish "polar caps" on Io, seen in many of these plates.

The few details that emerged over subsequent decades only deepened the mystery. The last half of the twentieth century revealed that Io was "dry," devoid of water ice seen on the other three satellites. A cloud of sodium ions (followed later by hydrogen, potassium, sulfur and oxygen)

Figure 5.3.7 Perspective view of a prominent double ridge, nicknamed "The Great Wall of Europa." The undulating topography of Europa's plains is evident. This three-dimensional visualization is based on unpublished topographic data produced by the author.

Figure 5.4.1 Prometheus in happier times. Prometheus brought fire to mankind, but he might have done better to steal it from Io than from Olympus. "Prometheus Recumbent," by Paul Manship, Rockefeller Plaza, New York. Photo by author.

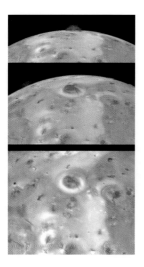

Figure 5.4.2 *Voyager*'s famous Prometheus plume sequence, showing three successive views of the 140-kilometer-high plume arcing above the surface, creating a circular ring around the volcano. The sequence shows that some arcs within the plume are dense with dust, others transparent. A second plume (at Maui) is also visible in the middle panel.

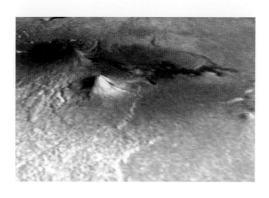

Figure 5.4.3 Perspective view of two shield volcanoes on Io. The reddish coated volcano in the background is Zamama Mons. The dark flows extending to the east (right in this view) are the source of a thermal hotspot and occasional plume. The volcanic cone in the foreground has apparently been dormant for some time. Both volcanoes are covered by numerous thin lava flows and rise 2 to 3 kilometers high. A small shield volcano is visible at far right. Zamama is unusual; most volcanoes on Io have little relief. See also Plate Ji8.4.

was discovered in Io's orbit, as was the timing of Jupiter's radio emissions to Io's orbital period. Sporadic reports of high temperatures on the surface were not taken very seriously, however. The *Pioneers*' modest glimpse of Io in 1973/1974 (Figure 1.2) provided few new insights, including a radio occultation of the ultra-thin atmosphere. One wonders what might have been revealed had *Pioneer* risked a much closer pass. Perhaps the most interesting concept of the time was that a global ocean had dried up, leaving only extensive evaporite salt beds, not unlike the Bonneville Salt Flats of Utah. Io remained a peculiarity as *Voyager* made its approach in early 1979, but few guessed at the surprise that awaited the spacecraft.

At the same time, Stan Peale, together with colleagues Cassen and Reynolds, were looking at the problem of Io's orbital resonance with Ganymede and Europa. They calculated that the eccentricity forced on Io's orbit by Europa and Ganymede raises surface tides so strong that they could melt large parts of the interior and trigger volcanism (as discussed in Chapter 1). Their prediction was published a week before the March 5th encounter.

Voyager changed our perception of Io, and with it the Solar System itself. As encounter week began, it was becoming clear that Moon-like Io was anything but Moon-like. There were dark and bright spots on the surface of all sizes, to be sure, but also no impact craters. The strange irregular markings began to look a lot like lava flows. As *Voyager* left Io behind, large umbrella-shaped plumes of gas and dust were seen arcing into space above the surface in global images (Figure 5.4.2). Prediction confirmed.

Volcanism on Io has a variety of flavors (Figure 5.4.3). Dark lava-flow fields, deep lava-covered calderas (called paterae), high plumes of gas and dust, and diffuse bright or dark deposits covering hundreds of square miles continually change the appearance of Io. The utter lack of impact craters on Io is consistent with very rapid volcanic resurfacing rates as high as a few centimeters per year (averaged globally).

Surprisingly little is known about Io's lavas. Sulfur compounds mask most of the surface, hiding much. Molecular sulfur is abundant and gives us the extensive yellow plains. Short-chain sulfurs are found in red deposits, whereas sulfur dioxide, SO_2, forms whitish deposits. SO_2 gives off a powerful rotten egg odor and makes Io the stinkiest place in the Solar System, if one could breathe there. A few volcanoes (Emakong Patera, for example, Plate Ji7.1) have had consistently lower temperatures of only a few hundred degrees, perhaps consistent with eruption of liquid sulfur, which may form reservoirs under the surface. Minimum temperatures for some volcanoes are in the range 1200 to 1600K, however, consistent with eruption of low-silica high-melting-point silicate lavas, perhaps similar to Hawai'ian or Icelandic basalts. (Temperature estimates from NIMS, SSI or Earth-based telescopes reported here require modeling and could be off by as much as a few hundred degrees.) Very high temperature peridotites or ultramafic komatiites are also possible in some cases.

At least 16 (and possibly 25) plumes have been observed on Io, and come in two types, dominated by either sulfur or sulfur dioxide. Most are intermittent, but some, such as Prometheus and Zamama, have been active since *Voyager*. Of Io's 500 plus volcanoes, at least 150 of them are "active" thermally, glowing fiercely in *Galileo*'s infrared-sensitive cameras (Figure 5.4.4). At least a few dozen are erupting at any given time. Some sites are in near-continuous eruption, many for only a few weeks or months. In fast-forward, Io would look like fireflies on a hot Louisiana summer's evening.

One consequence of *Galileo*'s malfunction was that volcano monitoring was not as consistent or complete as it could have been. During any given orbit, *Galileo* could image Io only a few times, and seldom over the entire surface. Hence, *Galileo* seldom witnessed major eruptions in progress or at high resolution (Tvashtar and Pillan, Plates Ji3.2 and Ji9.2, are prominent exceptions). *New Horizons* added immeasurably to our view by providing a complete view of Io in February 2007 to compare to our earlier piecemeal *Galileo* and *Voyager* views (a span of 28 years), including excellent views of the ongoing Tvashtar plume (Figure 5.4.5).

Underneath all this, the mantle of Io is likely partially molten and convecting. Although a large iron core was detected comprising ~20% of the total mass (Figure 5.4.6), *Galileo* was not able to detect a magnetic field, other than that induced directly by Jupiter's own powerful magnetic field. Apparently the core itself is not convecting or is molten. This curious absence is compounded by the existence of a magnetic field within Ganymede, a moon that has not been geologically active for at least a billion years.

It might seem strange to talk of Io as a tectonic planet, given the pervasive volcanism, but Io's 150 or so mountains provide an important

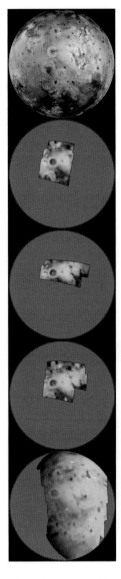

Figure 5.4.4 Io's volcanoes glow in the infrared like fireflies and *Galileo* was able to monitor these with some regularity. Its best views were of the anti-Jovian hemisphere dominated by Prometheus, the ring-shaped deposit left of center in the top view, an SSI color mosaic. *Galileo* NIMS looked at this region during the I24, I25, I27 and I32 orbits (second from top to bottom), spanning a two-year period from October 1999 to October 2001. The Prometheus, Culann, and Tupan hotspots near center glow reddish in these false-color renderings, showing Io at 1.31, 2.44, and 3.85 microns. Most of the thermal power is at 3.85 microns. Many of the hotspots are remarkably consistent from view to view. Subtle color variations in the surrounding plains are probably related to sulfur dioxide composition. All images are in orthographic projection centered on 15°S, 145°W. NIMS data are at 10 to 13 kilometers resolution.

Figure 5.4.5 Nearly simultaneous *New Horizons* imaging of Io, February 2007. LEISA image (left) of thermal emission shows active hotspots. MVIC color image (center) shows surface markings and a bluish-colored plume and reddish hotspot at top. LORRI image (right) shows details of a partially sunlit plume and tiny hotspot at top and lesser plumes elsewhere. The large plume and brightest hotspot are the Tvashtar eruption site (see Plate Ji3.2). Credit: NASA/Johns Hopkins University Applied Physics Laboratory/Southwest Research Institute.

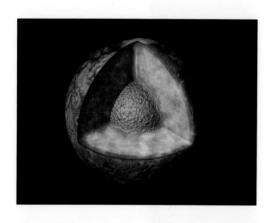

Figure 5.4.6 Cutaway view of likely interior structure of Io. Surrounding an iron core is a rocky mantle. The rigid outer lithosphere is probably 20 to 30 kilometers thick. Although the entire mantle could be involved, the upper part of the mantle is likely to be at least partly molten, as shown in orange. Modified by the author from a NASA/JPL diagram.

key to how Io works. Io's mountains are among the highest and steepest in the Solar System (Figure 5.4.7), much higher than anything on Europa, Ganymede, or Callisto. An average mountain rises 6 kilometers above the plains; the highest, Boösaule Montes (Figure 5.4.8), stands 17 kilometers high. Only a few resemble classic volcanoes (see Plates Ji8.4 and Ji14). Most appear to be mesas, rugged promontories, or large massifs formed by tectonic processes. Some are slowly deforming under gravity by surface creep, or are flanked by massive landslides or slumped blocks, such as at Euboea Montes (Plate Ji14) or Telegonis Mons (Plate Ji12.2).

While they may provide stunning vistas of surrounding country, Io's mountains are scattered and lonely promontories and do not form a global pattern like Earth's mountains. There are no Icelandic rift valleys, San Andreas Faults, Andes Mountains, or any of the tectonic zones that on Earth form the global pattern of plate tectonics. Plate tectonics involves the lateral movement of large crustal plates that pull apart, slide, and collide with each other due to mantle convection, creating volcanoes and earthquakes and recycling the crust back into the interior. Not so on Io.

The morphology and height of these mountains implies that compressive forces pushed these large blocks of crust above the surface seemingly out of nowhere, but why? Looking at landslides on Io in 1998 with Mark Bulmer, I developed a model that explains how global compression and mountain uplift are direct consequences of Io's

Figure 5.4.7 *Galileo*'s oblique view of the center of Tohil Mons (Plate Ji12.3). The steepest promontories rise 8 to 9.5 kilometers above the plains. Small discrete bright patches may be sulfur dioxide frosts. The dark-floored depression near center is Radagast Patera, which has cut steep cliffs into the flanks of the foreground range. The darkest flows within Radagast Patera are thermally hot.

ubiquitous volcanism. Over the millennia, Io's global volcanoes are continually resurfacing the surface. There is no place for older lavas to go as they are buried, except down (where they are eventually remelted). Think of Io as an anti-onion that grows from the outside in. Io's crust can be thought of as concentric shells of lava flows that are increasingly older with depth. As these older shells are forced downward, their radius decreases. The layers must shrink and compress, something they don't like to do. Eventually the crust ruptures along faults at random spots and large blocks of crust are forced upward, forming mountains. This cycle is ongoing as long as volcanism continues; Io's mountains will cease to form when volcanism does.

This volcanism-induced "spherical tectonics" may be unique in the Solar System. The story is certainly overly simplistic, but it serves as a framework for understanding much of what is going on inside Io and relates mountain and volcano formation. As seen in several locations in this *Atlas*, some volcanic features are located near mountains. Thrust faulting may occur near volcanic vents, or faults may provide conduits

Figure 5.4.8 The top of 17-kilometer-high Boösaule Montes pokes above Io's limb, backlit by the cold cloud decks of Jupiter. Limb profiling of this type is a common tool for identifying mountains, as are stereo and terminator images. Resolution of these images is ∼1.5 kilometers. Speckle on the left side of the mosaic is an example of radiation noise in the imaging system. The irregular dark feature in the foreground is the Pillan lava flows. The dark diffuse feature near the limb is the center of Pele.

for magma to ascend to the surface. *Galileo* looked at these inner workings but the radiation environment, antenna failure, and "antique" 1970s instrumentation conspired to severely limit *Galileo*'s peek. A return to Io with focused instruments and vastly improved data rates should be of the highest priority.

6
One big happy . . .

6.1 Why explore Jupiter?

The sheer complexity and dynamics of the Jupiter system set it apart. The planet itself contains ~71% of all the planetary mass in our Solar System. The dynamic, constantly changing, storm-tossed, and lightning-scorched atmosphere belies a turbulent interior. A vast radiation-charged magnetosphere spreads over a domain as large as the full Moon, as seen from Earth, and is the largest structure in the Solar System other than the Sun. A small but dynamic ring system points to a complex dusty environment close in. Jupiter also controls the dynamical state of the Asteroid Belt and small comets in the Inner Solar System. At least 63 moons swarm about the planet (I remember when only 10 were known!), most of which are captured asteroids or dead comets, but there is also a population of small comets in temporary orbit, one of which, Shoemaker-Levy 9, slammed into the giant planet in 1994 with much fanfare.

It is the four large planet-like moons that grab our attention here. Io and Europa are the stars, due in good measure to their very dynamic geologic histories, but as this *Atlas* attests Ganymede and Callisto are almost equally compelling, if not for youthful vigor. All feature an unearthly beauty that transcends their rich geologic histories. The Galilean satellites have revealed several fundamental truths. Four hundred years ago they tipped the balance in favor of the Sun-centered planetary system we now know to be true. In our own time they have demonstrated the importance of position and orbital dynamics in controlling the thermal history and fate of bodies in complex planetary systems. Europa, especially, has shown that potentially habitable environments might be found in diverse places far removed from the Sun's warming rays.

The four planet-like Galilean satellites are each glorious in isolation, but patterns emerge when we examine the lot. From Io out, satellite densities decrease, indicating increasing ice content by mass (see Appendix 4). From volcanic Io to resurfaced Europa and Ganymede to ancient Callisto

Figure 6.1 The four Galilean satellites at four scales; from left: Io, Europa, Ganymede, Callisto. These *Galileo* views show how the surface appearance changes as we zoom in on selected features at progressively higher resolution. Each row down represents an increase in resolution of roughly a factor of ten: the first two rows are at 10 kilometers, increasing to ~1 kilometer, ~100 meters, and finally ~10 meters (bottom). Top row shows our current best understanding of satellite interiors.

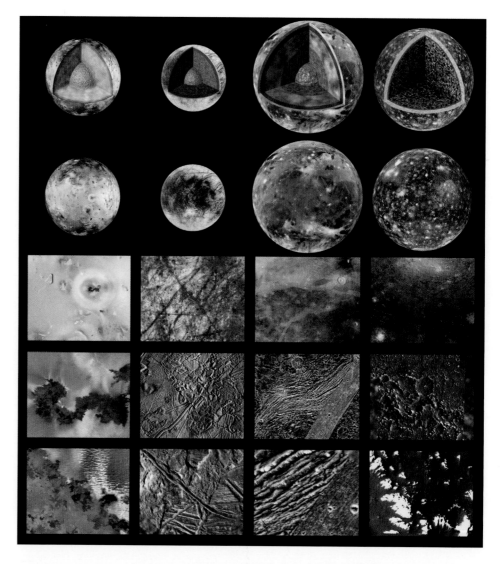

(Figure 6.1), the apparent surface age increases and the intensity of deformation also decreases outward. The satellites share some common geologic traits. Ganymede shares remnants of ancient cratered plains with Callisto and intensely deformed tectonic patterns with Europa. Yet Ganymede is not a hybrid Europa/Callisto. The desiccation of dark terrain on Callisto is not evident on Ganymede. Conversely, the double ridges, diapirs, and dilational bands of Europa are not to be seen on Ganymede, and the calderas of Ganymede are unknown on Europa.

The four satellites formed in the same Jovian planetary nursery and share a common compositional heritage. True, the rock-to-ice ratio changes with distance and other details with it. The key link binding the four siblings is location. The orbital resonance and tidal deformation driven by Jupiter's gravity are fundamental to their histories. Oscillating global daily tides, tens of meters high, power melting and volcanism on rocky Io and keep most of the subsurface water layers of Europa molten. They might have also provided one or more brief but intense periods of deformation on Ganymede. Alas poor Callisto. Without this mechanism it seems that these satellites might be a little less interesting than they are now.

With its active volcanic plumes, Io may command the stage, but Europa and its thick layer of water and ice stands apart. As a potential habitat for life, only Mars rivals Europa. No individual geologic feature or set of features demand it; rather it is the collective weight of evidence that demonstrates the probable existence of an ocean beneath the cold, brittle, icy surface. The most convincing evidence is the response of Europa's interior to Jupiter's magnetic field, indicating a conductive layer near the surface. Tidal heating is sufficient to keep a large portion of this water layer liquid, with the topmost layer frozen under the cold vacuum of space.

The thickness of this outer ice shell is a matter of intense and sometimes pointlessly acrimonious debate, one that is otherwise of some importance scientifically but potentially distracting from equally important issues. The weight of credible geologic evidence indicates the ice shell may be roughly 10 to 20 kilometers thick and does not allow sunlight to reach the water below, but the indirect tools we have from image analysis are imprecise. Geophysical models also suggest that the shell thickness varies considerably over time and may have been substantially thinner in the past. Since the formation of some features, such as diapirs, is related to shell thickness, global geologic mapping may well help determine the history of the shell and perhaps its thickness history as well. The fact remains that we do not yet know the exact thickness of the shell today, or in the past.

Speculation about the composition and biological potential of Europa's ocean, on the other hand, is unconstrained by data. If organic biology ever did or were ever to develop in the ocean, the thickness of the shell could control the form it takes. A thin shell (say only a few kilometers) could be easily disrupted, exposing liquid water to space and sunlight. Of course, just because this is an exciting possibility doesn't make it so. A thick shell (>10 kilometers thick) would effectively restrict what we call "ocean-surface exchange." A tethered submarine melting its way through the icy shell to cruise the ocean would have considerable difficulty if the shell were 10 or more kilometers thick. A thick shell would also exclude exposure of ocean waters to sunlight (and also to intense radiation) and prevent photosynthesis.

Critters require food and the question of the composition of Europa's ocean rises to the fore. Evidence for the ocean is robust but our information is indirect. We have no samples of the ocean. Although likely too thick to allow direct exposure of the ocean, geologic processes within the ice shell (such as fractures, diapirs, and large impact) may bring entrained bits of this ocean to the surface for us to sample. Europa's surface may include hydrated sulfates or salts, but is poorly sampled by *Galileo* and may have been altered even if it is related to ocean composition. Hydrothermal activity on the ocean floor is the most efficient way we know to alter ocean chemistry and supply nutrients – or caustic reagents – to extant life forms. Critters survive quite well on the sunless floor of Earth's ocean, but what is not known is whether they originated there in the dark long ago or invaded from the well-lit ocean surface. It's an important question. These critters survive because hydrothermal activity on the ocean floor allows them to use sulfur compounds through chemosynthesis. Whether hydrothermal or volcanic activity occurs on the floor of Europa's ocean is completely conjectural, but the presence of abundant sulfur on Io allows at least the possibility of this mechanism on Europa. Such arguments are of course pointless unless biology has gotten a start first on Europa.

We cannot focus on Europa in isolation, however. It is part of a complex planetary system and the evolution of the other satellites will bear directly on our understanding of Europa. The tides that heat Europa's ocean are directly evidenced on Io. Io's composition may hold clues to Europa's rocky interior. Ganymede's magnetic field, resurfacing, tidal and thermal history may help explain how and when it all started. Callisto's quiet history may have preserved clues to the composition and very formation of these satellites. Fortunately for us, the difficulties of getting to Europa require using their gravity to reduce the needed fuel to enter Europa orbit. A return to Europa should and most certainly will allow us to remap and closely study these other satellites.

6.2 The future

Post-*Galileo* exploration of the Jupiter system is coming in short infrequent bursts. Telescopes are more powerful each year, but observation time is

Figure 6.2 Best views of the four Galilean satellites from *New Horizons*, February 2007. Images acquired as the spacecraft passed on its way to Pluto. Surface resolutions range from 15 to 20 kilometers. Credit: NASA/Johns Hopkins University Applied Physics Laboratory/Southwest Research Institute.

at a premium. *Cassini* passed Jupiter in 2000 on its way to Saturn, but at 10 million kilometers away, it only acquired some new low-resolution thermal maps of Io and composition data of the satellites. *New Horizons* passed Jupiter for a few days in February 2007, close enough to monitor volcanoes and map recent changes on Io (see Figure 5.4.5), and peek at the other satellites (Figure 6.2). *Juno* will orbit Jupiter in the next decade, but its camera is not designed for satellite mapping. Encounters of this type add new knowledge but do not answer our more fundamental questions.

As of 2009, planning for future exploration of the Jupiter system focuses strongly on Europa and its potential subsurface habitat. This is rightly so, but Europa is not alone. *Voyager* showed us that the Outer Solar System is full of complex icy planetary bodies and beats Mars hands down for diversity and potential for habitability. After all, no liquid water has yet been found on Mars (though there was plenty 3 to 4 billion years ago). Io, Ganymede, and Callisto all have planetary attributes, as shown here in abundance. Triton, orbiting Neptune, is similar in size to Europa, has a geologically complex surface that may be younger than Europa's (Figure 6.3) and likely has an internal ocean of its own. Its atmospheric geysers may be related to internal heat, but Triton is a remote and difficult target and has been deliberately and unfairly demoted in importance because of this.

Figure 6.3 Perspective view of Triton's pitted volcanic plains. An 80-kilometer-wide caldera is visible in the upper-left corner. Rendering based on topographic data produced by the author.

The new discoveries at Saturn by *Cassini* are equally compelling. Small icy moons are normally cold and currently inactive but Enceladus, barely larger across than the state of Louisiana, is being resurfaced and is ejecting water ice and other particles into space. The source of this vapor is a mystery: molten liquid water beneath the surface is a possibility, but by

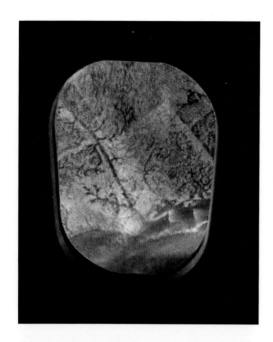

Figure 6.4 Huygens descent probe mosaic of Titan's channel networks, shown as if seen from some future winged aircraft on routine trans-Titan flight.

no means certain. Aside from Mars, Saturn's Callisto-sized moon Titan is the most Earth-like surface in the Solar System. Both Earth and Titan have thick atmospheres and see precipitation on the surface (perhaps, in the case of Titan), forming channel networks (Figure 6.4). On Titan, the rain is not water but ethane and methane. Saturn's only big moon, Titan is a sibling of Ganymede and Callisto, albeit from a different branch of the family tree. Its composition is therefore somewhat different, probably richer in the volatile ice phases. Titan's surface is awash in hydrocarbons, derived mostly from reactions of methane high in the dense atmosphere, but if Titan has a liquid ocean it is not near the surface like Europa's, and there may be no habitable zones. The link between the interior and atmosphere must be important, but evidence for geologic activity on Titan is cryptic at best. Titan's organic chemistry and precipitation cycle make it an important target for exploration and the atmosphere makes it easier to get to the surface.

What makes Europa so compelling amidst this family of fascinating moons? Many of the large icy worlds, including Ganymede and Callisto, and possibly Enceladus, Titan, and Triton (and perhaps Pluto) have oceans of liquid water deep inside. Only on Europa (and possibly Enceladus) is this water ocean near the surface and in direct contact with a geochemically complex (and perhaps warm) rocky interior. This potentially makes Europa's ocean chemistry more interesting. Its proximity to the surface also makes it possible to explore this ocean, at least indirectly through subsidiary geologic processes. Dark reddish areas such as Castalia Macula (Plate Je9.2) or Tyre (Plate Je3.1) have been put forth as sites where volcanism, diapirism or impact may have transferred oceanic – or lower crustal material – to the surface.

While *Voyager* and *Galileo* exposed the naked truth of these satellites, fundamental questions remain. What is the composition of Io's lavas? What roles do sulfur and other volatiles play in volcanism? Where do Io's sodium and potassium "clouds" come from? Is resurfacing truly global or restricted to caldera floors? How do Io's mountains relate to volcanism? Is Io's volcanism (and, by extension, tidal heating) variable or episodic? What are the mechanisms, intensity, and locations of tidal heating and melting inside Io (and Europa)? Is Io's mantle convecting and how much of it is molten? Have there been multiple episodes of nonsynchronous rotation or polar wander on Europa? How thick is the outer ice shell and is shell thickness (and geology) variable over time? What is the composition of the ice shell and the ocean and do either harbor organic molecules? Is there silicate volcanism on Europa's rocky core (that is, at the bottom of Europa's ocean)? Has life (or building blocks of life) ever evolved in the ocean? Is Europa geologically active today? Has volcanism ever occurred in ancient dark terrains of Ganymede or Callisto? What was the sequence of bright terrain formation? Has there ever been and, if so, where is the compressional deformation on Ganymede or Europa? How much did Ganymede expand? Did Ganymede ever rotate nonsynchronously (either rotationally or by polar wander) and for how long? When did resurfacing occur? Are bright terrain resurfacing, viscous relaxation, and possibly nonsynchronous rotation all linked with the history of the orbital resonance between Ganymede, Europa and Io? Why does Ganymede have a magnetic field, and Europa and Io do not? Why did Ganymede differentiate into a core and mantle and Callisto not? Why is Callisto's surface so sharply segregated into dark and bright materials and does this segregation occur at all latitudes and longitudes? What is dark material and is it primordial? Is Callisto's quiet geologic history related to orbital dynamics or a consequence of its original composition?

Thanks to its engineers, *Galileo*'s critical high-resolution data set achieved the important goal of revealing the true nature of the surfaces and something of the composition and interiors of these bodies. But, as cursory examination of the mapping quadrangles shows, *Galileo* did not achieve anything like the mapping success we now have for Venus, Mars and the Moon, where global imaging and topographic mapping at 100- to 300-meter scales has revolutionized our understanding of those planets in the past two decades. Only a few narrow longitudinal strips and a few isolated mosaics remain of what should have been near-global mapping coverage at 200- to 300-meter resolutions (see Appendix 3). Much of the lost *Galileo* imaging data would also have been in stereo, suitable for topographic mapping of 50 to 75% of the surface, instead of only 3 to 5% or so now possible. The near-infrared mapping camera (NIMS) designed to map surface composition suffered a similar loss. *The total volume of calibrated images from* Voyager *and* Galileo *combined for all four satellites fills less than one DVD*! Compare this to the hundreds of DVDs required to store the raw data files from the various imaging systems on the current Mars orbital space fleet. No other fact demonstrates the importance of a new mapping mission to the Galilean satellites. In practical terms, this *Atlas* should have been at 200-meter resolution or better, not 1 kilometer.

Figure 6.5 The author ice diving in June Lake, Sierra Nevada Mountains, 1990, training for possible future explorations of Europa. Those were cold fingers!

This is simply not sufficient to map the geologic history of any planetary bodies, and only a new mission to Jupiter will give us this.

Thanks to *Galileo* we can make educated decisions regarding which instruments to send to Jupiter and its moons. Global topographic, compositional, gravity field, magnetic field, subsurface, and high-resolution mapping of Europa (and similar mapping of the other moons) is critical. Measurement of the daily "tide" and libration (oscillation) of the surface as Europa completes its days in orbit around Jupiter, may answer the question of ice shell thickness. Advanced high-resolution orbiting spectrometers should help determine surface composition in areas where ocean or deep crustal material has been exposed; ultimately a lander may be required. Surface-penetrating radar should reveal exciting details of the workings of the ice shell and how the ocean communicates with the surface, but is not guaranteed to penetrate to an ocean, especially if it is deep. It should determine whether or not it is only a few kilometers thin, however. Global mapping will be critical for mapping out the evolution of the ice shell and selecting future landing sites.

A habitable ocean of liquid water: that is an irresistible target for human intelligence and imagination to swim about in. As a trained diver (Figure 6.5), I am ready to explore Europa's ocean, but for now the technology for direct exploration of this ocean is beyond our reach. A return to Jupiter with a Europa orbiter to map its geology and geophysics is the next logical step. The difficulty is in getting to Europa. The energy required to enter both Jupiter and Europa orbit is large (Europa has no atmosphere to provide free braking power, as does Mars or Titan). These difficulties do not make Europa less worth visiting. Surviving in the extreme Jovian radiation environment is also difficult, but electronics and instruments are more robust than they were when *Galileo* was built, which survived there for seven years. In February of 2009, NASA selected a Flagship mission to Europa as its next major priority. We can design and fund robust spacecraft to return to Europa and the Jupiter system today, it requires only the will to do so.

Atlas of the
Galilean Satellites

Callisto

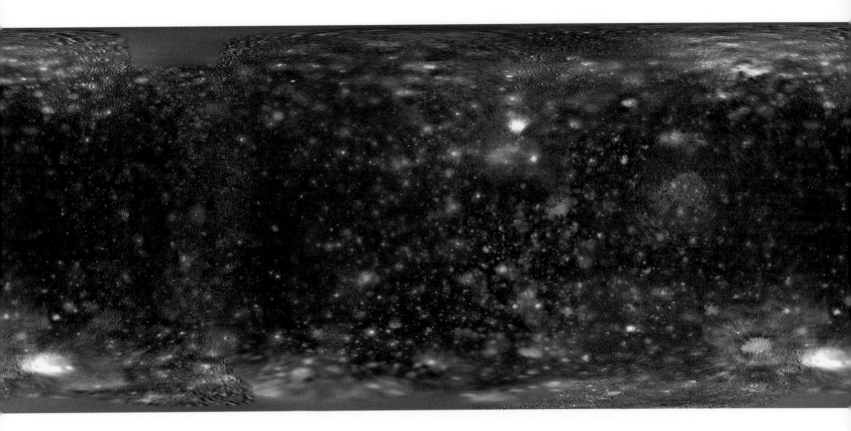

Plate Jc Global Map of Callisto
Simple Cylindrical Projection
Map Scale: 1 cm = 627 km
(valid at equator and along longitude lines)

Orthographic Projection
Center 15°N, 55°W

Plate Jc01
Global View 1: Valhalla

This view is centered on the Valhalla multi-ring impact system, the largest such feature and the dominant geologic feature on Callisto. At its widest, the ring system extends across 4000 kilometers of terrain. Several bright polar features are evident, including the bright crater Lofn to the far south and several bright-ray craters to the north. The bright patch to the west is likely an ancient impact scar.

Orthographic Projection
Center 55°S, 10°W

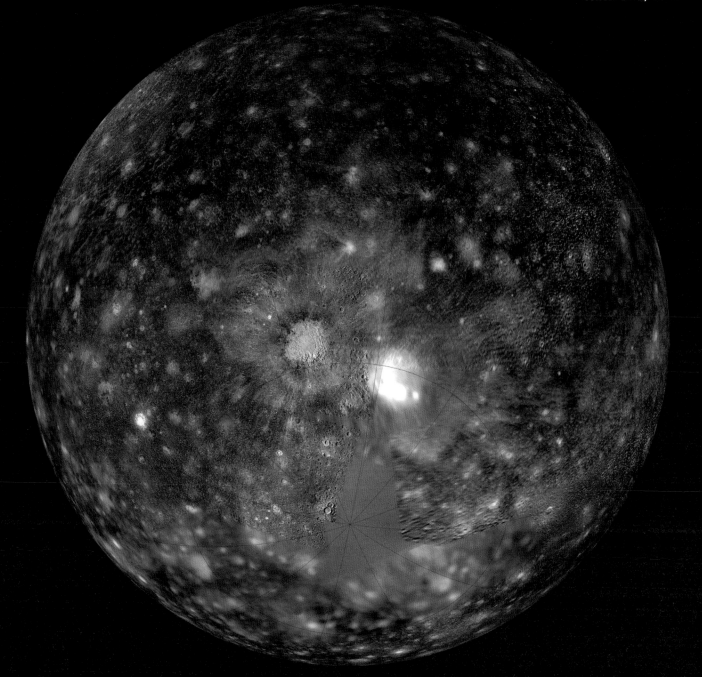

Plate Jc02
Global View 2: Large Impact Basins and South Polar Region

The south polar region is one of the most geologically complex on Callisto. There is an unusual clustering of large impact basins of increasing ages, including (from right to left) Heimdall, Lofn and Adlinda (see Plates Jc11 and Jc14). Areas south of 60 degrees latitude or so are brighter on average than equatorial terrains. Perhaps the segregation of dark non-icy and bright icy materials that occurs elsewhere on Callisto (for example, Plates Jc6.3 and Jc7.1) does not occur so extensively in colder polar regions. Bright frosts have been identified on pole-facing slopes on Callisto, similar to deposits on Ganymede (Plate Jg3.2). We also see several bright and dark polygonal patches hundreds of kilometers across. These are of unknown origin but some could be ancient impact scars, known as palimpsests. Parts of the Valhalla multi-ring system are on the limb to the upper right. The bland area at the pole has not been imaged.

Orthographic Projection
Center 48°N, 170°W

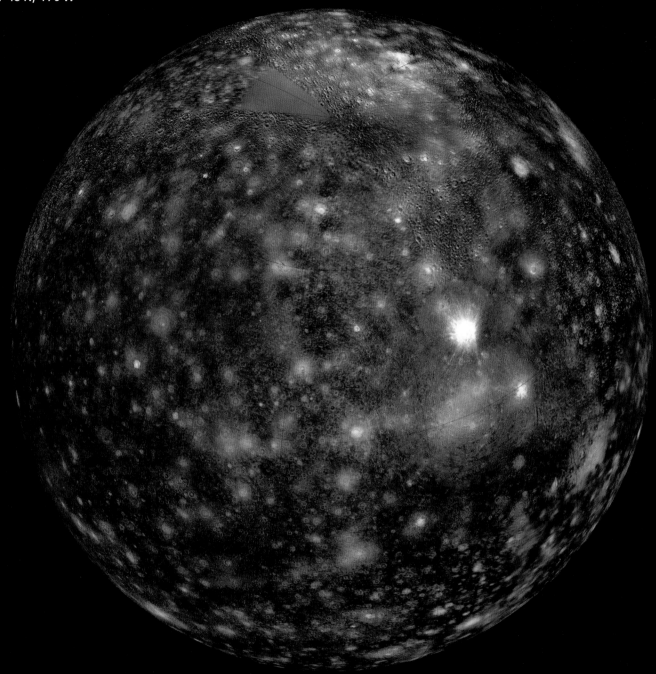

Plate Jc03
Global View 3: Asgard and North Polar Region

Although dominated by cratered plains, the large Asgard multi-ring structure at middle right and smaller impact features superposed on it stand out. The bright features of the north polar region are also apparent. Although Callisto does not seem to have bright polar caps like Ganymede, there does appear to be some enhancement of bright features in polar areas on Callisto. The small blank area at the pole has not been imaged.

Orthographic Projection
Center 0°N, 90°W

Plate JcHL
Global View: Leading Hemisphere

Most of the large multi-ring impact basins that characterize Callisto formed on this hemisphere. Numerous large bright patches are probably ancient impact scars, partially obscured by smaller younger bright and dark impact craters. Most of the crater chains formed by disrupted comets, which strike Callisto while outbound after a close passage to Jupiter, also formed on this hemisphere. This view, as with the global mosaic (Plate Jc) and the trailing hemisphere view, use the partial global color data shown in Chapter 3.4. Gaps in color coverage are filled with an averaged color. Color variations are subtle in these data.

Orthographic Projection
Center 0°N, 270°W

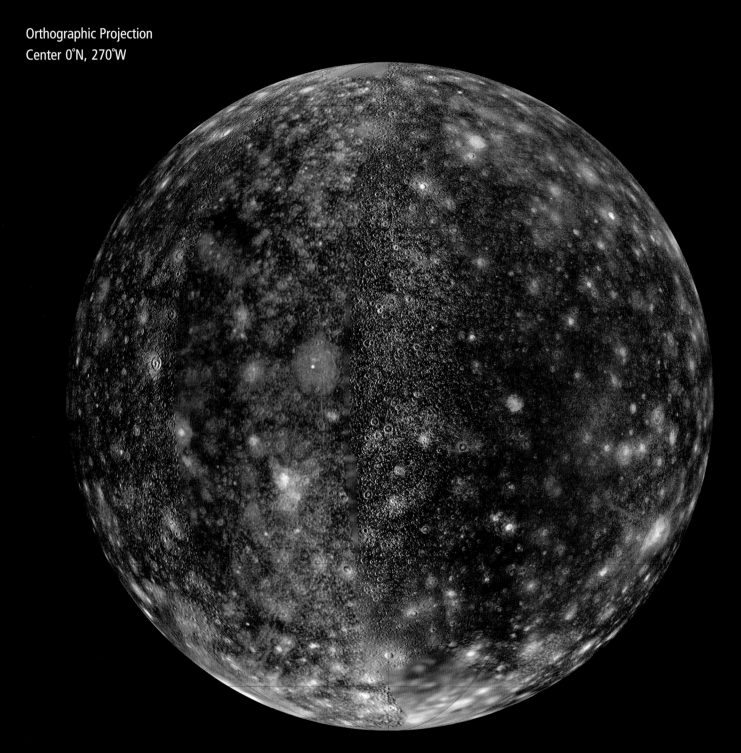

Plate JcHT
Global View: Trailing Hemisphere

This hemisphere appears heavily cratered, but less geologically variable than the leading hemisphere. Notable is the absence of large multi-ring impact basins and circular bright patches typical of those on the leading hemisphere. This is the only indication of a possible ancient cratering asymmetry on Callisto similar to that observed on Ganymede. There are highly degraded basins here, but they are few and very ancient. The south polar region appears somewhat brighter than the rest of the satellite, and dark and bright patterns are of unknown origin. In general, however, albedo variations are subdued compared with the leading hemisphere.

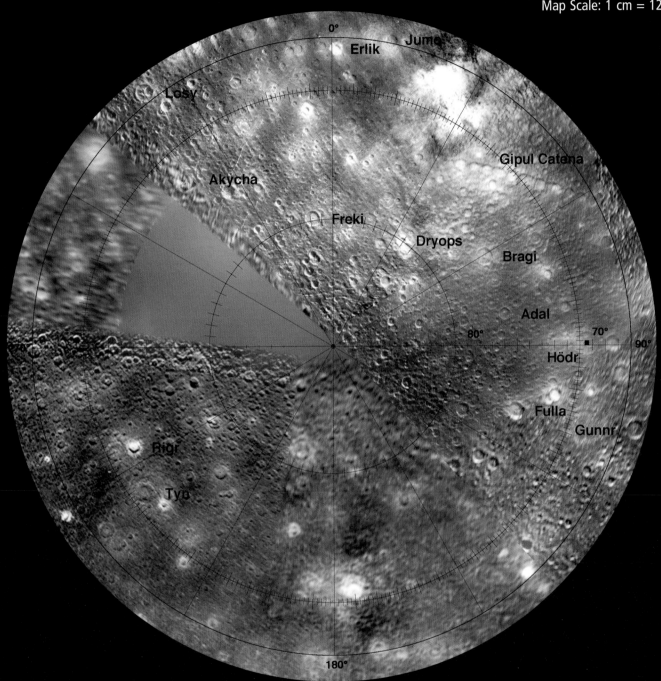

Plate Jc1 Gipul Catena Quadrangle

One of the highlights of this quadrangle is Gipul Catena, a 470-kilometer-long linear chain of craters. This and other similar chains are too large to be secondary craters formed by ejecta from large impact basins elsewhere on Callisto. Rather, this and similar chains (or catena) were formed by the impact of disrupted comets, broken into long chains of debris after close passage with Jupiter (see also Figure 5.1.5, and Plates Jc2.1 and Jg2.3). Most of these comets escape Jupiter, but over the eons a few have hit the Galilean satellites on their way out, forming long linear crater chains like Gipul Catena. The large bright patch nearby is a relatively young large bright impact feature. Although bright, the crater rim and structure are poorly defined.

There are few tectonic structures on Callisto, but we see here the highest concentration of such features. At least six narrow rimmed fractures can be identified in the upper left quadrant, all radiating from a point near 78°N, 285°W. Alas, this point of origin lies right in the center of the only unimaged part of Gipul Catena quadrangle! Most linear structures of this type on Callisto are associated with specific large impact craters (see Plates Jc3.1 and Jc8.3), but this system is by far the most extensive, suggesting that the crater at the center must be rather large, perhaps greater than 150 kilometers.

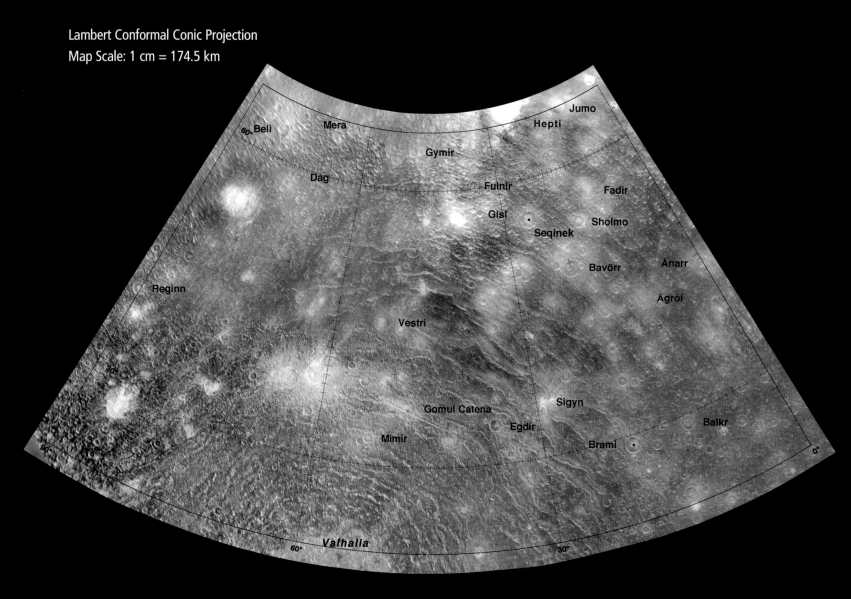

Lambert Conformal Conic Projection
Map Scale: 1 cm = 174.5 km

Plate Jc2 Vestri Quadrangle

Most of this quadrangle is dominated by the vast Valhalla multi-ring impact system, whose actual center lies to the south (see Plate Jc7). This area marks the widest extent of the system, approximately 2500 kilometers from the center. Here we see the transition from the ∆-shaped ridges in the inner rings to the scarps and graben that characterize the more widely spaced outer zones. Superposed on top of the Valhalla rings, at 35°N, 47°W, is Gomul Catena, a crater chain formed by a comet tidally disrupted by Jupiter (see also Plate Jc1).

Also evident in this region is the presence of bright pole-facing scarps, especially near 50°N, 20°W. Despite the southern location of the Sun at the time of imaging, the pole-facing southern rims of craters here are not dark but bright. These shadowed regions would be colder than other surfaces and act to trap water frosts, forming bright deposits. These deposits were never observed at high resolution, but a similar phenomenon was observed in polar regions on Ganymede (see Plate Jg3.2).

Callisto

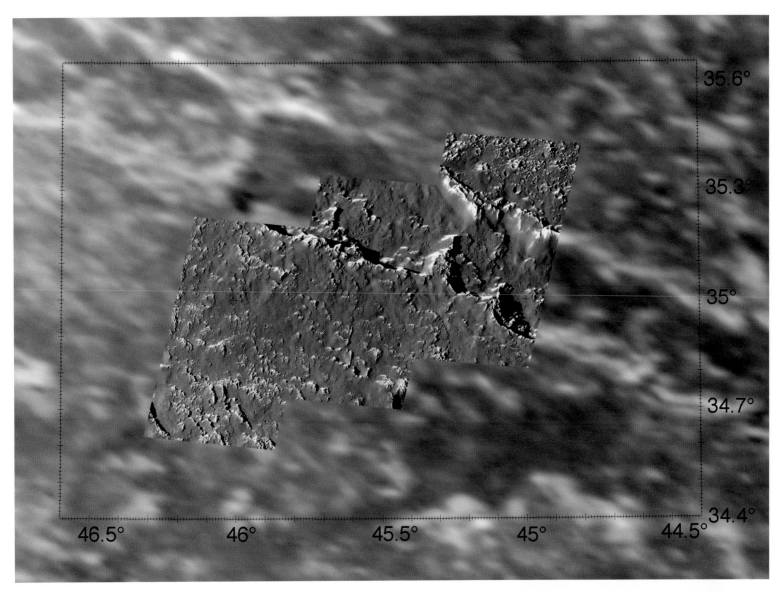

Plate Jc2.1 Gomul Catena
Encounter: *Galileo* C3   Resolution: 40 meters/pixel
Orthographic Projection   Map Scale: 1 cm = 4.54 km

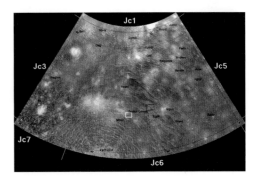

Galileo captured this high-resolution mosaic of the center of one of the eight or so crater chains formed on Callisto by disrupted comets. One such comet, Comet Shoemaker-Levy 9, broke apart in 1993 into two dozen or so fragments. The goal of this observation was to understand some of the dynamics of these impact events and the nature of the fragments that form the craters (were they "solid," unconsolidated rubble, or loose debris clouds?). The observation was somewhat frustrated by the unanticipated nature of Callisto's surface, which has been heavily modified by ice–rock segregation. This loss of water ice has lead to significant local erosion. Despite this, it appears that the bodies forming these craters were fairly coherent bodies, forming typical, if overlapping, impact craters.

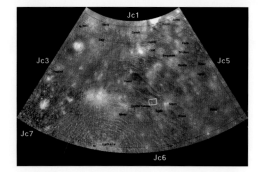

Plate Jc2.2 **Valhalla: Outer Ring Scarp**
Encounter: *Galileo* C3   Resolution: 46 meters/pixel
Orthographic Projection   Map Scale: 1 cm = 6.2 km

Our best observation of the Valhalla rings shows a clear sharp fault scarp at right. Several minor fault scarps and aligned knobs are also visible to the east of the large scarp, indicating that these terrains also experienced minor faulting. Most of the floor of the graben to the east is in shadow, however. The surrounding terrains are divided into the dark rolling plains and isolated icy knobs characteristic of eroded regions of Callisto. Several arcuate scarps are also apparent. These are deeply eroded impact craters. The large gaps in the first three frames are missing lines that were not transmitted to Earth.

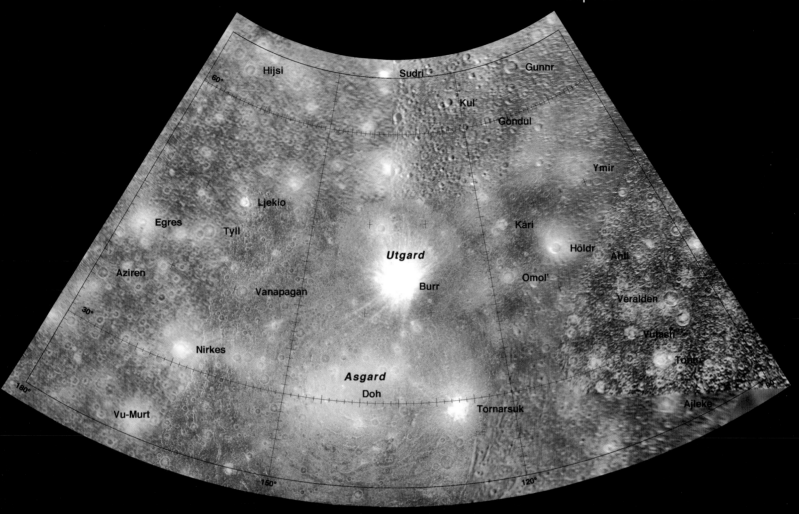

Lambert Conformal Conic Projection
Map Scale: 1 cm = 174.5 km

Plate Jc3 Asgard Quadrangle

The Asgard ring complex is the second largest of Callisto's prominent multi-ring impact systems. As in the case of Valhalla (Plate Jc2), we see the transition from concentric central ridges outward to concentric graben. Although it is not possible to identify the original crater rim from the numerous concentric scarps and graben, our best estimate of the original crater diameter is roughly 675 kilometers. Superposed on the northern portions of the Asgard system is the younger 625-kilometer-wide impact feature, Utgard (see Plate Jc3.1), which, although faint in this mosaic, can be recognized as a smooth-looking circular patch. On top of both is the very young bright-rayed crater, Burr.

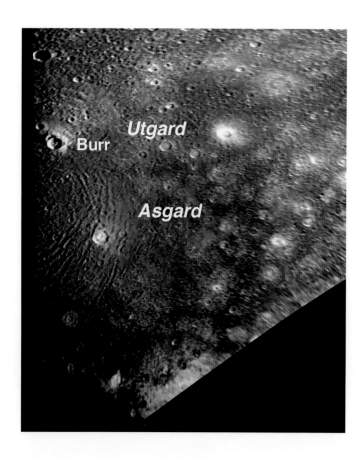

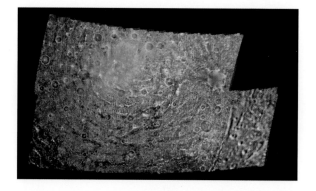

Plate Jc3.1 **Asgard Topography**
Encounter: *Voyager* 1   Resolution: 1000 meters/pixel
Cylindrical Projection   Map Scale: 1 cm = 238 km

This low-resolution low-Sun *Voyager* image of eastern Asgard is slightly smeared by camera motion. Nonetheless, the shadows highlight relief across the concentric ridges and graben comprising Asgard that is not apparent in the high-Sun *Galileo* views (Plate Jc3). The central area appears to be more chaotically developed. Some of the ridges and graben rim surrounding the central area are 2 to 3 kilometers high. A few concentric structures, as well as an outward-facing edge scarp, are also apparent at the large Utgard feature to the north. Finally, at bottom, we see part of a large circular bright patch, in which many of the highest-resolution images of Callisto were taken (see Plates Jc7 and Jc7.2). This patch may be an ancient impact scar.

Plate Jc3.2-N **Asgard: NIMS Infrared Map**
Encounter: *Voyager* 1   Resolution: 8.2 kilometers/pixel
Orthographic Projection   Map Scale: 1 cm = 216 km

NIMS spectral image of Asgard impact basin, Callisto. Here a 3-channel NIMS spectral map has been combined with the global mosaic. In this false-color version, red and brown tones represent non-ice materials. Cyan tones represent moderately ice-rich deposits, such as the older overlapping crater floor and impact ejecta deposits of Asgard and Doh (see Plate Jc3.3-X1). Bright blue spots are nearly 100% ice-rich ejecta deposits of young bright-ray craters. The strongest spectral ice bands are found in these craters, especially Tornarsuk, the large crater at right. CO_2 is often concentrated at such craters.

Callisto | 69

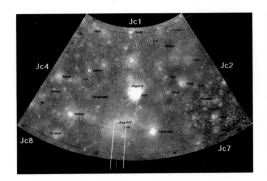

Plate Jc3.3 **Asgard Transect**
Encounter: *Galileo* C10   Resolution: 90 meters/pixel
Orthographic Projection   Map Scale: 1 cm = 54.5 km

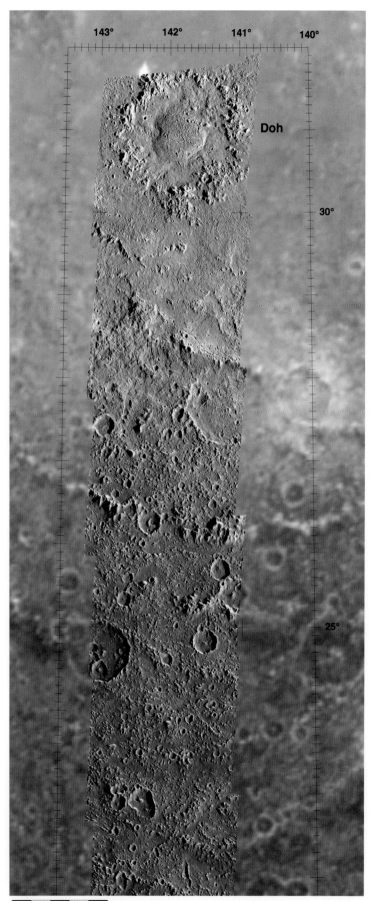

Plate Jc3.3-X1 **Asgard Transect 1**

Central Asgard: Doh, Inner Rings, and Landslides
Encounter: *Galileo* C10   Resolution: 90 meters/pixel
Orthographic Projection   Map Scale: 1 cm = 19.2 km

This is the first half of a spectacular radial swath showing the morphology of Asgard and its tectonic features from near center along a radial line due south (the southern half crosses into quadrangle Jc7). It shows the transition from Asgard floor and rim deposits and associated concentric ridges to ancient cratered plains, crossed by outer ring graben from Asgard.

The prominent 115-kilometer-wide penedome impact crater Doh obscures large areas in the center of Asgard. A large central dome complex dominates Doh. The dome itself is only 25 kilometers wide and is intensely fractured, a characteristic common to central domes seen elsewhere on Ganymede and Callisto (see Plates Jg2.1, Jg8.11, and Jc10.2). It formed during the cratering event when icy crust beneath the crater center rebounded upward, bringing bright soft ice up from several kilometers below the surface. The 60- to 80-kilometer-wide ring of broken massifs surrounding the central dome is also part of the vast central uplift complex, and comprises layers of icy crust deformed and uplifted as the central dome "punched" through to the surface. The crater rim, or what little remains of it, can be found at the short arcuate ridge to the southwest of the central dome at 29.5°N, 142.7°W. The low knobby plains between the central uplift and the dome are most likely a combination of impact melt (in the form of refrozen water) and severely fractured ice grains. Beyond this, ejecta can be seen partly obscuring Asgard ring structures down to a latitude of roughly 27°N.

The tall arcuate ridge near latitude 26°N is one of the inner rings of the Asgard structure. This ridge consists of closely spaced or overlapping peaks and is approximately 1500 meters high and is similar to inner rings of Valhalla. South of the ridge lie heavily cratered plains. Although partly obscured by shadows, several lobate landslides of debris can be seen at the base of inner crater rims in at least two craters southeast of the ridge, at 24.8°N, 141.4°W, and 25.7°N, 141.2°N. The fields of small knobs are part of the Asgard crater floor and ejecta deposit.

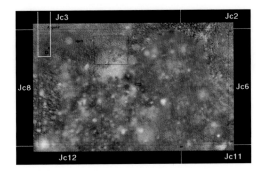

Plate Jc3.3-X2 **Asgard Transect 2**
Outer Asgard: Cratered Plains and Ring Graben
Encounter: *Galileo* C10   Resolution: 90 meters/pixel
Orthographic Projection   Map Scale: 1 cm = 19.2 km

This southern half of the Asgard transect shows two of the basin's major outer graben rings, at 15°N and 19°N latitude. The flanking graben walls appear to be degraded and partly eroded, much as the rest of Callisto. Nonetheless, the sinuous (snake-like) trace of the graben walls can be seen. This style is common for outer basin rings on Callisto. Although smaller impact craters can be seen in a variety of preservation states, few can be called fresh or pristine. Note that the fields of small knobs effectively disappear south of 15°N, roughly at the latitude of the southern graben ring. This may mark the furthest southern extent of Asgard ejecta deposits. The white rectangles in the southeastern corner were imaged later at 15 meters resolution (Plates Jc7.3 and Jc7.4). Shown at 50% scale to fit page.

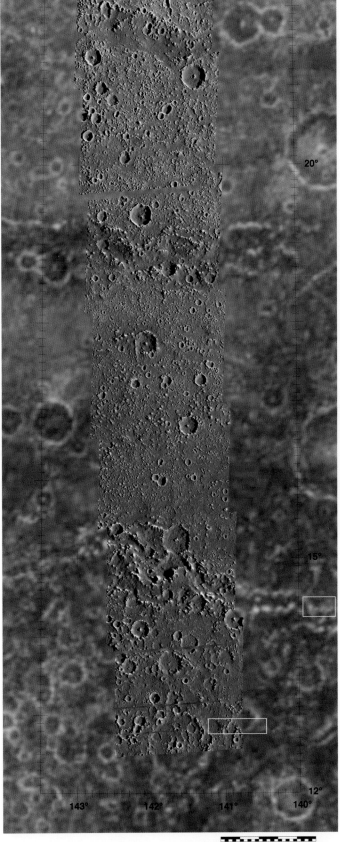

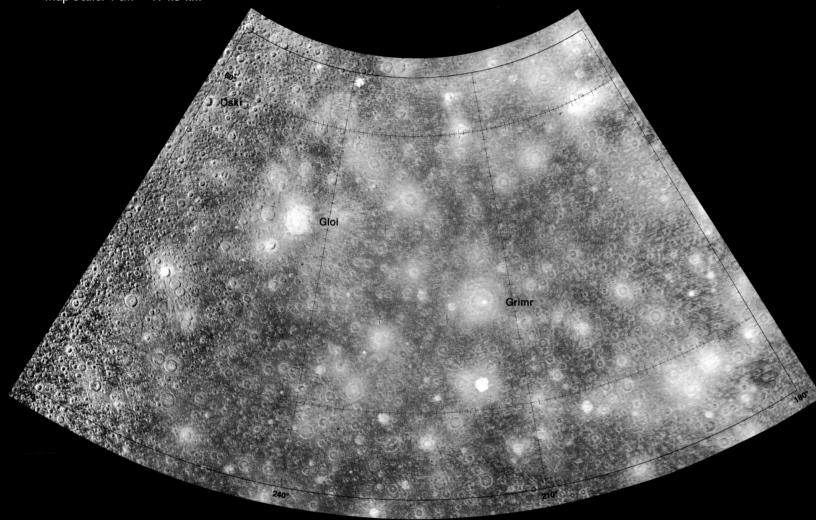

Lambert Conformal Conic Projection
Map Scale: 1 cm = 174.5 km

Plate Jc4 Gloi Quadrangle

This quadrangle is rather bland, even for Callisto. Despite this, a variety of impact morphologies can be seen. Crater morphology ranges from relatively fresh bright-floored craters (e.g., Gloi), to older degraded cryptic penedome craters (e.g., Grimr). *Voyager* 2 observed this terrain at only 2 kilometers resolution.

Lambert Conformal Conic Projection
Map Scale: 1 cm = 174.5 km

Plate Jc5 Askr Quadrangle

Here we see the uninterrupted cratered plains of Callisto. Even here, though, Callisto is not so simple. The origin of the irregular bright and dark patches remains unknown, but they are probably ancient. The very dark patch at 32°N, 308°W could be dark ejecta associated with the small crater at its center. Impact of a disrupted comet formed the short six-crater-long chain, Geirvimul Catena.

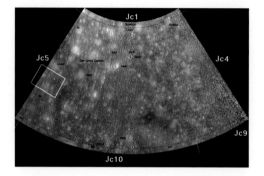

Plate Jc5.1 **Small Multi-ring Impact Structure**
Encounter: *Galileo* C10   Resolution: 430 meters/pixel
Orthographic Projection   Map Scale: 1 cm = 50.8 km

Callisto features several smaller ancient multi-ring structures (e.g., Plate Jc11), in addition to the large multi-ring impact structures such as Valhalla. This moderate-resolution view of a small multi-ring structure shows a relatively flat interior surrounded by several incomplete concentric ridges.

The region is heavily cratered, indicating that the ring structure is relatively old, probably more than 4 billion years in age. The large oval crater at lower right was probably formed by a very low-angle grazing impact or by a close binary comet or asteroid.

Map Scale at Equator: 1 cm = 163.6 km

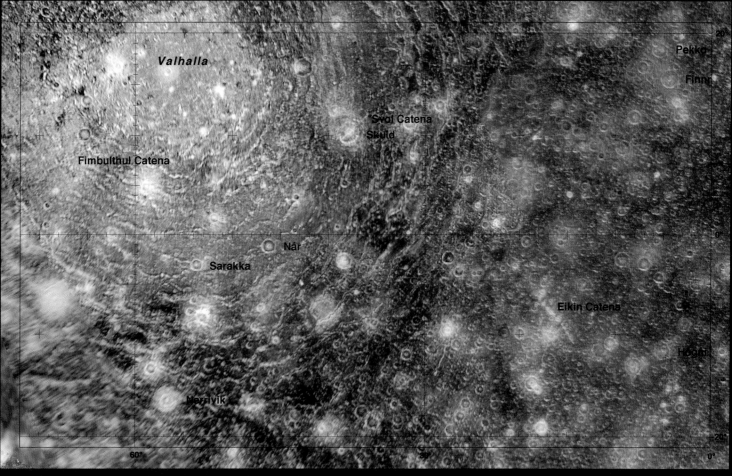

Plate Jc6 **Valhalla Quadrangle**

Here lies the heart of the Valhalla multi-ring impact system. The system is comprised of numerous concentric ridges, scarps and graben and is roughly 4000 kilometers across at its widest point. An irregular bright central patch is an ice-rich basin floor deposit composed of material excavated from the upper icy crust of Callisto. The zone surrounding Valhalla, out to ~900 kilometers from the center, is relatively bright.

This zone has a lower crater density and has been partly obscured by the Valhalla ejecta blanket. Valhalla formed from the impact of a large 100-kilometer-class asteroid or comet roughly 4 billion years ago or so. East of Valhalla center is a multiple-frame *Galileo* image mosaic designed to sample typical cratered plains (Plate Jc6.1).

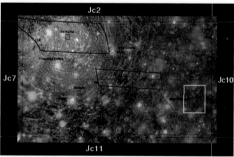

Plate Jc6.1 **Cratered Plains**

Encounter: *Galileo* C9   Resolution: 155 meters/pixel
Orthographic Projection   Map Scale: 1 cm = 26.5 km

This 6-frame mosaic is a representative high-resolution sample of cratered plains on Callisto and was used to determine cratering statistics of the surface at high resolution for surface age determination. Craters in a wide range of preservation state are apparent, indicating how surface modification processes continually operate to modify relief over time. The only features remaining in the large crater near mosaic center, for example, are the large central peak and two faint rings along the eastern crater rim. Contrast this with the slightly smaller but well-preserved central pit crater 43 kilometers across at 4°S, 5.5°W. The floors of several craters are deeply pitted, especially to the west and south, another indication of the erosion process desiccating the surface of Callisto. Small landslide deposits are evident on the floors of at least two smaller craters (at 7.5°S, 8.5°W, and 5.7°S, 6.5°W). Also visible are an elliptical grazing impact crater at 5°S, 4°W and several elongate craters and crater chains trending north–south along longitude 6°W. These latter craters may be secondary craters ejected from the Lofn or Heimdall impact basins located 2500 kilometers due south of this site.

Callisto

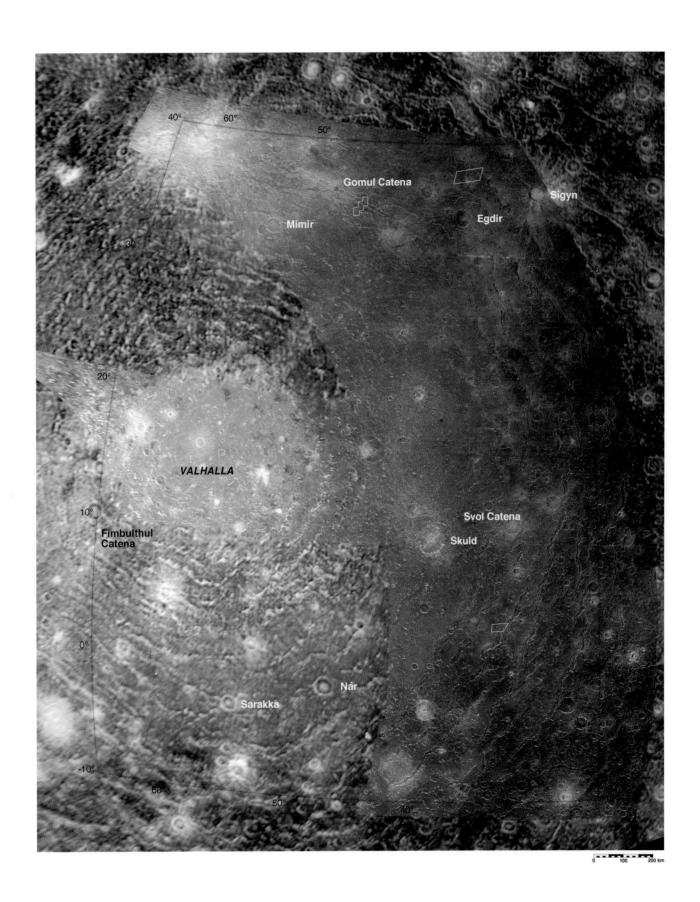

Plate Jc6.2.1 Valhalla: Regional Mosaic
Encounter: *Galileo* C9   Resolution: 410 meters/pixel
Orthographic Projection   Map Scale: 1 cm = 121.5 km

Galileo acquired a medium-resolution mosaic (shown at 50% scale) of the eastern half of Valhalla to further investigate how this global-scale event formed. The central zone is a relatively bright but rugged circular patch, probably formed by refrozen impact deposits on the floor of the impact. Unlike lunar basins there is no deep central depression here. Outward ~1000 kilometers from the center extends a zone intermediate in brightness between the central feature and dark-cratered plains beyond. This is similar to the zone of knobs seen at Asgard (Plate Jc3.2) and is probably part of the Valhalla ejecta deposit. This area features numerous concentric ridges, many of which retain relief of 2 to 3 kilometers. Beyond this zone lie graben and scarps of the Valhalla system, superposed on cratered plains. Although the original crater rim can no longer be identified, the size of the original impact crater is estimated to be approximately 1000 kilometers. The original rim, if it ever existed, is now obscure. Small white rectangles mark the locations of high-resolution images (Plates Jc6.2.2, Jc6.2.3, Jc2.1 and Jc2.2).

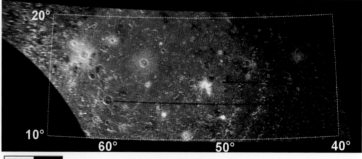

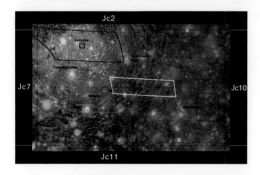

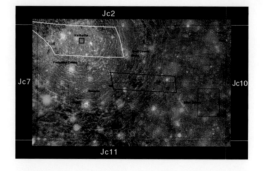

Plate Jc6.2.2 Valhalla: Regional Color Mosaic
Encounter: *Galileo* G8   Resolution: 1150 meters/pixel
Orthographic Projection   Map Scale: 1 cm = 135.8 km

One of the few color images taken of Callisto, this set shows a part of the transition zone from inner ridge zone to the outer graben zone in the Valhalla impact structure.

Plate Jc6.2.3 Valhalla: Basin-floor Color Mosaic
Encounter: *Galileo* C9   Resolution: 1100 meters/pixel
Orthographic Projection   Map Scale: 1 cm = 130 km

One of the few color images taken of Callisto, this set (part of the larger Jc6.2.1 mosaic) shows the central bright basin-floor deposit. A few patches of dark brown material poke through the bright deposit. Bright craters of differing brightness and color are also visible. Other than these, no major color variations are apparent.

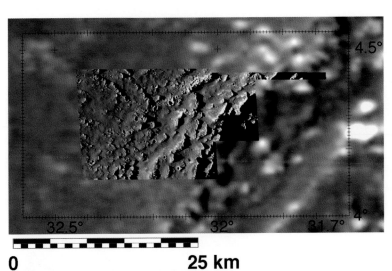

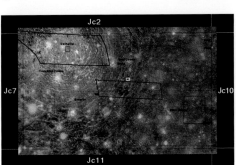
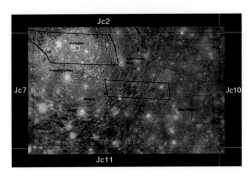

Plate Jc6.2.4 **Valhalla: Outer Graben**
Encounter: *Galileo* C3   Resolution: 42 meters/pixel
Orthographic Projection   Map Scale: 1 cm = 5.0 km

Galileo observed the western flank of this Valhalla ring graben at high resolution while the floor of the graben itself was in shadow (see also Plate Jc2.2). This structure is a classic Valhalla graben, formed in response to the main impact event 1000 kilometers away. Several parallel ridges cut the outer flank of the graben wall, indicating that the surrounding region was also faulted during graben formation.

Plate Jc6.2.5 **Valhalla: Center Palimpsest**
Encounter: *Galileo* C3   Resolution: 30 meters/pixel
Orthographic Projection   Map Scale: 1 cm = 5.9 km

This single image is our best view of the bright central floor of Valhalla. Numerous isolated knobs of various size and undetermined origin are visible, indicating that the central floor of Valhalla is quite rugged. Several small craters are also visible. The terrain has been eroded in a manner seen elsewhere on Callisto, making it difficult to resolve original textures or deposits.

Map Scale at Equator: 1 cm = 163.6 km

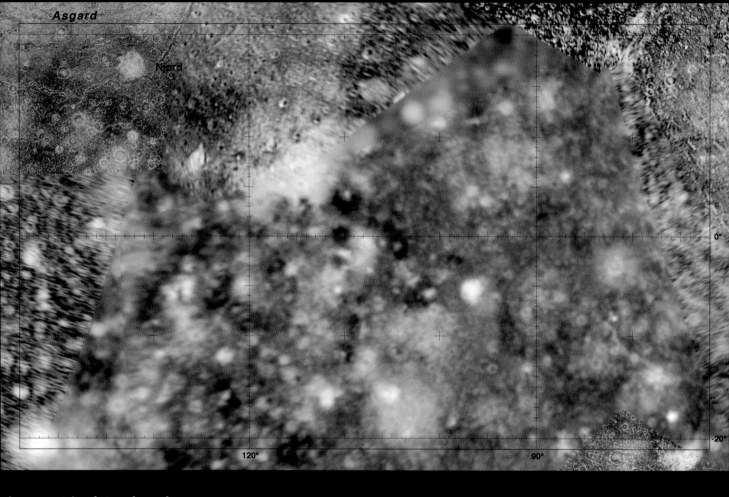

Plate Jc7 **Njord Quadrangle**

Although this mosaic is composed of a patchwork of images of different resolution, it is evident that this quadrangle has a complex history. Several irregular dark and bright patches intermingle. Ring graben from both the Asgard and Valhalla multi-ring impact structures intrude from the upper left and upper right, respectively. The 600-kilometer-wide bright patch at 8°N, 115°W is probably a large impact scar, such as a palimpsest. Numerous dark splotches that interrupt its surface are probably younger small impacts that excavated through the bright material to dark material beneath (see also Plate Jc8.3.2).

This giant feature would be an excellent target for high-resolution mapping. Several of the very high resolution images in Plate Jc7.2.1 lie on the northern edge of this bright patch. Another possible ancient multi-ring structure is centered in the poorly imaged area at 2°N, 134°W.

Some of our highest-resolution views of Callisto (Plates Jc7.1, Jc7.2, Jc7.3, Jc7.4, and Jc8.1) are located within the central parts of this quadrangle. With some exceptions, the resolution of this base map (2 to 7 kilometers) is far too poor to allow us to pinpoint their exact locations.

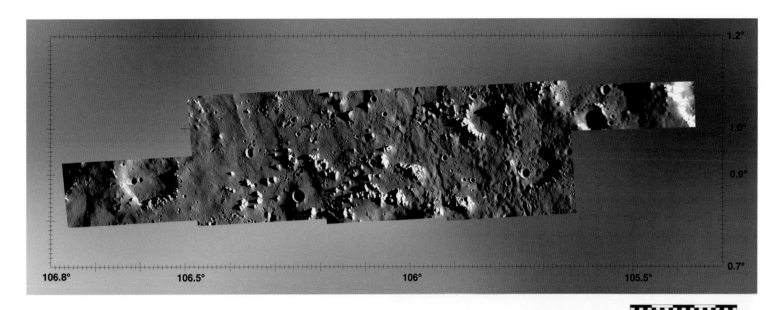

Plate Jc7.1 **Cratered Plains: Very High Resolution**
Encounter: *Galileo* C21   Resolution: 15 meters/pixel
Orthographic Projection   Map Scale: 1 cm = 3.45 km

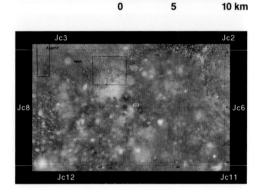

Despite the lack of context imaging for this site, quite a few interesting features are apparent. The isolated bright peaks break the rolling dark plains. Some of these peaks form the rims of degraded impact craters. Compare these with fresh craters in Plate Jc6.1 and on Ganymede (Plate Jg2.2). The high crater density of the dark material here and elsewhere on Callisto suggests that the segregation of dark and bright materials occurred a long time ago and that the process may no longer be occurring.

The extensive dark plains are complex in morphology. This terrain is located in a relatively dark region, consistent with the relatively low density of bright knobs. In addition to a peppering of small craters, we see several north–south trending scarps and ridges. These indicate that Callisto may not have been completely dead, but could have undergone a period where local or global-scale compressional tectonism deformed the surface after dark plains formed. Without a lot more high-resolution mapping, however, it will be impossible to determine how extensive this deformation is across Callisto. Shown at 60% scale to fit page.

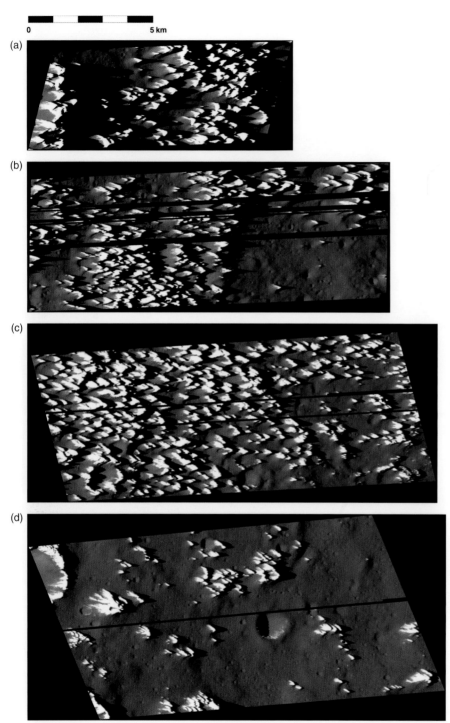

Plate Jc7.2 (a–d) **Highest Resolution (1): Four Views**

Encounter: *Galileo* C30   Resolution: 5–10 meters/pixel
Orthographic Projection   Map Scale: 1 cm = 1.48 km

These are our highest-resolution views of Callisto. They feature essentially random views of Callisto's surface within a hundred kilometers or so of each other. Despite their proximity, they are very different in appearance. Bright peaks largely dominate the first view (a), dark plains the last (d). The dark plains are not flat but rolling and are heavily cratered, suggesting that they formed long ago. The bright peaks are often saturated in these images and few details can be resolved. They are very sharp crested, however, as indicated by their shadows, and are typically several hundred meters high. The views are oblique, accounting for the distorted shapes of the bright peaks. Shown at 40% scale to fit page. Frame (d) is featured in Figure 5.1.3.

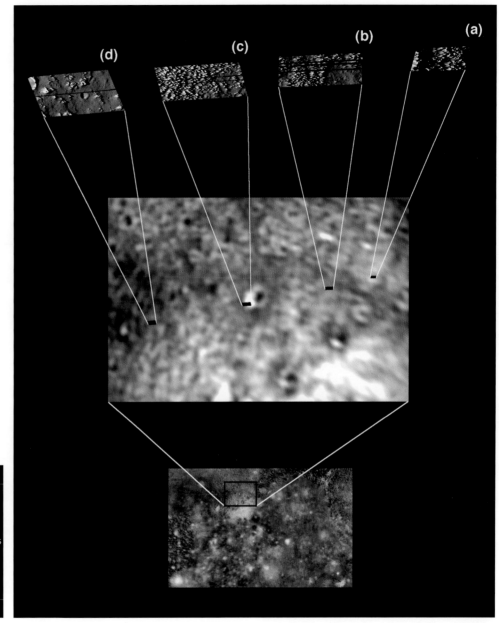
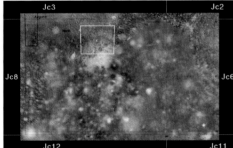

Plate Jc7.2.1 **Highest Resolution (1): Four Views Context Image**
Orthographic Projection

This context view is our best guess as to the location of the four frames of Plate Jc7.2. No context imaging was obtained of these sites except this slightly smeared *Voyager* image at 1.5 kilometers resolution. It suggests that most of these frames lie on the large circular bright patch centered on 16°N, 115°W. This patch resembles impact palimpsests seen elsewhere on Ganymede and Callisto. The third frame may lie near the rim of a 25-kilometer-wide dark ejecta crater, perhaps accounting for the abundant knobs. Without better context imaging the causes of these variations will likely remain shrouded in mystery.

Atlas of the Galilean Satellites

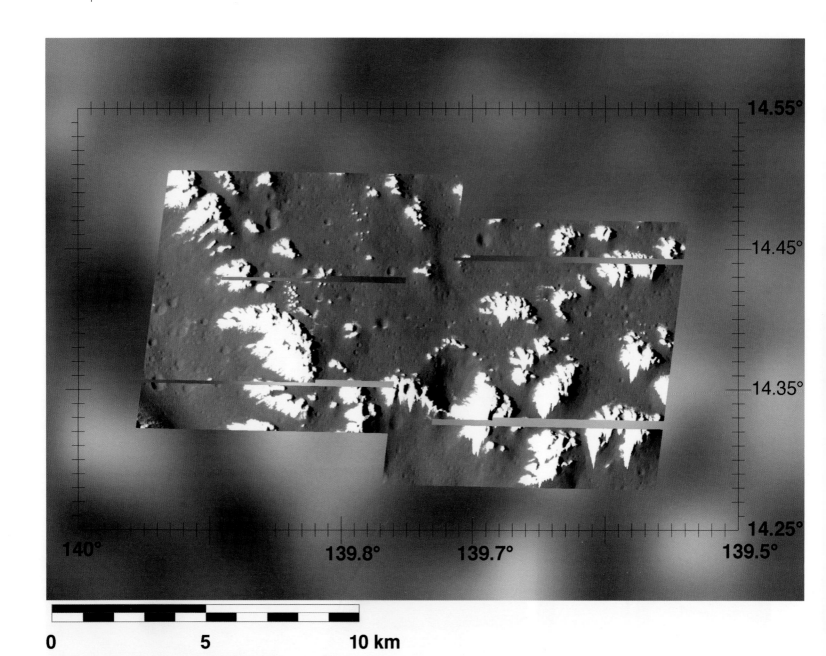

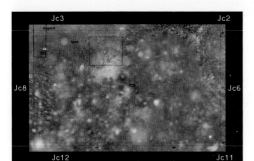

Plate Jc7.3 Highest Resolution (2): Asgard Ring Graben
Encounter: *Galileo* C30   Resolution: 10 meters/pixel
Orthographic Projection   Map Scale: 1 cm = 1.2 km

This two-frame mosaic includes the northern flank of one of the outer ring graben of Asgard and a small impact crater, both of which are highly eroded. In this mosaic, the northern wall of the graben extends along the bottom of the mosaic and then bends toward the north at longitude 139.8°W. The graben wall is not continuous, but has been eroded into a number of large bright massifs. Like other high-resolution frames (e.g., Plates Jc7.2), the bright peaks are often saturated and few details can be resolved. Otherwise, the rolling dark plains have been moderately cratered since their formation. Some of the craters in the plains appear to be rounded due to ongoing micrometeorite bombardment, similar to what is seen on some lunar terrains. (See Plate Jc3.3-X2 for additional context.)

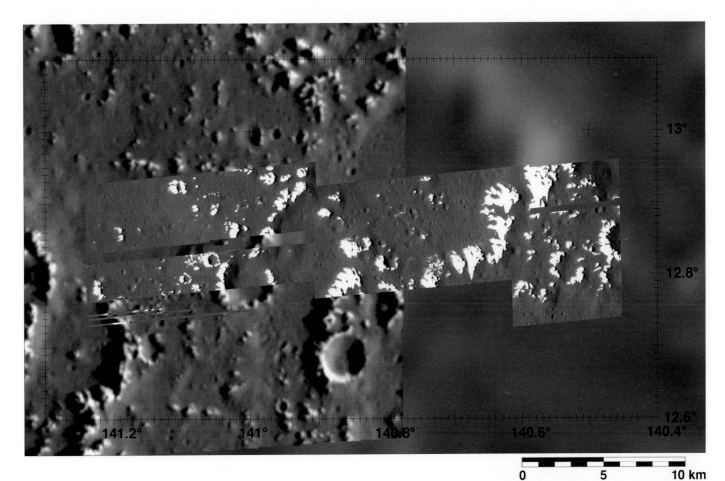

Plate Jc7.4 Highest Resolution (3): Ancient Craters

Encounter: *Galileo* C30   Resolution: 14 meters/pixel
Orthographic Projection   Map Scale: 1 cm = 2.2 km

The western half of this incomplete 3-frame mosaic overlaps part of the 90-meter-resolution Asgard transect mosaic (see Plate Jc3.3-X2). This may lead to some confusion because the Sun was located in the opposite part of the sky in that mosaic, compared to when this very high resolution mosaic was acquired. Nonetheless, the Asgard transect is included to provide additional context. The mosaic includes several old craters. The largest of these, at 12.9°N, 140.75°W, is only 12 kilometers across. The pervasive effects of landform degradation are apparent. The rims of these craters have all been partially eroded, leaving behind rings of discontinuous knobs. Some of the very small craters in the dark plains here have a rounded appearance, consistent with a great age for these dark deposits.

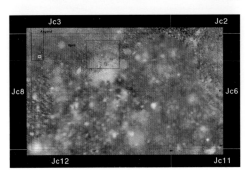

Map Scale at Equator: 1 cm = 163.6 km

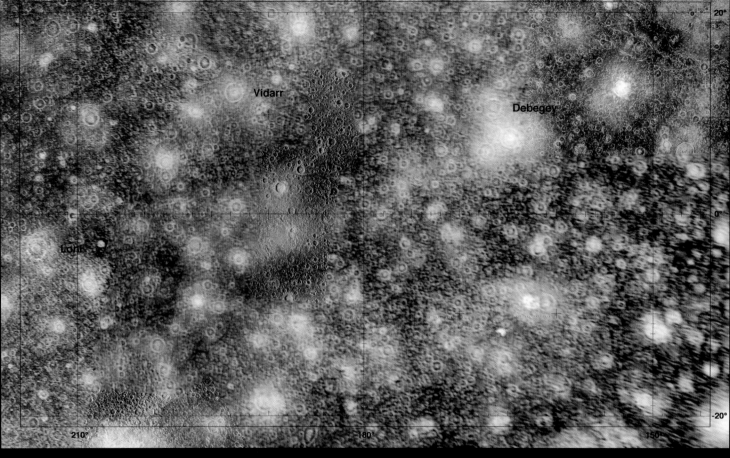

Plate Jc8 Vidarr Quadrangle

Endless tracts of cratered plains typify much of Callisto. Valhalla rings creep in from the upper right, but the two penedome craters Loni and Debegey are signs of smaller ancient impact events. Debegey was a *Galileo* imaging target (Plate Jc8.2). *Voyager* 2 covered most of this quadrangle at 2-kilometer or worse resolution. The central area was well imaged by *Galileo*, showing that this regional view can be misleading. Targeted features also include a dark irregular flow-like deposit near the ancient large crater centered at 2°S, 188°W.

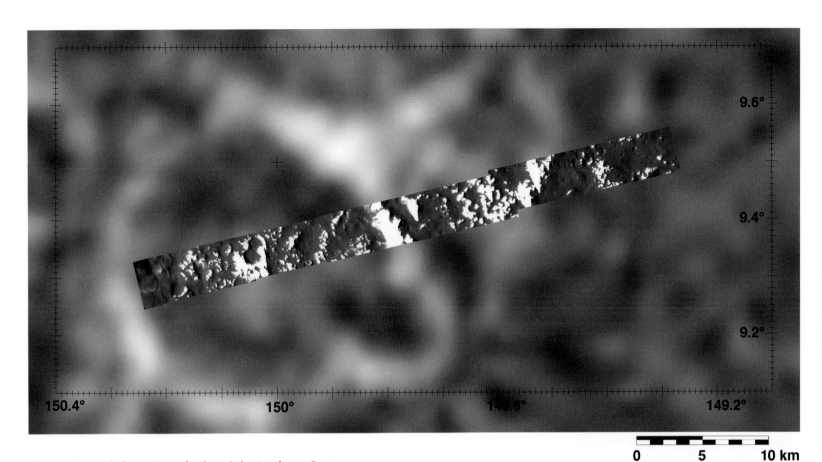

Plate Jc8.1 Highest Resolution (4): Ancient Crater
Encounter: *Galileo* C30 Resolution: 18 meters/pixel
Orthographic Projection Map Scale: 1 cm = 2.76 km

Only fragments of this planned 3-frame mosaic were transmitted to Earth. They cut across the center of a 21-kilometer-wide degraded impact crater at 9.3°N, 150°W. At far left can be seen the toes of several lobate landslide deposits that slid from the western rim.

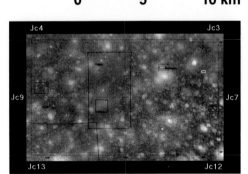

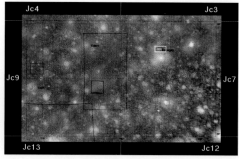

Plate Jc8.2.1 **Debegey (Penedome Crater): Context Image**
Encounter: *Galileo* C30   Resolution: 120 meters/pixel
Orthographic Projection   Map Scale: 1 cm = 14.2 km

These high-Sun images show the central and eastern sector of this ancient impact crater. The western edge of the mosaic clips part of the 35-kilometer-wide central dome, the bright circular feature centered at 10.6°N, 166.6°W (see Plate Jc8). The steep-sided arcuate cliffs to the east of the dome (at 165.6°W longitude) are part of the 70-kilometer-wide central ring complex that surrounds the dome. This crater is similar to Neith (Plate Jg2.1) and Doh (Plate Jc3.3-X1), although it appears to be older. The crater rim is actually located along longitude 164.8°, giving an estimated crater diameter of 150 kilometers. Although no rim scarp is recognized, the location is known based on established scaling relationships between the diameters of the central dome and crater rim.

Plate Jc8.2.2 Debegey (Penedome Crater): High Resolution
Encounter: *Galileo* C30   Resolution: 24 meters/pixel
Orthographic Projection   Map Scale: 1 cm = 4.9 km

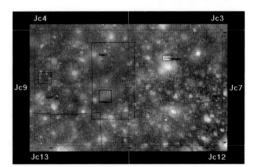

Despite the high-Sun illumination, there is a surprising amount of detail in this high-resolution strip across the center of the context image in Plate Jc8.2.1. Here we see a bit more of the fractured central dome (far left: 166.0° to 166.6°W) and the arcuate central ring complex that surrounds it (longitude 165.6°W). A lobate landslide deposit lies at the base of one of these steep-sided peaks, at 10.9°N, 165.65°W. Both the dome and ring surrounding it have relief (from stereo data) of ~1000 meters. A crater rim should occur at longitude 164.8°W but is virtually nonexistent. This is a fundamental characteristic of large craters of this age (see Plates Jc3.3-X1 and Jg2.1), however.

The lowest areas appear to have smooth dark deposits within them, consistent with landform erosion and degradation as seen elsewhere on Callisto. This is especially true at the old central peak crater at 11°N, 163.8°W. Debegey lies within the secondary crater field of a large bright-rimmed crater 100–200 kilometers to the south, which may account for many of the small craters in this scene. Shown at 60% scale to fit page.

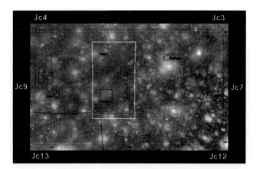

Plate Jc8.3.1 Cratered Plains

Encounter: *Galileo* C20   Resolution: 260–430 meters/pixel
Orthographic Projection   Map Scale: 1 cm = 47.2 km

This 4-frame mosaic of cratered plains was intended for crater statistics and for geologic mapping. A cursory examination confirms that cratered plains are not necessarily uniform. A large irregular bright patch is centered at 3°S, 188°W. The cryptic concentric arcs and ridges associated with this dark patch clearly indicate that it is a degraded impact basin, possibly a palimpsest. Note the linear structure to the east of this bright patch, which is probably a secondary crater chain or linear fracture. Smooth dark patches can be seen scattered across the mosaic, including extensive dark material along the southern edge of the large degraded impact basin. These dark deposits are featured in high resolution in Plate Jc8.3.2. A second linear feature located at 6°N, 189°W is probably a linear fracture or crater chain associated with the young bright-floored crater at 2.5°N 188.7°W (see also Plate Jc1 for similar crater chains).

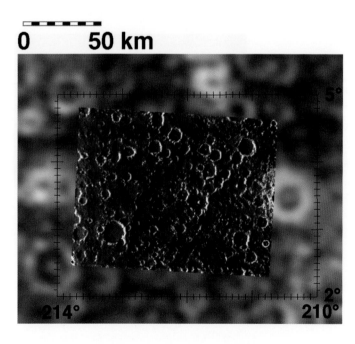

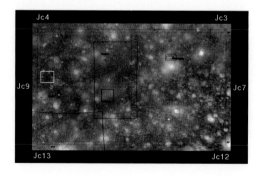

Plate Jc8.4 Medium-resolution Color

Encounter: *Galileo* C20   Resolution: 250 meters/pixel
Orthographic Projection   Map Scale: 1 cm = 23.6 km

This image is part of a color sequence intended to search for local color variations on Callisto. This frame shows typical cratered plains. The bright knobs at center right are part of the ejecta blanket of a 20-kilometer-wide crater just outside the scene. No obvious color variations are apparent. This is the only color imaging of dark-cratered terrain we have on either Ganymede or Callisto at better than 1-kilometer resolution.

Plate Jc8.3.1

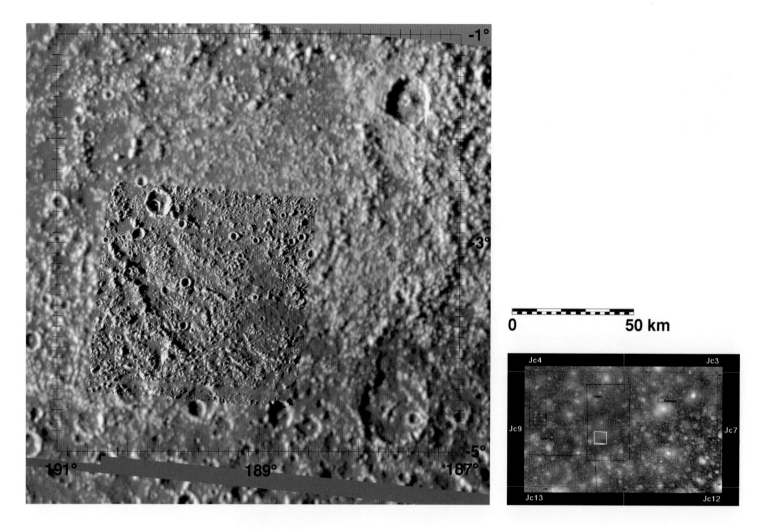

Plate Jc8.3.2 **Cratered Plains: Dark Material**
Encounter: *Galileo* C30   Resolution: 110 meters/pixel
Orthographic Projection   Map Scale: 1 cm = 15.3 km

This image (which combines with part of Plate Jc8.3.1 to form a stereo mosaic) was designed to investigate the origins of irregular dark deposits located across Callisto. It includes the southwestern quarter of the large bright spot, described in the Plate Jc8.3.1 as a large degraded impact feature. This view includes part of the southwestern rim segment. The dark deposit is seen here as a relatively smooth unit emplaced on top of the older crater floor and clearly fills topographic lows. While this could be evidence of some form of volcanism, it may be ejecta related to one or more craters at the edge of the larger basin in the southeast corner of this view. Alternatively, it could simply be an area where erosion has removed more ice than other areas.

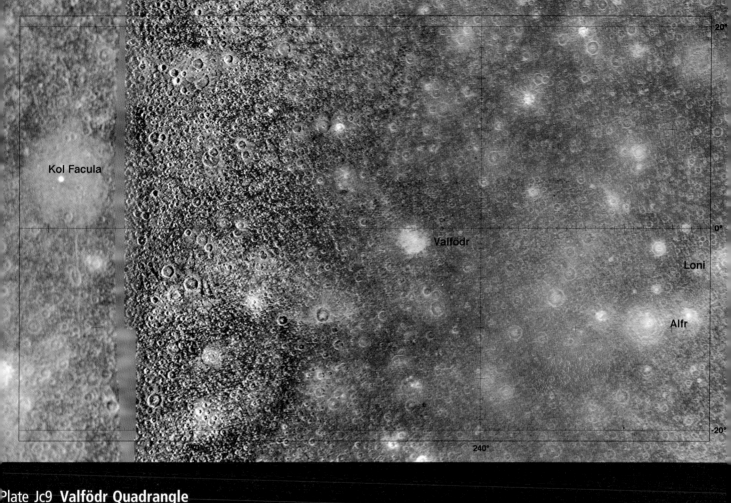

Plate Jc9 **Valfödr Quadrangle**

Although much older and smaller, the large 600-kilometer-wide ancient multi-ring impact structure at 3°S, 230°W, is a close cousin to Valhalla and Asgard. The large bright patch (Kol Facula) at 5°N, 282°W, on the other hand, is probably a palimpsest similar to Memphis Facula on Ganymede (Plate Jg7). Several linear features, possibly similar to those at the north pole (Plate Jc1) radiate from this bright patch. These are either radial fractures or secondary crater chains.

Intact features of this type are rare on Callisto compared to Ganymede. Valfödr is an ordinary large bright-floored crater, but the ring-like features Loni and Alfr are probably close cousins to the penedome craters seen in Plates Jc8 and Jc11. The change in surface characteristics at longitude 282°W corresponds to the edge of *Voyager* 2 imaging in this area. The fuzzy terrains at far west are part of low 4-kilometer-resolution *Galileo* mapping coverage.

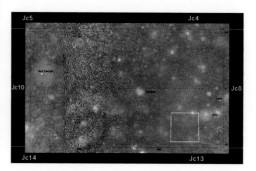

Plate Jc9.1 **Valhalla Antipodal Terrain**
Encounter: *Galileo* C30   Resolution: 340 meters/pixel
Orthographic Projection   Map Scale: 1 cm = 40.2 km

This region is located almost exactly antipodal to (the point on the opposite side of Callisto from) the giant Valhalla impact basin. On Mercury, the location antipodal to the Caloris impact basin is disrupted and fractured due to seismic shaking during the impact. This moderate-resolution image reveals no major features that could be associated with the Valhalla impact. If such disruption occurred on Callisto, it may be below the resolution of this image.

Mercator Projection
Map Scale at Equator: 1 cm = 163.6 km

Plate Jc10 **Vali Quadrangle**

The cratered plains seen in this quadrangle are typical of Callisto. Numerous vague irregular bright and dark patches are seen throughout. The bright patch at 15°N, 351°W was chosen for high-resolution imaging by *Galileo* to evaluate whether these features were formed by impact or volcanism (see Plate Ic10.1).

96 Atlas of the Galilean Satellites

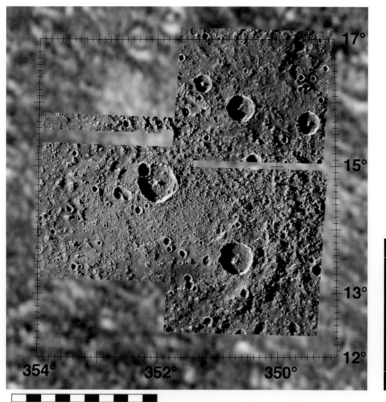

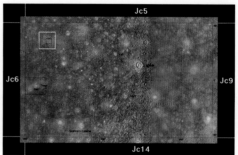

Plate Jc10.1.1 **Bright Patch: Medium Resolution**
Encounter: *Galileo* C10   Resolution: 265 meters/pixel
Orthographic Projection   Map Scale: 1 cm = 25.0 km

This 3-frame mosaic was targeted to investigate the origins of quasi-circular bright patches observed by *Voyager* at low resolution. The motivating idea was that these patches might be ancient ice volcanic deposits, although the general shape also suggested they are eroded ancient impact scars similar to palimpsests (see Plate Jg3). Despite the evident degradation, a number of arcuate features appear to be centered on the bright patch, consistent with the impact hypothesis, especially when compared to similar structures seen elsewhere on Callisto (see Plate Jc8.3.1). The large numbers of smaller craters on this feature attest to its great age. These craters are on the order of 1 kilometer deep, some of which appear to be relatively pristine. Either they are relatively young or this deposit does not erode as easily as other terrains.

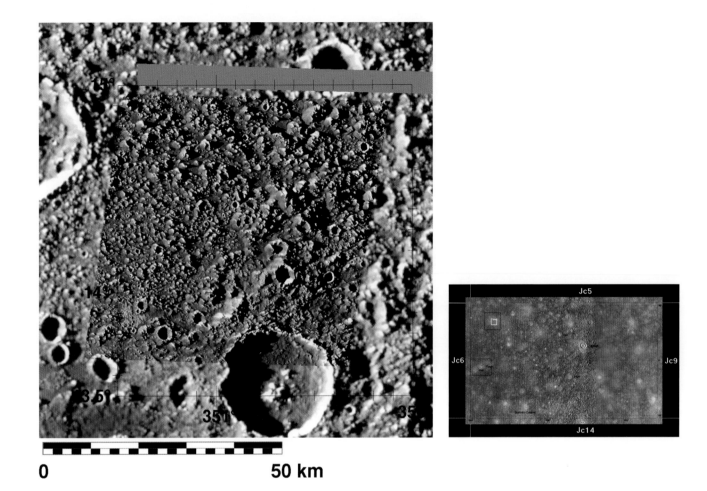

Plate Jc10.1.2 **Bright Patch: High Resolution**
Encounter: *Galileo* C10   Resolution: 65 meters/pixel
Orthographic Projection   Map Scale: 1 cm = 7.8 km

This image of the center of the bright spot in Plate Jc10.1.1 shows the dense eroded angular hills that cover the interior of this feature. The hills are several hundred meters high and are evidence of slope degradation. Parts of the interior concentric scarps are also visible at 13.8°N, 351°W, just north of the young 23-kilometer-wide, 1-kilometer-deep crater at lower right.

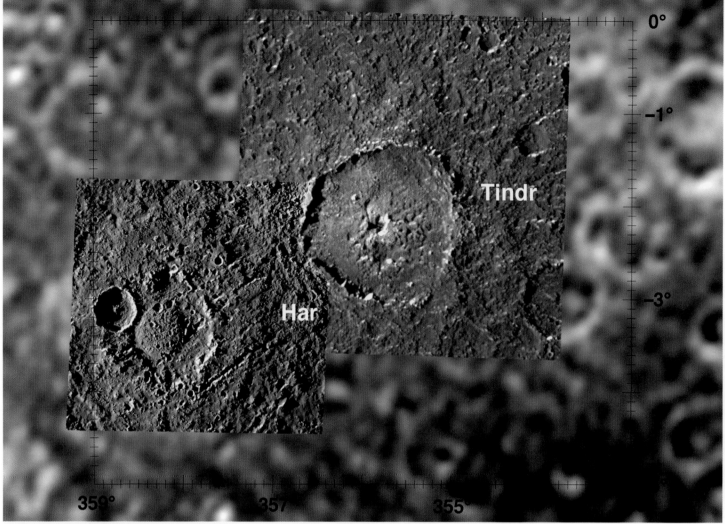

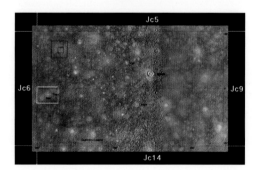

Plate Jc10.2 Hár and Tindr

Encounter: *Galileo* C10   Resolution: 400 meters/pixel (Tindr)
Encounter: *Galileo* C9   Resolution: 145 meters/pixel (Hár)
Orthographic Projection Map Scale: 1 cm = 17.1 km

This combined pair of discrete observations compares craters of ancient and intermediate ages. Tindr, the younger large central pit crater to the east is 70 kilometers wide. Quasi-radial textures beyond the rim to the southwest are due to ejecta blasted out during impact that scour the surface nearby, including penedome crater Hár. Small secondary craters from Tindr can just be resolved in the upper right corner of the high-resolution image. The central pit is actually a complex structure with fissures and valleys extending outwards. Central structures in craters of this size include material uplifted rapidly from depths of a few kilometers. Central pits may form in smaller craters where uplifted central dome material does not quite make it to the surface, as they do in larger central dome craters (Plate Jc3.3-X1).

The ancient impact scar Hár is the degraded concentric set of topographic features to the southwest of Tindr. Surrounding the 25-kilometer-wide smooth rounded central dome at 3.5°S, 358.2°W is a 15-kilometer-wide ring depression and an annular plateau. The dome and ring depression of Hár are part of the central uplift complex and the annular plateau is the remains of the crater floor impact deposit. A younger 19-kilometer-wide central peak crater lies astride the annular depression, just west of the central dome. The true rim on such ancient craters is a very subtle poorly formed feature.

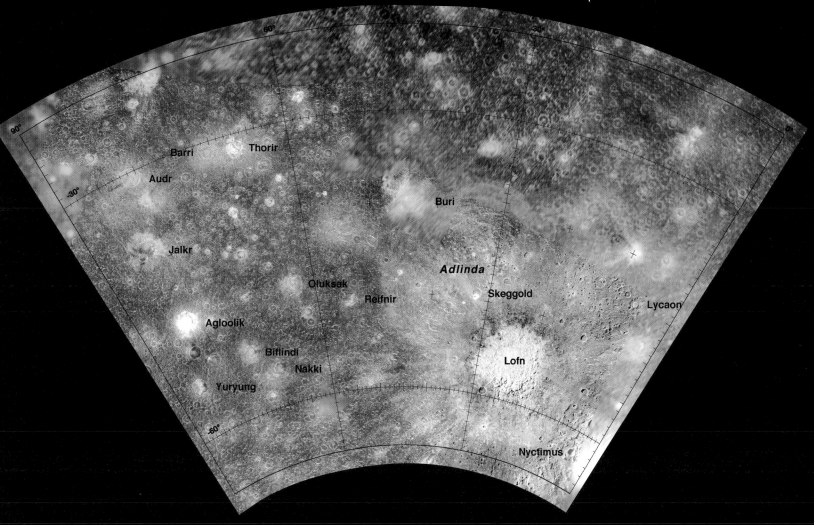

Plate Jc11 Adlinda Quadrangle

The southern temperate zone sampled in this quadrangle (and extending eastward into quadrangle Plate Jc14), includes perhaps the richest concentration of unusual geologic features on the surface of Callisto. In addition to the numerous central dome and penedome craters in the western sector (despite a gap in the *Galileo* images, the crater Buri appears to be a penedome crater very similar to Jalkr) are the two large impact basins, Lofn and Adlinda. A third multi-ring feature may be just gleaned from amidst the overlapping craters in the western areas. This very ancient and barely preserved structure consists of a few concentric graben centered near 35°S, 75°W. The bright patch and rugged terrain along the southeastern edge of the quadrangle is part of a fourth large impact feature in this area, the Heimdall impact complex (see Plate Jc14 and JcO2).

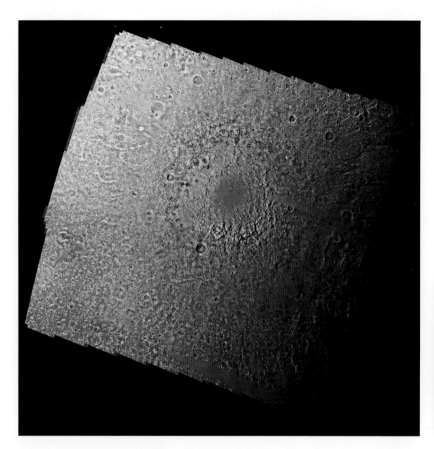

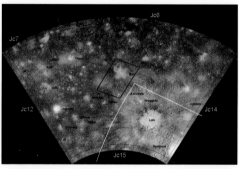

Plate Jc11.1-N Lofn: NIMS Infrared Map
Encounter: *Galileo* G8 Resolution: 8.9 kilometers/pixel
Orthographic Projection Map Scale: 1 cm = 118 km

Lofn impact basin is one of the most distinct on Callisto. This view combines the global mosaic with low-resolution NIMS infrared imaging, and can be referenced to the monochromatic quadrangle view. In this false-color rendition, reds and blues indicate higher non-ice or ice content, respectively. Lofn, a relatively young 350-kilometer-wide impact basin, is the largest known impact feature on Callisto that is not ancient and not multi-ringed. This structure may be only 2 billion years old. The large central bright patch, shown here in blue, has relatively high ice content. Beyond the central bright patch is a relatively crater-free zone extending out ~400 kilometers from basin center. The outer part of this zone has a pale blue color indicating enhanced ice content, forming a faint ring. Large secondary craters from Lofn are found just beyond the edge of this crater-free zone. Lofn's original crater rim is not apparent, but scaling of these ejecta deposits tells us that it lies just outside the large central bright patch.

Lofn is superposed on top of the older Adlinda structure to the northwest (visible in Plate Jc11), an ancient multi-ring impact feature related to Asgard and Valhalla. The difference in morphology between Lofn and Adlinda, two similar-sized impact basins, is an indication that the cold outer layers of Callisto thickened with time, changing the response of the icy crust to large impact events. Adlinda formed when the lithosphere was warmer and thinner and easier to fracture. Lofn formed when it was colder and thicker, and hence we see far fewer concentric fractures. The browns of the cratered plains in this region give way to a greener shade in the vicinity of Adlinda. Whether this is related to Adlinda ejecta is unknown. Small bright-rayed craters also have enhanced ice content. Prominent concentric rings from Heimdall are visible to the southeast.

Plate Jc11.2-N Buri: NIMS Infrared Map

Encounter: *Galileo* G8   Resolution: 2 kilometers/pixel
Orthographic Projection   Map Scale: 1 cm = 94.5 km

At 240 kilometers diameter, Buri is one of the largest penedome craters on Callisto. This false-color combined NIMS–SSI map shows the prominent central dome complex is ice-rich, consistent with its bright albedo. The crater floor is marked by several radiating bright spoke-like features. The greenish material at left has a mixed ice and non-ice composition and is the western edge of the Adlinda impact feature (see also Plates Jc11 and Jc11.1-N). The greenish material at left is part of another but smaller ancient impact scar. Scene is centered at 42°S, 48°W.

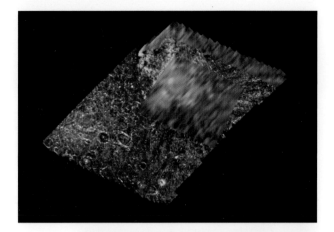

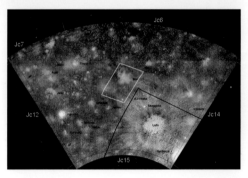

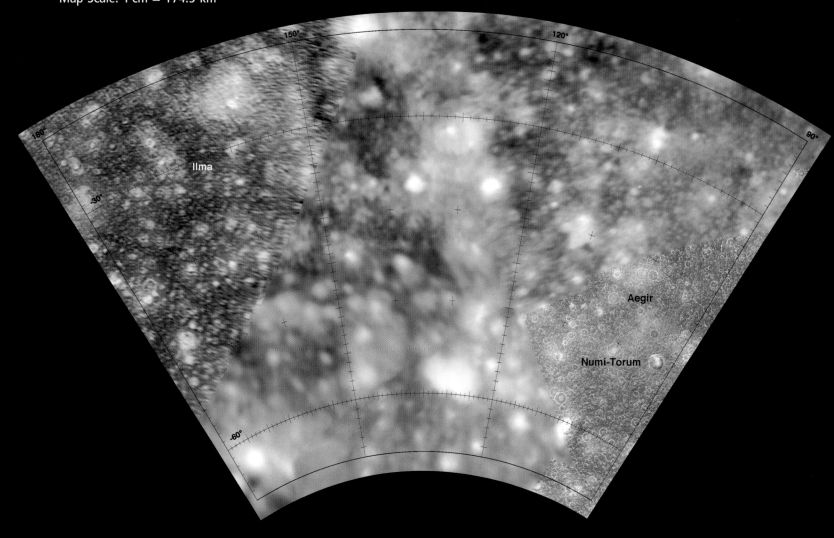

Plate Jc12 Ilma Quadrangle

Typical of Callisto, the cratered plains here feature many irregular amoeba-like dark and bright patches of unknown origin. Similar patterns are strongly evident in quadrangle Jc7. The two circular bright patches at 32°S, 130°W and 24°S, 158°W are most probably large impact features and may be palimpsests analogous to those observed on Ganymede or on Callisto (quadrangle Jc9). Image resolution varies from ~10 to 0.8 kilometers across the quadrangle.

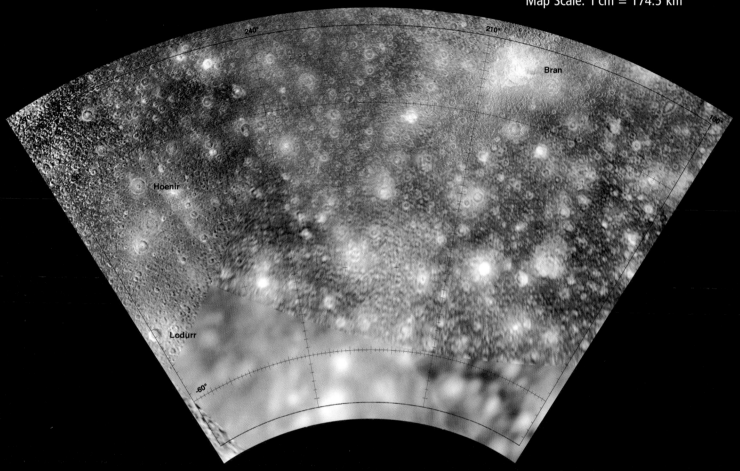

Lambert Conformal Conic Projection
Map Scale: 1 cm = 174.5 km

Plate Jc13 Hoenir Quadrangle

Standing out amid the endless cratered plains is Bran. Bran is relatively young and is one of the largest "normal" Moon-like impact features on Callisto (see Plate Jc13.1). An apparent highly degraded ancient multi-ring impact basin is centered at 50°S, 225°W, although it may be artifacts in the oblique *Voyager* imaging. *Voyager* 2 observed most of this terrain at resolutions of only 2 to 4 kilometers.

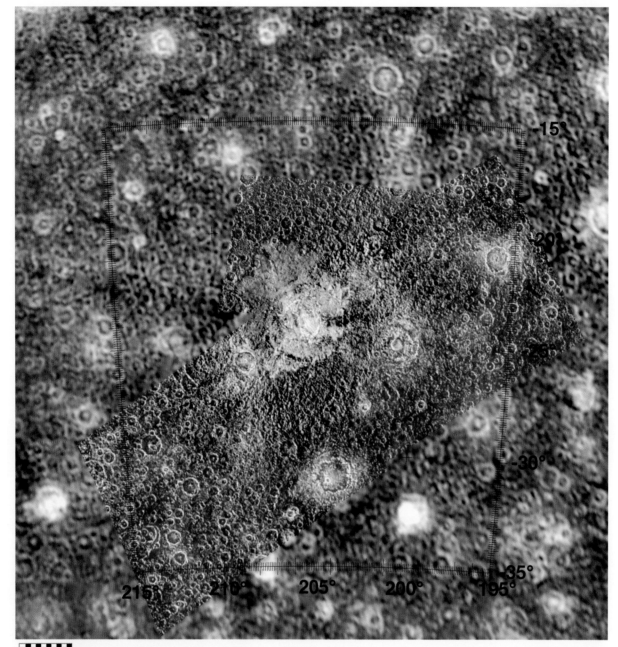

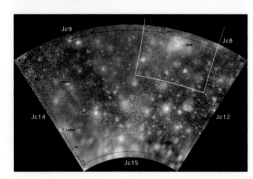

Plate Jc13.1 **Bran**

Encounter: *Galileo* C20   Resolution: 600 meters/pixel (northern image)
Encounter: *Galileo* C30   Resolution: 630 meters/pixel (southern images)
Orthographic Projection   Map Scale: 1 cm = 70.9 km

Bran, at 148 kilometers diameter, is one of the largest intact "regular" impact craters on Callisto. The brightness of the floor and rim indicates it is relatively young in age, possibly less than 1 billion years old or so. Although it has a sharp rim scarp and a central dome and central ring complex similar to other large impact craters seen on Ganymede and Callisto, parts of the northern rim are obscured or missing. A large relatively smooth bright grayish deposit extends to the north and west from this "missing rim." This smooth deposit is probably part of the impact deposit that usually forms in the floors of large craters, except that here it appears to have breached the rim and flowed or was thrown onto surrounding terrain during impact. Surrounding areas include cratered plains and large impact features of various morphologies and ages, including a partly buried older penedome crater just to the east at 25°S, 200°W. This is a 3-frame mosaic of images obtained during two different *Galileo* orbits.

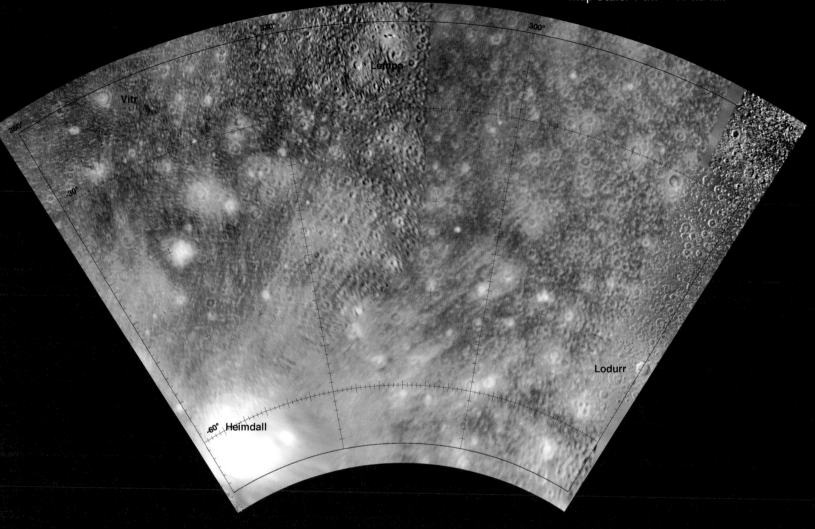

Plate Jc14 Lempo Quadrangle

Although poorly imaged, the bright impact basin Heimdall dominates this region (see also Plate Jc11). High-Sun imaging precludes an assessment of its morphology. Unfortunately, the only portions of this structure that are resolved are several rugged ridges along the western margin shown in the Adlinda quadrangle (Plate Jc11), indicating that there is a lot of relief within this structure. The age of Heimdall is unknown, but the high albedo, lack of bright rays and the bright-rimmed craters superposed on Heimdall indicate that it is relatively young but not recently formed. It appears to be somewhat younger than either Adlinda or Lofn. Extensive radial textures appear to extend from Heimdall to the north and east. Whether these are tectonic grooves or ejecta rays is unknown. Also evident within these cratered plains are several irregular bright patches of enigmatic origin, such as the patch at 42°S, 325°W.

Polar Stereo Projection
Map Scale: 1 cm = 124 km

Plate Jc15 Keelut Quadrangle

The south polar region is typical of Callisto: heavily cratered terrain separated into relatively bright and dark patches. Here, close inspection suggests that some of the brighter patches include smooth areas. These could be attributable to volcanic resurfacing, but no direct evidence has as yet been found to confirm or refute that volcanism ever occurred on Callisto. Bright frost-covered pole-facing slopes similar to those in north polar terrains should be present here, but if they exist they are not obvious in these mosaics.

This quadrangle is unusual in that it is the only region in the *Atlas* to include NIMS imaging data in any of the global mapping mosaics. Although low in resolution (only ~9 kilometers), NIMS imaging of much of the region between 115° and 170°W longitude is better than any available *Voyager* or *Galileo* imaging data. The NIMS channel used here is in the visible band of the spectrum. These data allow us to map the extent of a large polygonal patch of bright material centered near 80°S, 90°W. We are not much closer to understanding the origin or these bright (and dark) regions, other than to confirm that darker areas are less ice-rich than brighter areas.

Ganymede

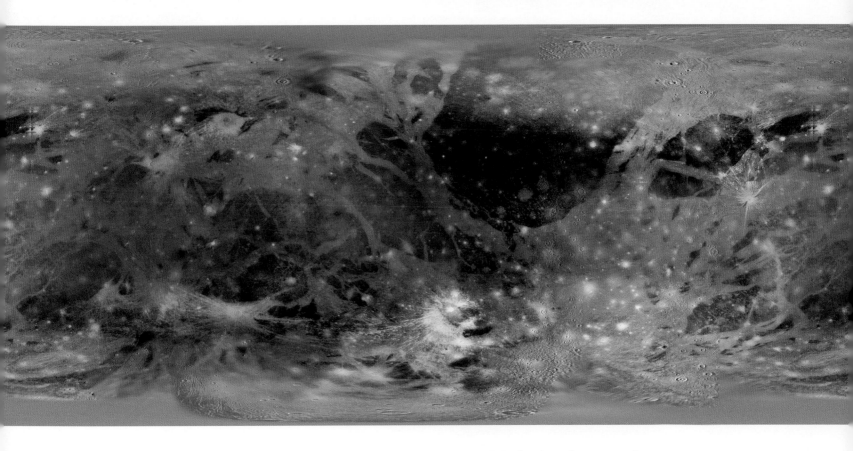

Plate Jg **Global Map of Ganymede**
Simple Cylindrical Projection
Map Scale: 1 cm = 676 km
(valid on equator and along longitude lines)

Orthographic Projection: Center 42°N, 130°W

Plate Jg01
Global View 1: Galileo Regio and North Polar Cap

Two major features dominate Ganymede when viewed from great distances. The huge dark feature in the center, Galileo Regio, is the largest contiguous geologic feature on Ganymede. Astronomers may have seen Galileo Regio from Earth and *Pioneer* definitely saw but did not resolve it. It is one of numerous remnants of dark terrain, which predate the younger bright terrains. Arcuate furrows cross Galileo Regio. The other global feature is the bright polar frost cap covering much of Galileo Regio, a small part of which was resolved by *Galileo* at high resolution (Plate Jg3.2). The polar caps are believed due to charged-particle sputtering and redistribution of water ice controlled by Ganymede's magnetic field.

Orthographic Projection: Center 20°S, 175°W

Plate Jg02
Global View 2: Large Impacts, Old and Young

This view highlights three major impact events and three major epochs in Ganymede's history. The most obvious is the prominent bright-ray crater, Osiris (Plate Jg12.1), which may be the youngest geologic feature on the surface. To the southeast lies Gilgamesh (Plate Jg12.2), a somewhat older impact basin 600 kilometers across. It is the largest fully preserved impact feature on Ganymede and formed not long after bright terrain. This view is centered on the ancient dark-terrain furrow system, the concentric sets of arcuate graben that formed as a result of a very large impact early in Ganymede's history. Many of these furrows can be seen in Galileo Regio at upper left.

Orthographic Projection: Center 30°S, 355°W

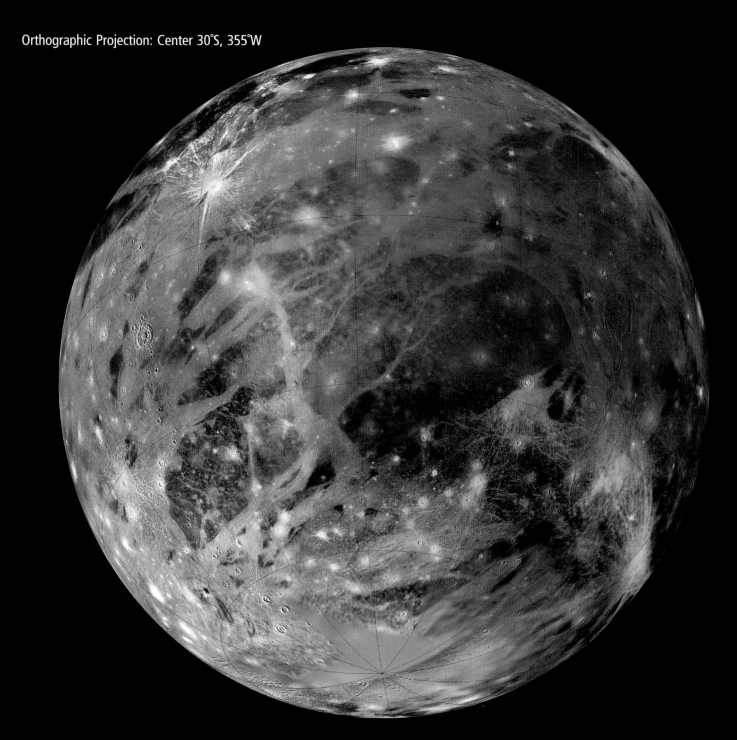

Plate Jg03
Global View 3: Nicholson Regio

This region contrasts with and is nearly antipodal to Galileo Regio. Like the Galileo Regio hemisphere (Plate Jg01), dark terrain dominates this side. Dark terrain is considerably more fragmented on this hemisphere, however. The largest block, Nicholson Regio, lies near center. Long bright rays extend from the very young central dome crater Tros at upper left. The color contrast between equatorial and polar regions is also evident in this view.

Orthographic Projection: Center 0°N, 90°W

Plate JgHL
Global View: Leading Hemisphere

Large tracts of contiguous bright and dark terrain dominate the "forward"-facing side of Ganymede. The large block of dark terrain is Galileo Regio. The contrast between dark and bright terrains is greater on this hemisphere than on the trailing hemisphere. The bright polar frost caps are dramatically evident. Also noticeable is a color contrast between equatorial latitudes, which tend to be subtly redder than higher latitudes in these enhanced-color mosaics. The boundary between this color change is ~25° latitude on this side, and is much closer to the equator than on the trailing hemisphere.

Orthographic Projection: Center 0°N, 270°W

Plate JgHT
Global View: Trailing Hemisphere

The trailing face of Ganymede has considerably fewer (~25%) impact craters than the leading hemisphere (compare quadrangles Jg7 and Jg9). This asymmetry is a consequence of orbital geometry, as more comets are expected to strike the leading hemisphere. Dark-ray craters (Plate Jg9) are more prevalent on this hemisphere, especially on the eastern side where *Voyager* obtained high-Sun images.

The center longitudes (265–300°W) also have the lowest-resolution coverage (3.5 kilometers) on Ganymede. The color difference between reddish equatorial and bluer higher latitudes is especially evident in this view, although here the boundary between the two regions lies at ~40° latitude. The origin of this color contrast is related to Ganymede's magnetic field and plasma bombardment of the surface.

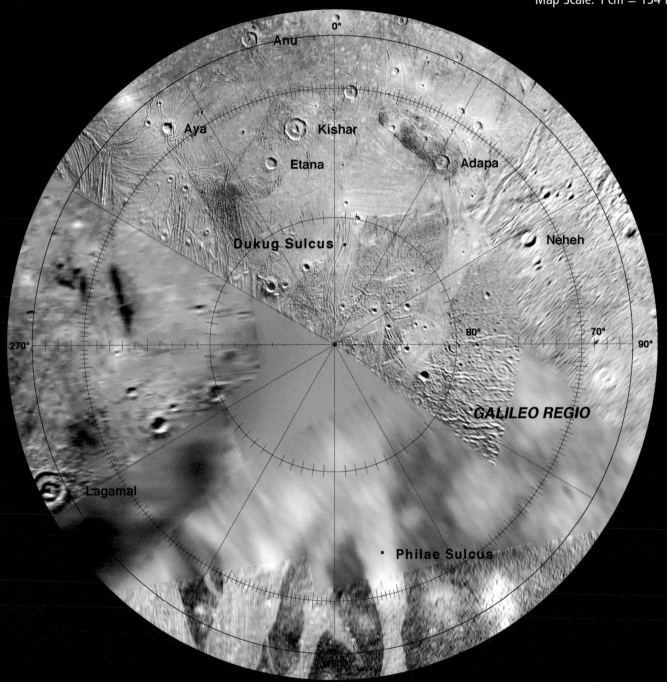

Plate Jg1 **Etana Quadrangle**

Ganymede's polar regions are covered in a thin water-ice frost, forming thin polar "caps." This deposit extends down to mid-latitudes (approximately 45°), and covers this entire quadrangle, but is not thick enough to obscure geologic features (see Plate Jg3.2). Ancient furrows and circular impact scars called palimpsests can be seen on the dark terrain of Galileo Regio (see Plate Jg3).

Lambert Conformal Conic Projection
Map Scale: 1 cm = 185 km

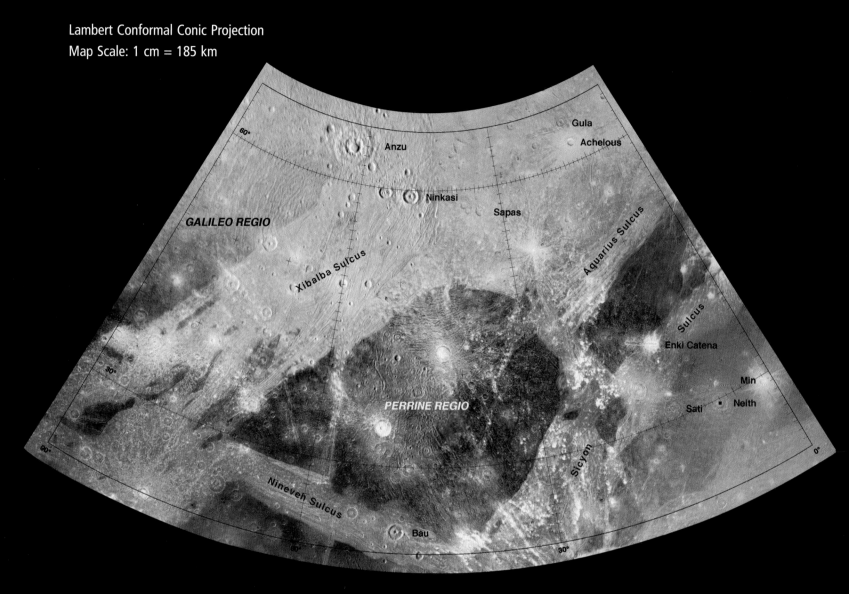

Plate Jg2 **Perrine Quadrangle**

Perrine Regio, a remnant of ancient dark terrain, is one of the darkest regions of Ganymede. At least two sets of parallel curved furrows (ancient graben) cross Perrine Regio. Part of Galileo Regio can be seen to the north and west. The remnants of a small multi-ring furrow system, similar to the much larger system that extends over most of Galileo Regio, is centered at 57°N, 68°W. Xibalba Sulcus, a broad lane of younger bright terrain, separates the two regios. At least two large penedome craters, Neith (Plate Jg2.1) and Anzu are present. These older craters are actually elevated above surrounding terrain, the opposite of normal craters.

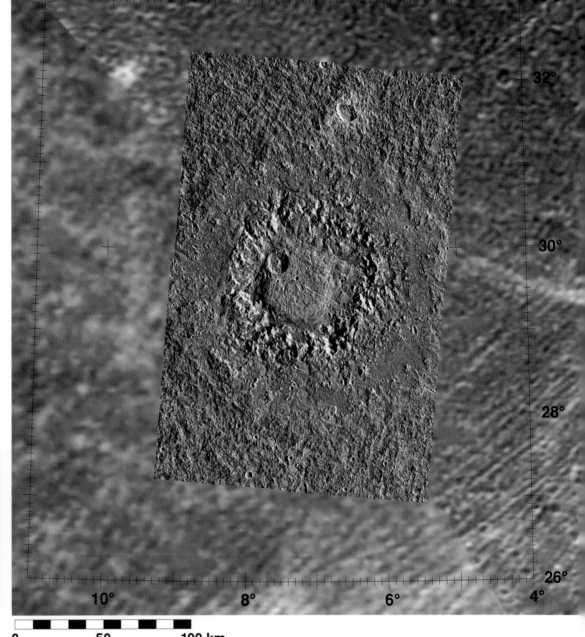

Plate Jg2.1 **Neith**   Encounter: *Galileo* G7   Resolution: 140 meters/pixel   Orthographic Projection   Map Scale: 1 cm = 20.8 km

Neith is a type example of a penedome crater on Ganymede. Although craters of this type are broadly similar to central dome craters such as Eshmun or Melkart (see Plates Jg8.10 and Jg8.11), key features are missing. Neith is dominated by a central dome and inner ring of massifs. Like Melkart, this dome is formed of warm soft ice uplifted from depth. The rim, however, which should have a diameter about twice as large as the inner massif ring and a kilometer or more of relief, is mostly absent. The zone between the inner massif ring and the putative rim is mostly smooth, possibly due to water ice partly melted during impact ponding in topographic lows. Craters like Neith are somewhat younger than bright terrain but no longer form today. The unusual morphology of Neith is related in some way to higher heat flow in the ancient past of Ganymede and Callisto. Compare with similar craters Menhit (Plate Jg12.3) on Ganymede and Doh on Callisto (Plate Jc3.3-X1).

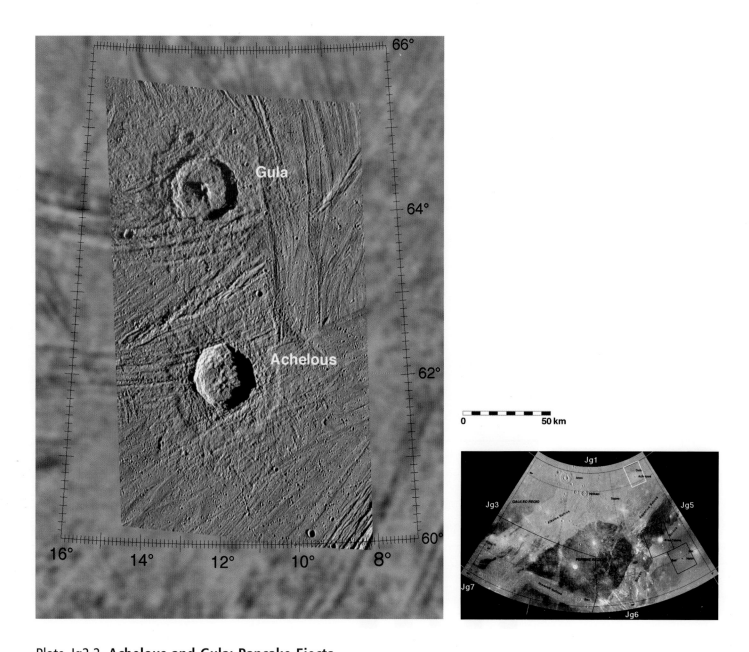

Plate Jg2.2 Achelous and Gula: Pancake Ejecta

Encounter: *Galileo* G7   Resolution: 180 meters/pixel   Orthographic Projection   Map Scale: 1 cm = 20.1 km

Galileo targeted the large craters Achelous and Gula, 35 to 38 kilometers across, respectively, in order to image pancake ejecta deposits. Once termed pedestal ejecta, they form an annular plateau outside the crater rim (see Plate Jg10.6.1). Their similarity to pancake ejecta craters on Mars led to speculation that deposits on both might be related to liquid water in the ejecta. At high resolution, these deposits are anything but smooth. These images reveal no direct evidence that liquid water was part of the ejecta deposit during formation. Rather, the deposit can be thought of a type of large circular landslide formed during impact.

The mosaic also reveals that, as can be seen by the lengths of the shadows, Gula is about half as deep as Achelous, despite the fact they are the same size. This difference in depth may be related to the fact that Gula has a prominent central peak. Achelous does not have a central peak and floor collapse and uplift was less severe, making the floor relatively deep. In the background are several grooved terrain units, which are peppered with small secondary craters from both craters.

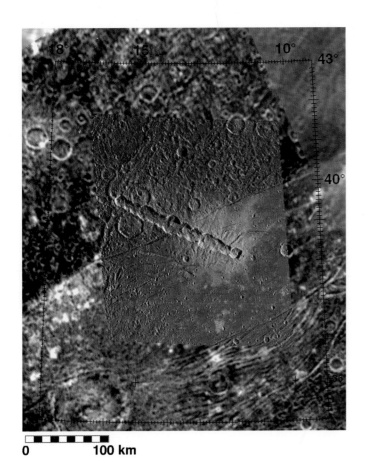

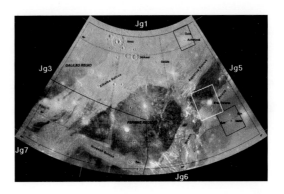

Plate Jg2.3 Enki Catena

Encounter: *Galileo* G7   Resolution: 550 meters/pixel
Orthographic Projection   Map Scale: 1 cm = 43.3 km

Enki Catena is a relatively young crater chain that most likely formed from the impact of a tidally disrupted comet similar to Comet Shoemaker-Levy 9. This comet broke into at least 13 coherent fragments after passing very close to Jupiter before hitting Ganymede. Similar chains can be found elsewhere on Ganymede and Callisto (see Plates Jc1, Jc2.1, and Jg10.2). The lack of walls or scarps between the individual craters indicates that the impacts formed almost simultaneously. It is interesting that bright impact ejecta are visible only on the southern half of the crater chain, which are on bright terrain. Dark terrain is inherently less able to create or retain bright crater ejecta due to greater amounts of dark particulates. Note also the relaxed topography of all but a few of the impact craters on dark terrain.

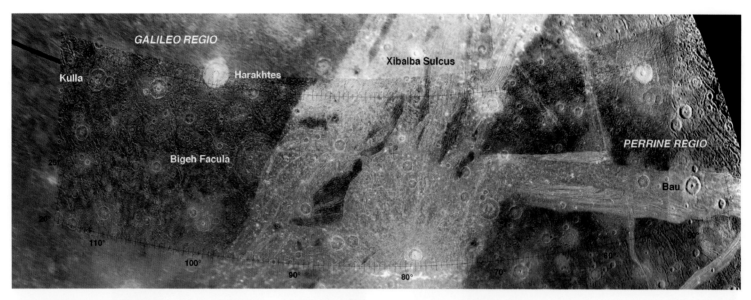

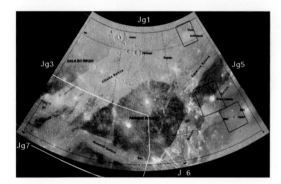

0 150 300 km

Plate Jg2.4 From Perrine to Galileo Regio

Encounter: *Galileo* C9   Resolution: 850 meters/pixel
Orthographic Projection   Map Scale: 1 cm = 154.5 km

This four-frame mosaic across Perrine Regio, Xibalba Sulcus, and Galileo Regio provides our most detailed imaging of these longitudes and the bright terrain between. Due to the *Galileo* antenna failure, what should have been a hemisphere-wide mosaic at our best resolution was reduced to this four-image strip. Sections of bright terrain near 26°N, 40°W, and 28°N, 92°W, have spotted rather than grooved textures. This type of bright terrain is relatively rare, and was imaged once at very high resolution in Plate Jg3.1. Shown at 75% resolution to fit page.

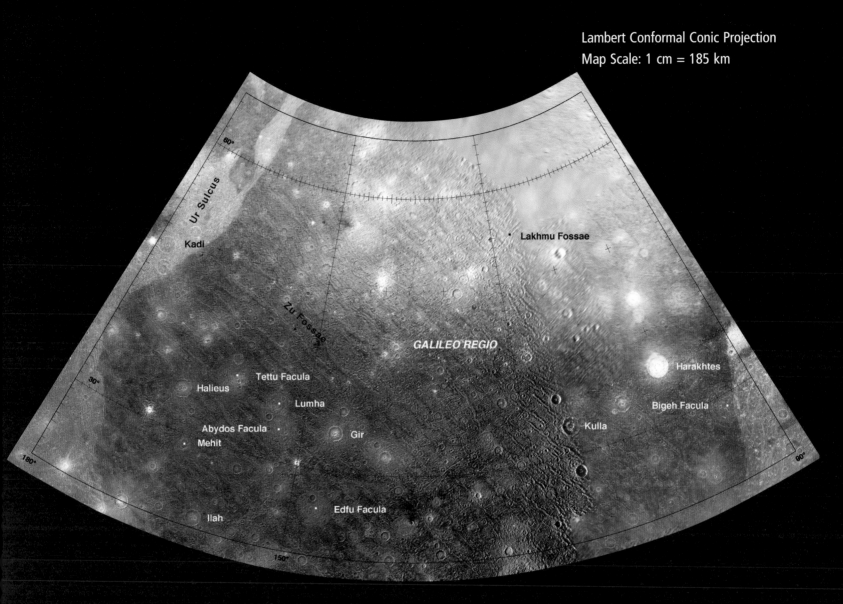

Lambert Conformal Conic Projection
Map Scale: 1 cm = 185 km

Plate Jg3 Galileo Regio Quadrangle

Galileo Regio, which includes almost all of this quadrangle and parts of several others, is by far the largest block of ancient terrains preserved on Ganymede. The bright parallel lineations (Zu Fossae) crossing the terrain are ancient graben-like fractures known as furrows. They probably formed during a giant impact event, the original center of which lies well to the south (at 20°S, 178°W) and has since been destroyed by bright-terrain formation. Most of these furrows are concentric, but several are quasi-radial to the center of the structure. The bright region across the northern part of Galileo Regio is part of the northern polar frost cap of Ganymede (see Plate Jg3.2). Several cryptic quasi-circular ancient palimpsests are also visible.

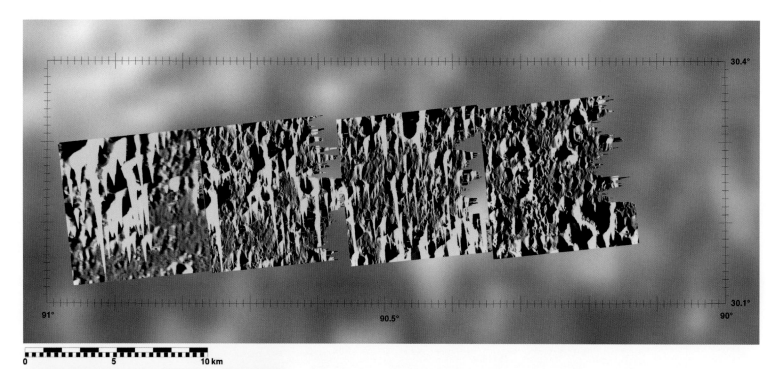

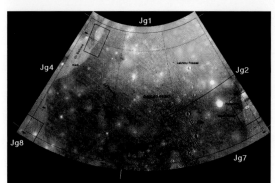

Plate Jg3.1 Xibalba Sulci: Highest-resolution Imaging of Ganymede
Encounter: *Galileo* G1   Resolution: 11 meters/pixel
Orthographic Projection   Map Scale: 1 cm = 1.98 km

This four-frame mosaic, of which the first is smeared, is our highest-resolution image of Ganymede. The mosaic is located in an area of bright terrain that may not be typical of bright terrain elsewhere (see Plate Jg3.1.1). The grooves so prevalent on bright terrain are absent in this mosaic. Instead we see a landscape of craters scattered amidst rounded and broken hills of various sizes, most of which trend north–south. There is also a more subtle north–northeast fabric of lineation, parallel to the general groove direction of nearby grooves seen in the context mosaic. Details can often be seen in shadows due to backlighting. The mosaic is afflicted with an excessive amount of bleeding, a phenomenon due to overexposure and saturation of the pixels on the *Galileo* CCD. The saw teeth along the right-hand edges of each frame are due to a type of image compression that truncates image lines depending on content. Shown at 60% resolution to fit page.

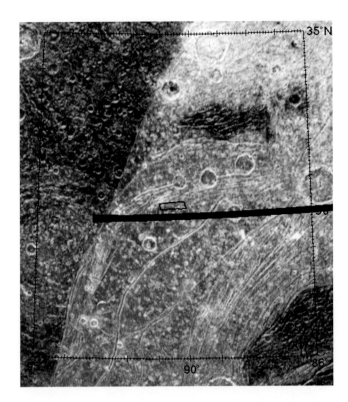

Plate Jg3.1.1 Xibalba Sulci: Highest-resolution Imaging of Ganymede, Context Map

The G8 context imaging for Plate Jg3.1 provides surprisingly little insight because it was planned at too low a resolution (850 meters) to fully understand the features seen in the small ultra-high-resolution mosaic. These first *Galileo* highest-resolution views were targeted in the blind based on even poorer resolution *Voyager* images. In fact, the precise location of the mosaic was unknown for some years, but the largest craters in the mosaic easily correlate with bright spots in the low-resolution context imaging. These are likely small secondary craters from nearby large craters.

Plate Jg3.2-X1 Northern Polar Frost Cap: Close-up 1
Encounter: *Galileo* G2   Resolution: 45 meters/pixel
Orthographic Projection   Map Scale: 1 cm = 5.3 km

These enlargements of Plate Jg3.2 reveal bright frosts in the southwestern corners of small impact craters at high latitudes. These pole-facing slopes are relatively cold during the Ganymedean day and water frosts are more stable here. Slopes that face south and are warmer through the day are relatively frost-free. Examples of such deposits can be seen in the southern hemisphere as well (Plate Jg14.1.2). The northern scene (Plate Jg3.2-X3) also shows the crater-scarred boundary between bright (grooved) terrain to the left and dark (cratered) terrain to the right.

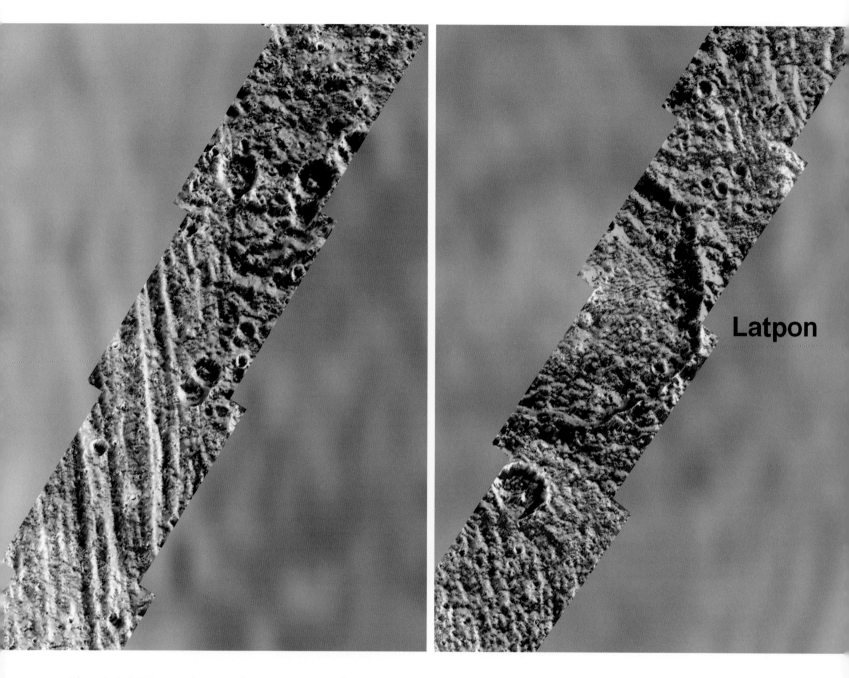

Plate Jg3.2-X2 **Northern Polar Frost Cap: Close-up 2**

Plate Jg3.2-X3 **Northern Polar Frost Cap: Close-up 3**

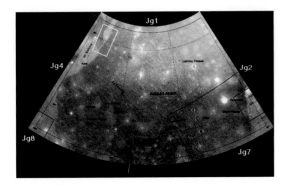

Plate Jg3.2 **Northern Polar Frost Cap**
Encounter: *Galileo* G2
Resolution: 45 meters/pixel
Orthographic Projection
Map Scale: 1 cm = 21.3 km

This long narrow 16-frame mosaic crosses four geologic units and one large crater, Latpon. The craters at the bottom of the mosaic are secondary craters from Epigeus (Plate Jg4.5). This is the most polar of *Galileo*'s Ganymede observations and was designed to elucidate the nature of Ganymede's polar frost caps (Plates Jg01, JgHL, and Jg3). These images were over-compressed by the onboard software, resulting in box-like compression artifacts throughout the mosaic. Despite this, details of the frost distribution are revealed. These images show that the polar caps formed from discrete local deposits of frost locally concentrated in pole-facing slopes (see enlargements). This frost distribution is due to plasma bombardment in polar regions, controlled by Ganymede's intrinsic magnetic field, redistributing water-ice molecules to colder pole-facing slopes. Shown at 25% resolution to fit page.

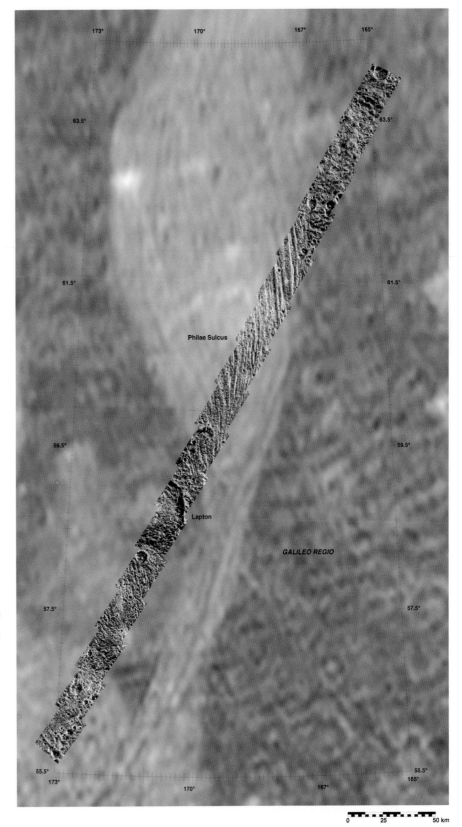

Lambert Conformal Conic Projection
Map Scale: 1 cm = 185 km

Plate Jg4 Philus Sulcus Quadrangle

This quadrangle is typical of Ganymede. The large blocks of ancient dark terrain, Marius Regio, have been partially broken apart by narrow bands of younger bright terrain. *Galileo* obtained several high-resolution observations of this process. Also, the large impact basin Epigeus is visible in the southeast corner and much of this area is influenced by ejecta, including large secondary craters, from this large basin.

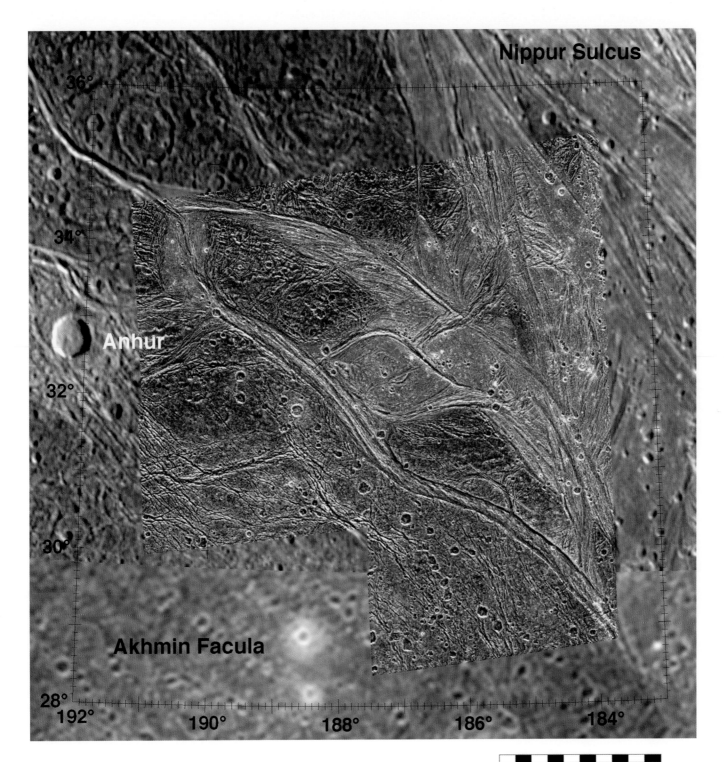

Plate Jg4.1 **North Marius Regio**
Encounter: *Galileo* G2   Resolution: 190 meters/pixel
Orthographic Projection   Map Scale: 1 cm = 22.4 km

Dark terrain in this area is broken into small blocks by narrow lanes and patches of bright terrain in this 4-frame mosaic. Both smooth and grooved bright terrain is evident. Dark terrain here is extensively fractured, especially toward the south. These are likely relics of bright-terrain formation. The medium-sized craters toward the south and east are secondaries from Epigeus, located just out of view to the southeast (Plate Jg4.5).

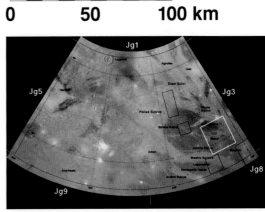

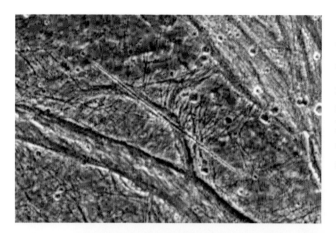

Plate Jg4.1-X North Marius Regio: Close-up of Crater Chain
Orthographic Projection   Map Scale: 1 cm = 9 km

This narrow needle-like feature, shown in this enlargement from Jg4.1, is radial to the large nearby impact basin, Epigeus (Plate Jg4.5), as is a similar radial feature located to the south on Buto (see Plate Jg8.3). Both are secondary crater chains from Epigeus. They differ from disrupted comet crater chains (Plate Jg2.3) in that they are spatially associated with a specific impact basin.

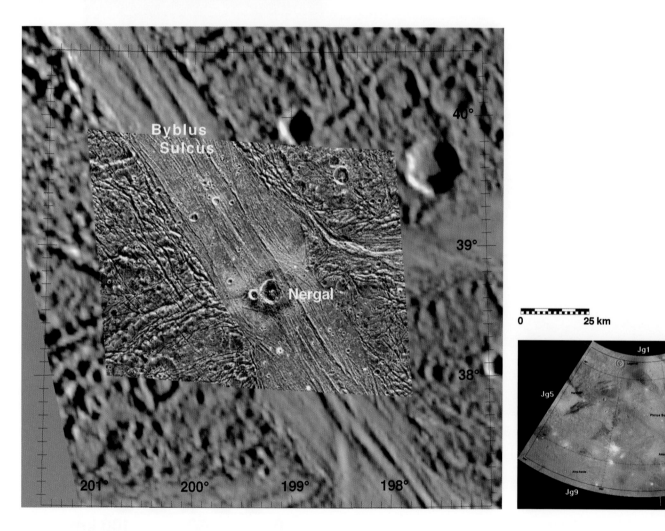

Plate Jg4.2 Byblus Sulcus
Encounter: *Galileo* G2   Resolution: 85 meters/pixel
Orthographic Projection   Map Scale: 1 cm = 13.4 km

This two-frame mosaic shows the intersection of two narrow lanes of bright terrain. The background image is a low-Sun *Galileo* image at 900 meters resolution that gives us a sense of the relief of these terrains. At the center lies the 8.5-kilometer-wide dark-floored crater, Nergal (see also Plate Jg8.7). The block of dark terrain to the west includes a large degraded and truncated furrow at 38.5°N, 200.5°W, similar to those in Galileo Regio (Plates Jg3 and Jg8.1). Numerous narrow fractures similar to those in Plates Jg4.1 and Jg4.3 and elsewhere occur on both sides of Byblus Sulcus, indicating that a pervasive episode of extensional deformation occurred across Marius Regio prior to or in association with emplacement of bands of the bright terrain.

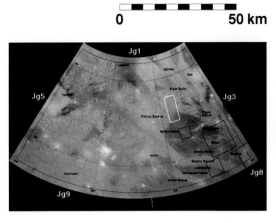

Plate Jg4.3 **Nippur Sulcus**

Encounter: *Galileo* G2   Resolution: 100 meters/pixel
Orthographic Projection   Map Scale: 1 cm = 15.8 km

This three-frame *Galileo* mosaic shows crosscutting bands of bright terrain units that have been deformed to different degrees. The narrow diagonal band Nippur Sulcus (at 48°N, 204°W) is smooth and is the least deformed. Only a few faint lineations betray minor deformation. The smoothness is consistent with volcanic resurfacing by water lavas. This unit truncates a lightly grooved packet of bright terrain to the south. To the north of this narrow band lies a wide zone of intensely grooved and fractured bright terrain. This zone can be seen partly fracturing and destroying a zone of older lightly grooved bright terrain at the top of the mosaic near 52°N, 203°W.

The dark terrain to the south is itself heavily deformed by numerous narrow grooves and fissures. The ancient impact crater at 45.5°N, 203°W is relaxed and distorted, indicative of higher heat flow in the distant past.

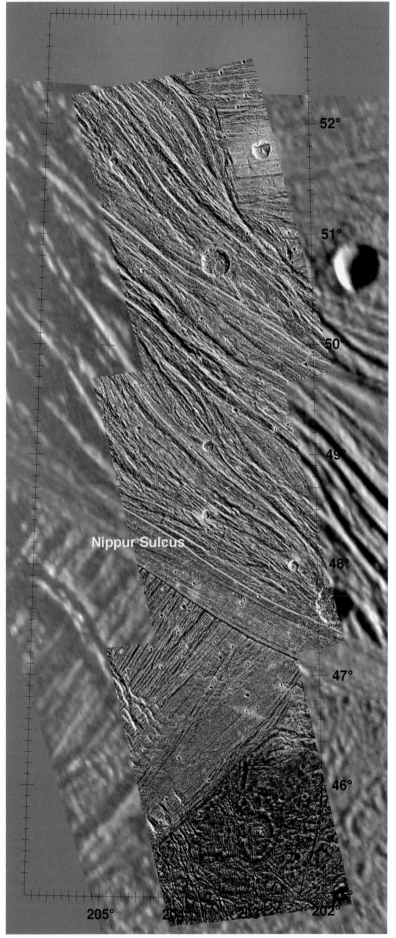

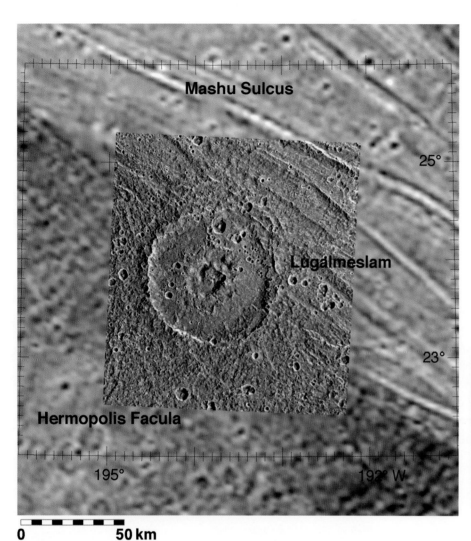
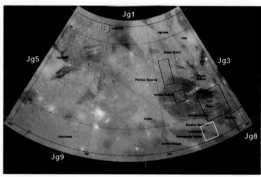

Plate Jg4.4 **Lugalmeslam**
Encounter: *Galileo* G8   Resolution: 145 meters/pixel
Orthographic Projection   Map Scale: 1 cm = 17.7 km

Lugalmeslam is a typical, if somewhat older and deformed, crater 63 km across. It is transitional from central pit to central dome crater, in that a small central plug appears to partly fill the deformed central pit. The crater straddles the contact between bright and dark terrain to the south. Although the northeast rim is elongate, it is not clear if groove formation is responsible. Many of the larger craters superposed on top of Lugalmeslam and surrounding areas are secondary craters from the Epigeus impact basin (Plate Jg4.5) to the east.

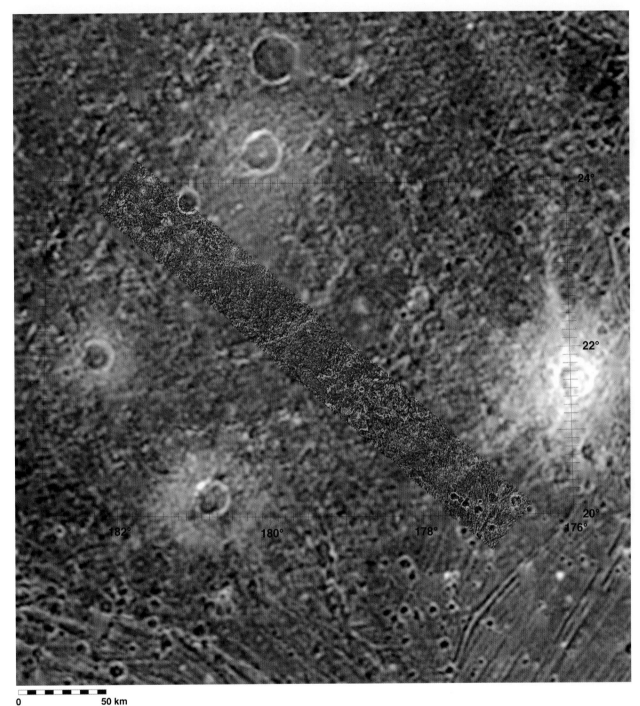

Plate Jg4.5 **Epigeus**

Encounter: *Galileo* G2   Resolution: 90 meters/pixel
Orthographic Projection   Map Scale: 1 cm = 20.8 km

This post-bright-terrain impact basin is one of the largest and best preserved of its type. Both *Voyager* global mapping and this narrow radial *Galileo* mosaic transect were obtained at local noon, however, washing out many topographic details. Nonetheless, abundant details can be seen. Two concentric rings are apparent in this southwestern quadrant. These form two concentric zones of ridges and massifs, separated by (superficially) smooth units.

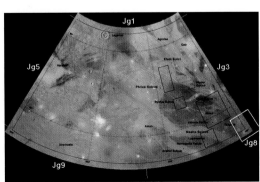

Plate Jg4.5-X1 Epigeus: Close-up 1: Central Unit and Inner Ring
Orthographic Projection   Map Scale: 1 cm = 10.4 km

The otherwise smooth central area of Epigeus at upper left is intensely fissured with a web of very narrow fractures. The inner ring is a narrow ridge several hundred meters high, which appears here as a bright arcuate scarp at frame center. The outer ring (in the southeast corner) is a complex zone corresponding to the crater diameter (210 kilometers).

The zone between the two rings is relatively smooth and is covered in small knobs. Several lobes of this smooth unit appear to have flowed into valleys within this outer ring zone. The smooth units in the center and between the inner and outer rings may be a solidified impact-melt sheet.

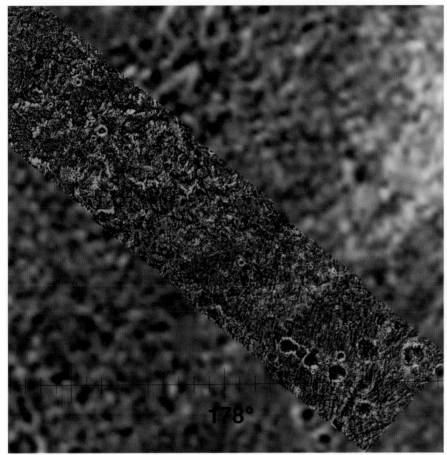

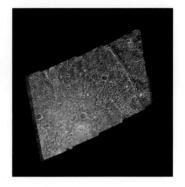

Plate Jg4.5-N **Epigeus: NIMS Infrared Map**

Encounter: *Galileo* G2
Resolution: 6.5 kilometers/pixel

This map of the infrared spectra of the northern Epigeus area clearly shows the high ice content of the band of bright terrain and the high non-ice component of the dark terrain to either side. The Epigeus impact basin has a greenish color in this version, indicating that the basin deposits are ice-rich, but also have a distinctly unique composition or texture.

Plate Jg4.5-X2 **Epigeus: Close-up 2: Outer Ring and Ejecta**

Orthographic Projection   Map Scale: 1 cm = 10.4 km

In this view we see the outer ring at upper left, a wide zone of swirling bright scarps and smooth-floored valleys. These are analogous to the crater rim deposits. The large craters at lower right are secondary craters. Between these and the rim lie relatively smooth textured ejecta deposits.

Lambert Conformal Conic Projection
Map Scale: 1 cm = 185 km

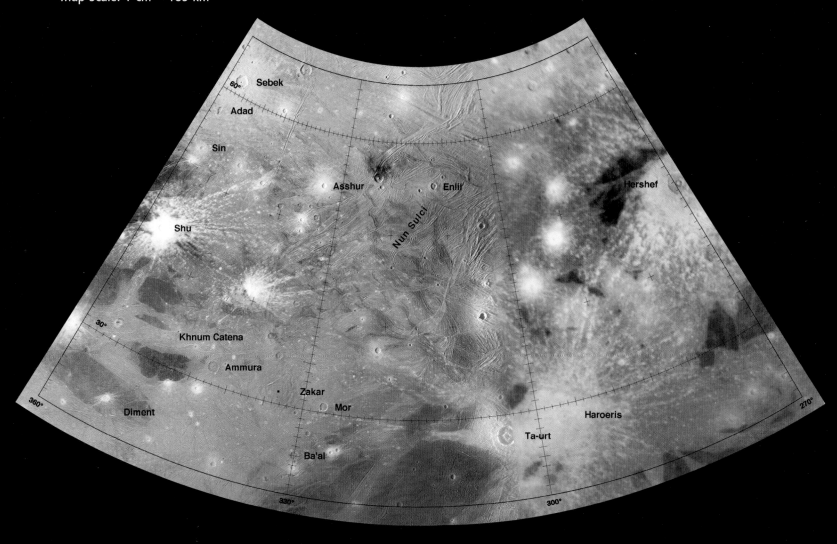

Plate Jg5 Nun Sulci Quadrangle

Extensive tracts of smooth and grooved bright terrain cross this quadrangle. The crosscutting lanes of Nun Sulci form a complex sequence of deformation episodes that were the focus of a *Galileo* observation (Plate Jg5.1). Several prominent craters populate this region, including the large penepalimpsest Zakar. Bright ejecta from several prominent young bright-rayed craters are spread across much of this region. Asymmetric dark ray and bright craters are located at 56°N, 323°W and 41°N, 342°W.

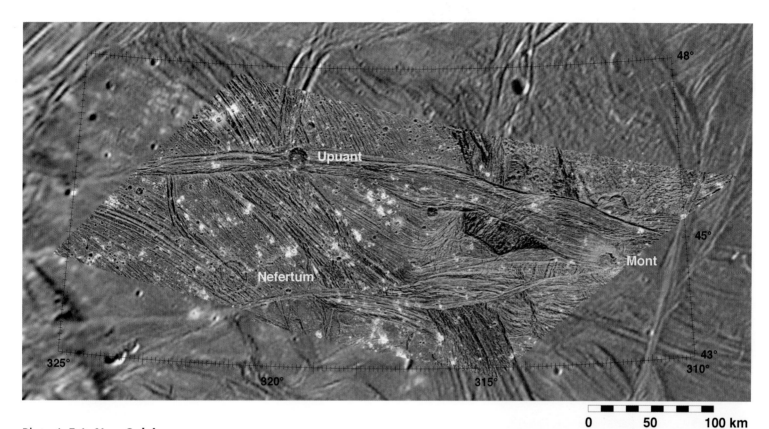

Plate Jg5.1 **Nun Sulci**

Encounter: *Galileo* G7   Resolution: 170 meters/pixel
Orthographic Projection   Map Scale: 1 cm = 29 km

This 2-frame mosaic shows a complex set of intersecting and crossing bands of bright terrain. The older crater in the southwest corner (44°N, 321°W) has been extensively fractured and elongated as groove formation stretched the icy crust. The scallop-shaped features in the eastern corner (45.5°N, 312°W) and at 43°N, 324°W are volcanic calderas similar to those in Sippar Sulcus (Plate Jg13.1). The bright splotches are secondary craters from the bright-ray craters Tros (Plate Jg6) and Haroeris located to the southeast and the bright-ray craters to the west.

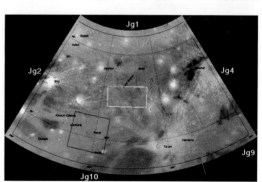

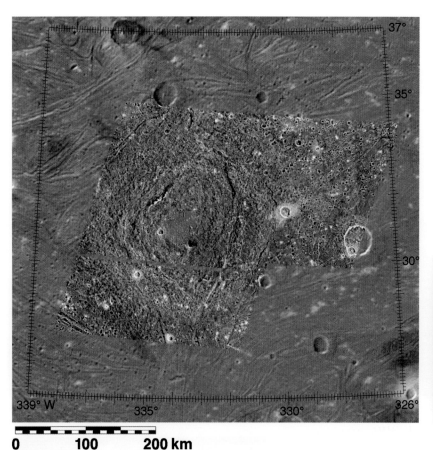
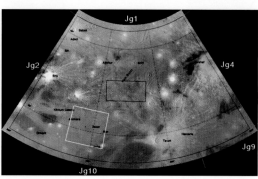

Plate Jg5.2 Zakar
Encounter: *Galileo* G7   Resolution: 440 meters/pixel
Orthographic Projection   Map Scale: 1 cm = 52 km

Zakar, a typical penepalimpsest, does not have a sharply defined rim scarp but consists rather a set of low concentric ridges. There is no deep basin. Rather the entire floor may be elevated 1 or more kilometers above the surrounding terrains. Beyond the concentric ridges lies a mottled and etched zone (the ejecta blanket) surrounded by a large field of secondary craters. The original crater diameter is approximately 162 kilometers, corresponding with one of the outer concentric ridges. This crater and others like it probably formed at a time during or just after bright-terrain formation (due to the fact that the density of superposed craters is the same as on bright terrain), when the icy lithosphere of Ganymede was very warm, soft and less able to support the steep-sided topographic features characteristic of normal impact craters.

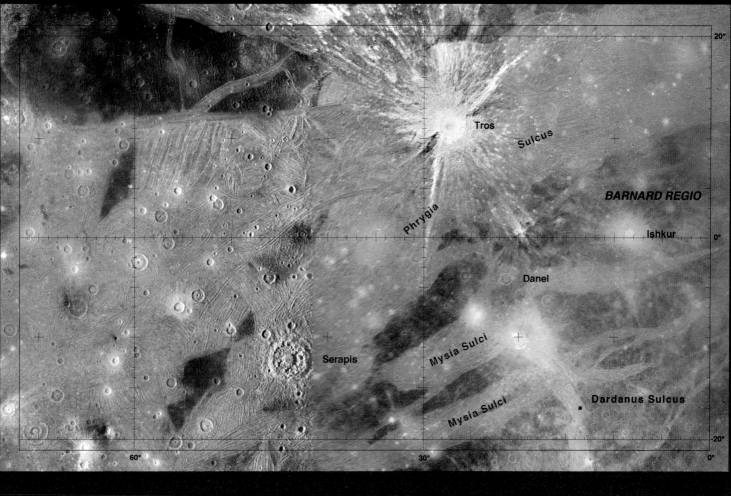

Plate Jg6 Dardanus Sulcus Quadrangle

Extensive tracks of bright terrain formed in this region, but it is the large impact craters that stand out. Tros, a 94-kilometer-wide bright-rayed central dome crater, is one of the youngest on Ganymede, and may be only slightly younger than Osiris (Plate Jg12.1). This type crater contrasts sharply with the older 175-kilometer-diameter penedome crater, Serapis. This type of crater formed several billion years ago, not long after bright terrain during a period when Ganymede's heat flow was higher than it is now. Large impacts into warmer ice apparently form these modified impact craters (e.g., Plate Jg2.1) due to the enhanced flow properties of ice.

Map Scale at Equator: 1 cm = 175 km

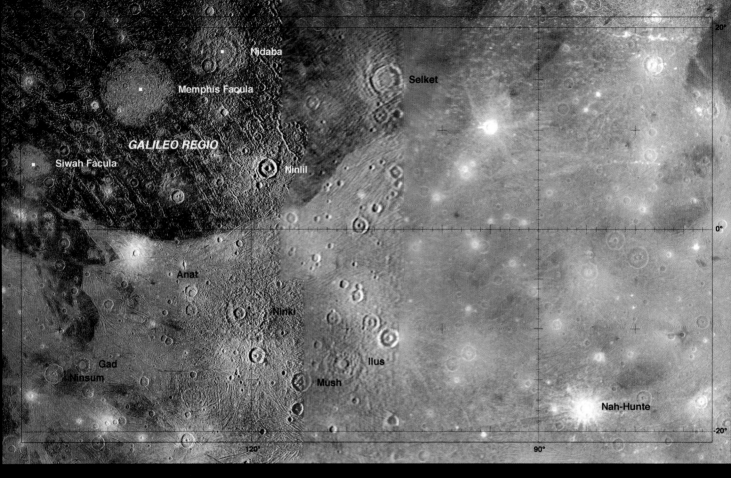

Plate Jg7 Memphis Facula Quadrangle

The apex area centered on 0°, 90°W is the region that faces forward as Ganymede orbits Jupiter. As a result, it is the most heavily cratered region on Ganymede. This region can be compared with the more sparsely cratered antapex region in Plates Jg9 and Jg10. The dark terrain is the southeast portion of Galileo Regio (Plate Jg3). The most striking features here are the two large bright circular spots, Memphis Facula and Nidaba. These are type examples of palimpsests and penepalimpsests, respectively. Both are ancient impact scars that have been flattened due to viscair flow or creep of water ice when internal temperatures were higher than they are now. Nidaba preserves more of its original features, including parts of the rim and central complex, and is likely younger than Memphis Facula. Penedome craters Ilus and Gad and at 12°N, 72°W, also stand out. The highly elliptical crater (10°N, 84°W) may have formed by a highly oblique (grazing) impactor or by a disrupted comet. The high-Sun illumination of the *Galileo* images in the eastern half of the mosaic washes out much of the surface morphology.

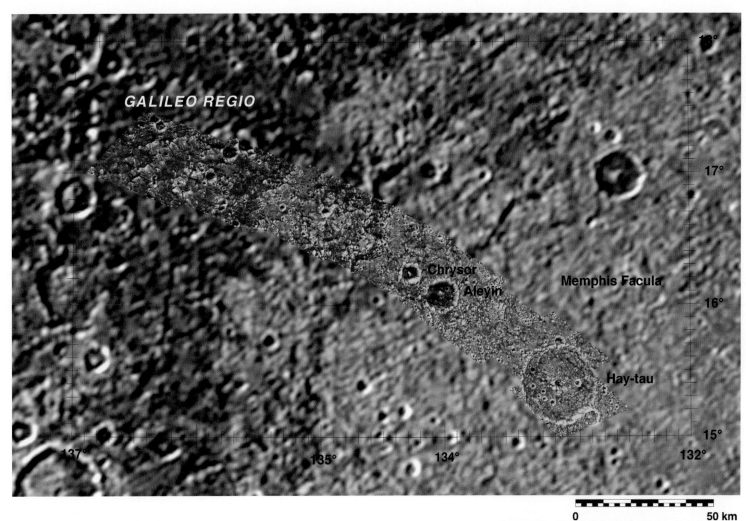

Plate Jg7.1 **Memphis Facula**
Encounter: *Galileo* G1   Resolution: 65 meters/pixel
Orthographic Projection   Map Scale: 1 cm = 13.5 km

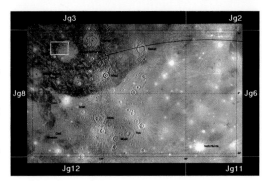

Memphis Facula is one of the largest and best-preserved palimpsests, and the only one observed at high resolution. This 355-kilometer-wide impact scar and others like it are ancient and formed before bright terrain. Although some faint concentric features can be seen in *Voyager* images, little original topographic expression is preserved, and the crater rim is not recognizable. *Galileo* completed a narrow radial transect across the western section of Memphis Facula, although it missed the central area. The outer zones of Memphis Facula consist of rounded knobby hills and massifs of various sizes, and are highly modified, consistent with its greater age. The outer edge of the bright deposit is a low scarp a few hundred meters high at most. The three named craters are interesting in that Chrysor and Aleyin are dark floored and 25-kilometer-wide Hay-tau is not. Dark-floored craters in the outer zones are believed to have excavated through a thin layer of bright material into buried dark material. By comparison with other large impact features, such as Nidaba, the original crater diameter of Memphis is probably 245 kilometers, corresponding roughly with the location of Chrysor or just inside Aleyin. Shown at 65% scale to fit page.

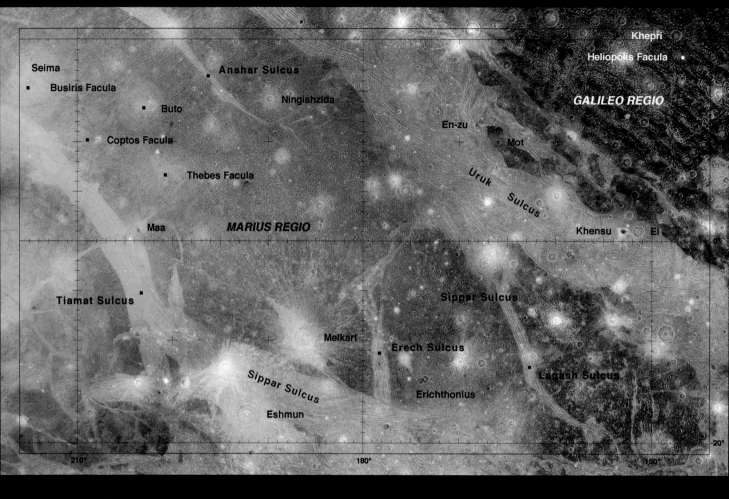

Map Scale at Equator: 1 cm = 175 km

Plate Jg8 Uruk Sulcus Quadrangle

One of the highlights of this quadrangle, aside from the rich abundance of higher-resolution images shown in the following plates, is the abundance of bright palimpsests in Marius Regio, especially to the west. These ancient impact scars are found only in dark terrain (e.g., Plate Jg3) and have been flattened due to high heat flow and creep of water ice. The chaotic center of the vast ancient furrow system (e.g., Plates Jg3 and Jg8.1) lies at 20°S, 178°W and covers most of Marius Regio. More coherent arcuate furrows can be seen to the northeast and northwest, a part of the vast Galileo Regio furrow system.

Other prominent features are the long swaths of Tiamat Sulcus and Uruk Sulcus bright terrain. Noteworthy in Uruk Sulcus are several smaller craters surrounded by dark halos (2°S, 160°W for example). These dark rings are believed to be due to excavation into dark terrain buried at shallow depth beneath bright terrain. Several small dark-rayed craters also stand out in random locations within the quadrangle (see Plate Jg8.7). These randomly distributed craters are probably due to asteroid or cometary impactor contamination of the ejecta deposit.

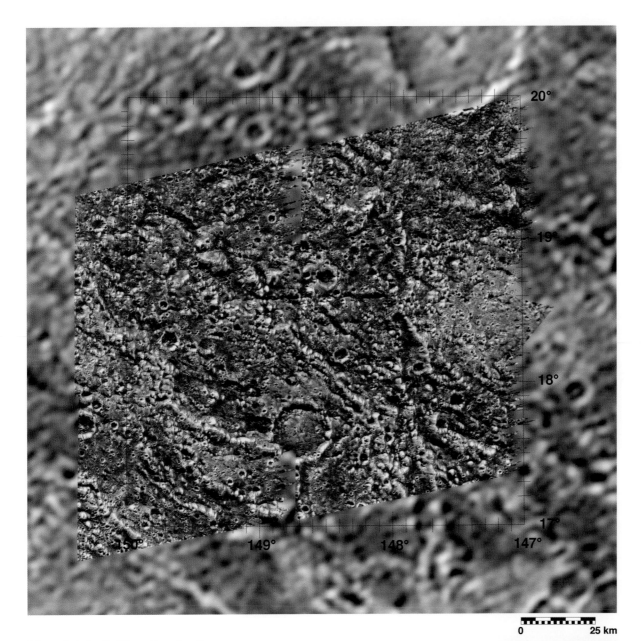

Plate Jg8.1 Galileo Regio: Furrows
Encounter: *Galileo* G1   Resolution: 77 meters/pixel
Encounter: *Galileo* G2   Resolution: 90 meters/pixel
Orthographic Projection   Map Scale: 1 cm = 12.1 km

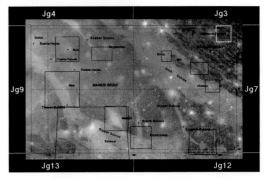

This 5-frame mosaic (four from orbit G1, one from G2) is among our highest-resolution observations of dark terrain. They are centered on the intersection of several sets of giant furrows that arc across most of Galileo and Marius Regios. The furrows are 1- to 2-kilometer-deep arcuate trenches, graben, flanked by parallel raised rims, and were formed as part of an ancient giant impact. They were originally deeper and have been modified by viscous creep and landform degradation. Relaxation of impact craters is common throughout dark terrain due to the slow creep of ice at elevated temperatures. Bright and dark materials are partly segregated into bright slopes and peaks and darker valleys and low areas. This process appears to involve down-slope transport of dark material, leaving bright ridge crests, but not the radical segregation of ice and rock as seen on Callisto (e.g., Plate Jc7.1).

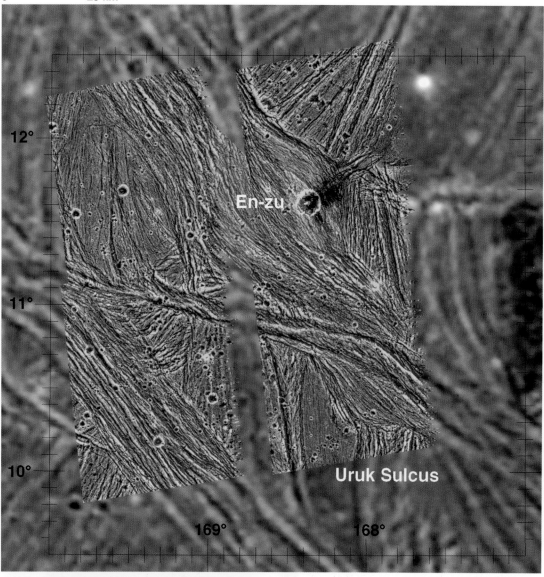

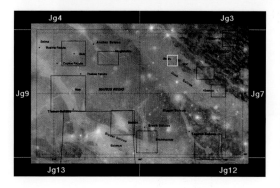

Plate Jg8.2 Uruk Sulcus: Grooves

Encounter: *Galileo* G1   Resolution: 75 meters/pixel
Encounter: *Galileo* G2   Resolution: 40 meters/pixel
Orthographic Projection   Map Scale: 1 cm = 10.4 km

This 5-frame mosaic (four images from G1, one from G2) shows several intersecting bands of bright terrain. In the northeast corner (12°N, 168.7°W), one band of bright terrain is cutting into an older section of bright terrain, partially cutting and destroying an older impact crater (12.2°N, 168.6°W). In this area, smoother areas are the remains of older terrains. Here as elsewhere on Ganymede we see the segregation of darker material to topographic lows, such as the floors of craters and the troughs between the ridge crests. The area is lightly peppered with small craters, the largest of which are secondaries from nearby Epigeus impact basin (see Plate Jg4.5). A dark-floored crater (En-zu) is also present.

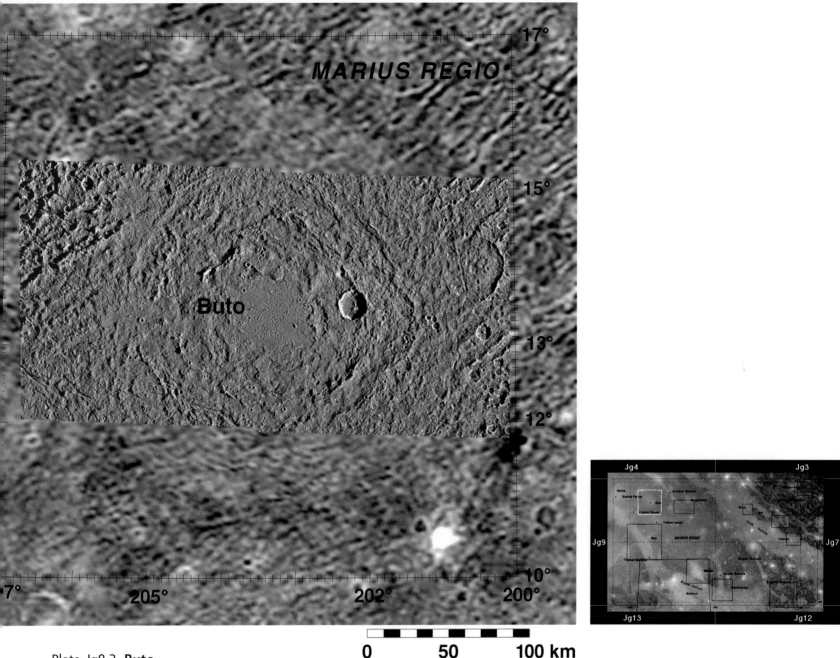

Plate Jg8.3 **Buto**
Encounter: *Galileo* G8   Resolution: 180 meters/pixel
Orthographic Projection   Map Scale: 1 cm = 22.4 km

This large ancient impact structure, or penepalimpsest, consists of a smooth central plain surrounded by a wide zone of concentric scarps and ridges. Here we do not see the steep rim scarp and high-standing central domes common in more recently formed craters such as Melkart (Plate Jg8.11), indicating that penepalimpsests are relatively old. In fact, they appear to be about as old as most of the bright terrains on Ganymede. The edge of Buto buries all older geologic features, including numerous furrow ridges on both sides and part of an older impact crater to the east (at 14°N, 200.5°W). These features have been covered or obliterated by ejecta and basin floor deposits. The original crater rim probably corresponds to the outermost concentric scarp and indicates an original rim diameter of 157 kilometers.

Note also the narrow fractures within dark terrain to the west. Some of these are buried by Buto material, whereas some continue across. Evidently dark-terrain fracturing as seen here and elsewhere continued over a protracted span of time.

The narrow linear feature at 12.3°N, 202.9°W is in fact a chain of secondary craters radial to Epigeus and is one of two similar such features (Plate Jg4.1-X).

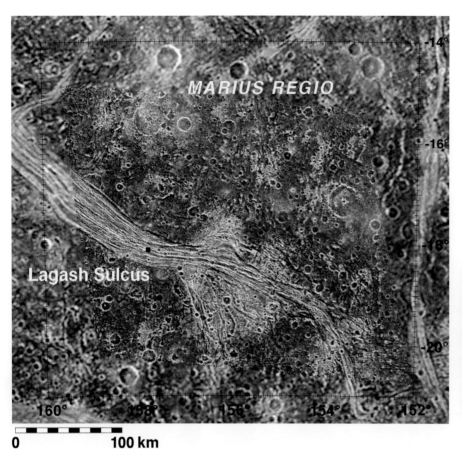
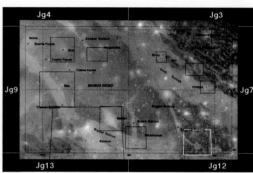

Plate Jg8.4 **Lagash Sulcus**
Encounter: *Galileo* G8   Resolution: 290 meters/pixel
Orthographic Projection   Map Scale: 1 cm = 34.3 km

Although heavily cratered, the ancient dark terrain in this region is complex, segregated into irregularly shaped patches of brighter and darker material. Some of this dark material is smoother and may be more lightly cratered. Segregation into darker and lighter patches may be part of the erosional processes afflicting ancient terrains on both Ganymede and Callisto (see Plate Jc7.1). This region is near the center of the vast furrow system seen in Marius and Galileo Regios (see Plate Jg3), however, and much of these textures could be volcanic, ejecta or floor deposits created by this vast basin.

Ganymede

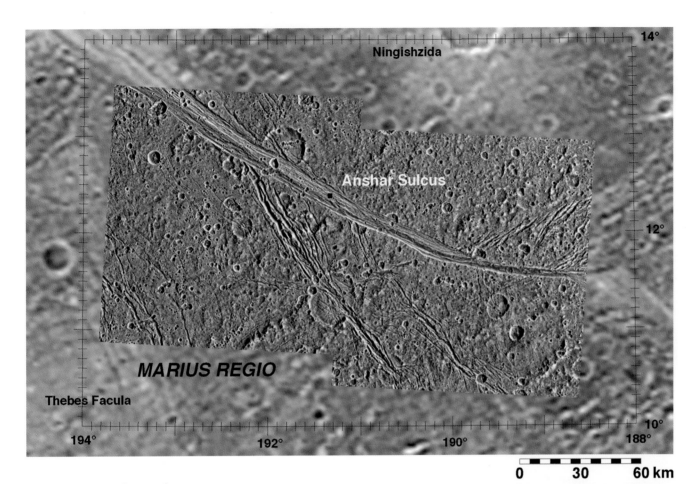

Plate Jg8.5 **Anshar Sulcus**
Encounter: *Galileo* G8   Resolution: 145 meters/pixel
Orthographic Projection   Map Scale: 1 cm = 18 km

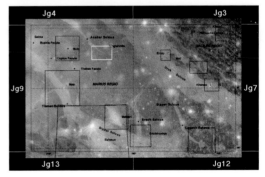

This mosaic features a narrow band of bright terrain. This lane is of special interest because of the truncated craters in the dark terrain either side of the band. If the band formed by the simple fracturing and pulling apart of blocks of ancient dark-terrain crust (such as we see at dilational bands on Europa: Plate Je8.3), then each crater would have matching halves on the opposite side of the band. This is not the case. Rather, a large slice of dark terrain appears to be "missing." This lane of bright terrain formed when the icy crust fractured, stretched and thinned to form a walled depression. A thin layer of icy lavas later flooded the depression and buried the deformed dark-terrain material. Hence the missing crater halves are buried beneath bright-terrain material. Continued tectonic deformation produced the extensive groove patterns.

Extensive fracturing can also be seen throughout dark terrain as well. The prominent set of fractures diagonally crossing the scene south of Anshar Sulcus have broken and pulled apart two large relaxed impact craters, the largest of which is ~22 kilometers across. These fractures never proceeded to full bright-terrain development. The fractures in the southwest corner appear to be concentric around Thebes Facula, an ancient palimpsest on the east edge of the mosaic. Also visible in the dark terrain are several old degraded furrows trending northeast–southwest.

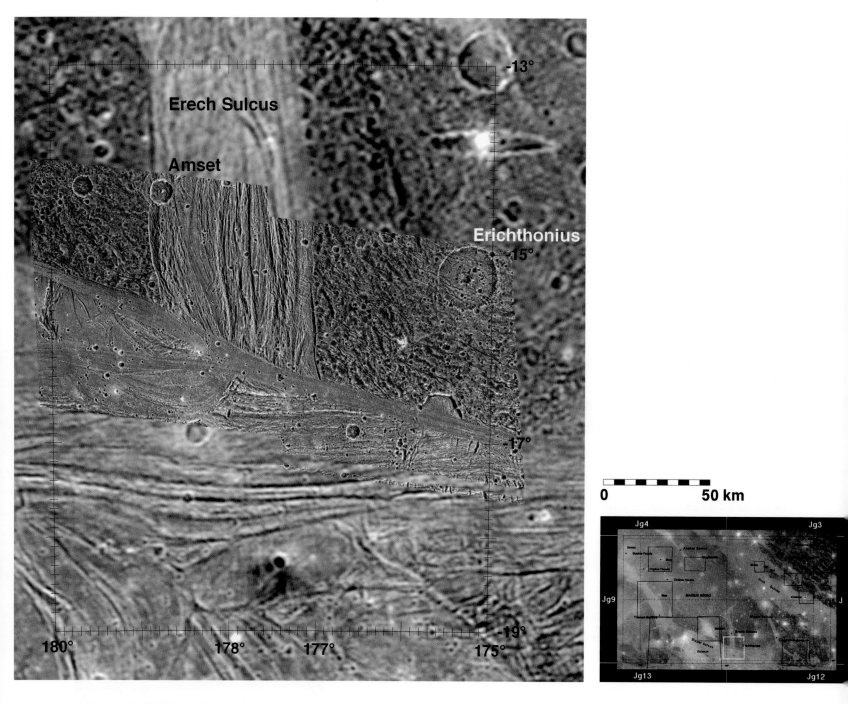

Plate Jg8.6 Erech Sulcus

Encounter: *Galileo* G8   Resolution: 145 meters/pixel
Orthographic Projection   Map Scale: 1 cm = 20.2 km

This mosaic shows the variable degree of deformation seen in bright terrain. The older north–south trending band in the center of the mosaic has been intensely fractured and has considerable relief. The younger east–west trending bands to the south are smooth to lightly grooved. The smoothest band along the northern margin of Sippar Sulcus may also be associated with the half-moon-shaped, caldera-like feature at 16.8°S, 175.4°W. Smooth units such as these are among the best evidence that bright terrains formed in association with volcanic flooding of low-lying troughs. Most of these bright terrain units were later deformed by groove formation, however. The large worm-like crater at 15.5°S, 175.2°W is a crater chain of unknown origin.

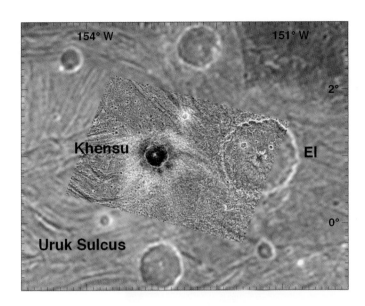 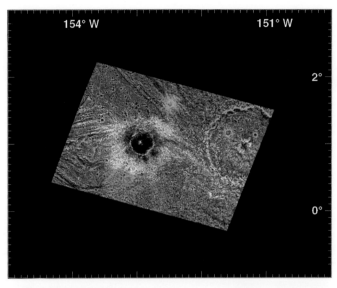

Plate Jg8.7a Khensu: Dark-floored Crater
Encounter: *Galileo* G2   Resolution: 220 meters/pixel
Orthographic Projection   Map Scale: 1 cm = 26 km

On occasion, craters on Ganymede are dark, not bright. This 15-kilometer-wide example, although the Sun was high in the sky at the time of this observation, shows that the dark material fills the crater floor and extends a short distance from the rim, beyond which the ejecta and ray system becomes bright. Like other examples (see Plate Jg10.1), the dark material at Khensu is believed to be due to concentrations of dark cometary or asteroidal material in the crater deposits. Why only some craters should behave this way is unknown. The small pits to the north and west of this dark-floored crater are secondary craters from El, the 61-kilometer-wide large crater to the east.

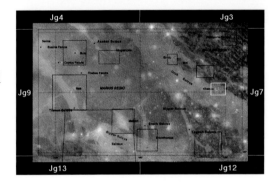

Plate Jg8.7b Khensu: Color
Encounter: *Galileo* G2   Resolution: 440 meters/pixel
Orthographic Projection   Map Scale: 1 cm = 26 km

Simultaneous SSI color observations of Khensu confirm that it is unusually reddish in color for a Ganymede crater, as well as dark. This is consistent with contamination of the ejecta by an object of unusual composition.

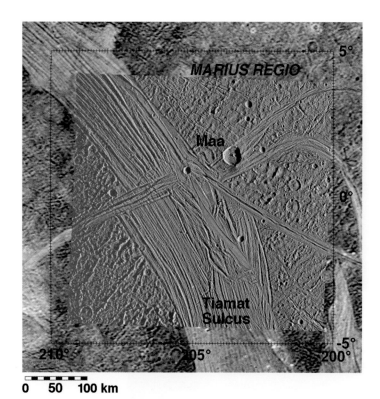
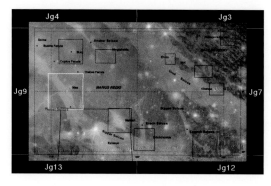

Plate Jg8.8 **Tiamat Sulcus**
Encounter: *Galileo* G8   Resolution: 500 meters/pixel
Orthographic Projection   Map Scale: 1 cm = 59.1 km

Tiamat Sulcus is peculiar because it is significantly wider on the southern side of the narrow band of bright terrain that cuts it in half. One of the major questions raised by this is whether there was more extension on the southern side than on the northern, pushing the blocks of dark terrain past each other. The region has a complex history of successive crosscutting formation of bands of bright terrain. There is no obvious evidence for lateral movement between the blocks of dark terrain.

Note the number of flattened or relaxed impact craters within dark terrain. Only a few younger craters retain their original depths. We also see again the evidence for parallel fine-scale fracturing across much of dark terrain. These fractures are usually parallel to the margins of nearby bands of bright terrain, suggesting they formed at the same time but were not involved in resurfacing. The contrast between the *Galileo* image and the background *Voyager* context image is due to the very different illumination conditions during the two encounters.

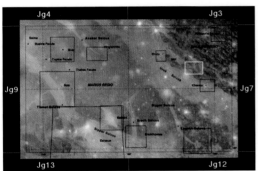
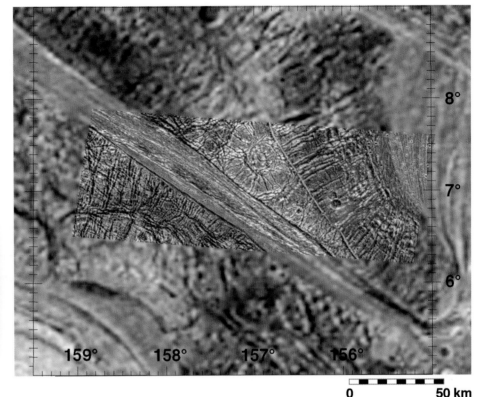

Plate Jg8.9 Reticulate Terrain

Encounter: *Galileo* G8   Resolution: 160 meters/pixel
Orthographic Projection   Map Scale: 1 cm = 19.5 km

This view of the southern margin of Galileo Regio features a variety of resurfaced units of different ages. The diagonal band of smooth terrain has probably been resurfaced by water or icy lavas but has also been weakly faulted. The highly dissected blocks of dark terrain have experienced multiple episodes of fracturing and are known as reticulate terrain.

There are surprisingly few impact craters at this site, suggesting that this terrain is not ancient. The major exception is the partially truncated 36-km-wide concentric bright circular spot at 7.3°S, 157°W. The original relief (~1 kilometer) of this severely degraded ancient central dome crater has been almost completely erased and the original crater floor intensely fractured by viscous relaxation, creating the wagon-wheel pattern of fractures on the floor.

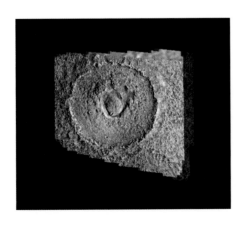

Plate Jg8.11.2-N Melkart: NIMS Infrared Map

Encounter: *Galileo* G8   Resolution: 2.7 kilometers/pixel
Orthographic Projection   Map Scale: 1 cm = 42.6 km

This NIMS infrared spectral map of Melkart (Plate Jg8.11.1) shows the compositional contrast between ice-rich bright terrain (to the west) and ice-poor dark terrain to the east. The ice-rich areas appear bluish in this false-color view. It also shows that this same compositional contrast continues across the crater floor uninterrupted, indicating that lateral mixing of impact debris within the crater floor is not very extensive. The peculiar greenish patch on the west side of the crater floor is of unknown origin, and does not correlate with any observed features in Melkart.

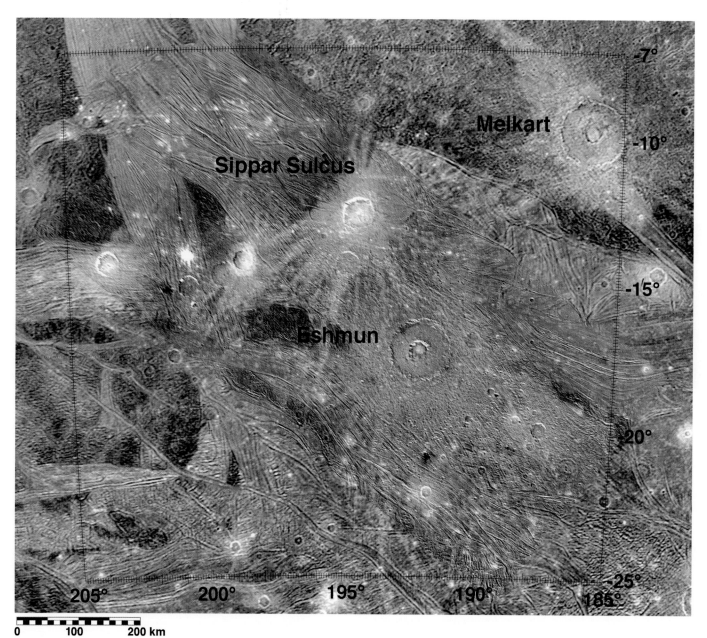

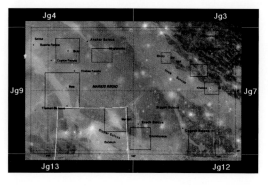

Plate Jg8.10 **Eshmun**

Encounter: *Voyager* 2   Resolution: 500 meters/pixel
Orthographic Projection   Map Scale: 1 cm = 59.1 km

Eshmun is a typical central dome crater. Instead of a central peak or central pit, craters of this size feature a rounded central dome surrounded by a ring scarp or ring of massifs. A field of densely packed small secondary craters surrounds Eshmun. This *Voyager* observation features a bright-ray crater (12.5°S, 195°W) and a number of crisscrossing lanes of grooved bright terrain. An area of multiply deformed reticulate terrain (see also Plate Jg8.9) can be found in the southeast corner. The large central dome crater to the northeast is Melkart (featured in Plate Jg8.11).

Plate Jg8.11.1 Melkart
Encounter: *Galileo* G8   Resolution: 180 meters/pixel
Orthographic Projection   Map Scale: 1 cm = 59.1 km

Melkart, at 107 kilometers diameter, was selected as the type example of a central dome crater. Domes are formed of warm soft ice uplifted from several kilometers below the surface, and offer us a peek at the lower crust of Ganymede. Numerous small fissures and lineations score the surface of the 20-kilometer-wide central dome. Such fracturing is usually due to stretching of the outer skin of a domical deposit during dome growth or later spreading. The dome is surrounded by an irregular moat depression. The floor of the crater itself is both knobby and smooth. The smooth areas may be refrozen impact melt. The knobby and craggy zone beyond the short narrow rim scarp is ejecta blasted out of the crater and deposited beyond the rim at high speed. Sets of small secondary craters can be identified approximately one crater diameter from the rim scarp, in both the *Galileo* and *Voyager* images. See also Plate Jg8.11-N for NIMS map.

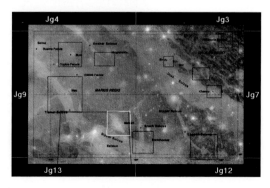

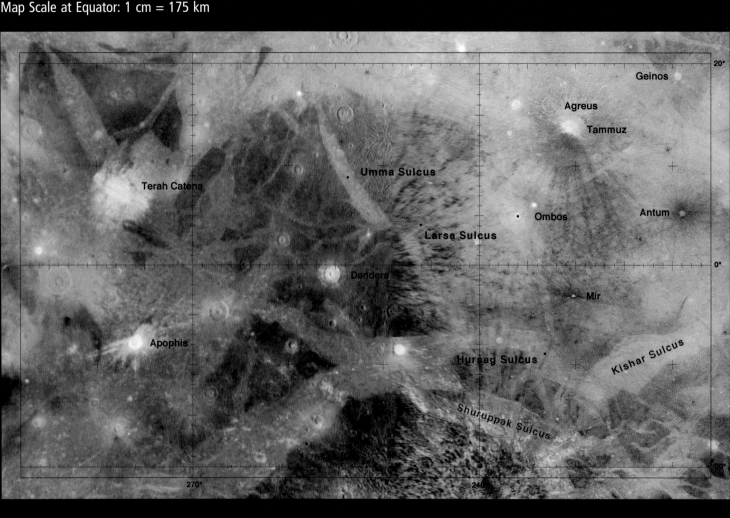

Plate Jg9 Tiamat Sulcus Quadrangle

This quadrangle includes the antapex region that faces backward as Ganymede orbits Jupiter (centered on 0°, 270°W), but it has a split personality. The eastern half is mapped by *Voyager* images obtained under noontime solar conditions (highlighting albedo variations). The western half is mapped by low-Sun evening *Galileo* images, which highlight relief. *Voyager* images show very prominent dark rays centered on impact craters. Similar rays are seen in *Voyager* images of the western portion too, but are not visible in the *Galileo* images due to the shadowing. For example, dark rays radiating from Dendera are visible only in the high-Sun *Voyager* but not the *Galileo* images.

Most crater rays on Ganymede and Callisto (and the Moon) are bright, due to exposure of ground-up water ice. The formation of these dark rays is attributed to enhanced concentrations of dark material from the comets that formed the impact. This is possible because the antapex is also the area of Ganymede that is least heavily cratered by smaller impacts and most heavily bombarded by radiation from Jupiter's magnetic field, which would preferentially remove water ice and leave surface deposits of rock-rich material.

The other prominent feature is the bright-rayed crater chain, Terah Catena. This is one of the largest of the tidally disrupted comet crater chains on Ganymede (see Plate Jg2.3), although it was not observed at high resolution. Numerous impact craters pepper the scene, including Ombos, which appears to be a bright spot, but actually consists of three concentric bright zones and is most likely a central dome crater.

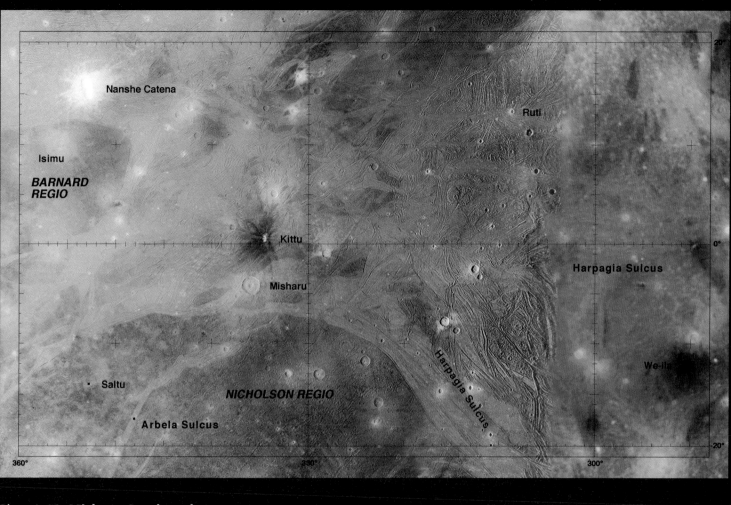

Plate Jg10 **Misharu Quadrangle**

The prominent dark splotch is the dark-ray crater Kittu (Plate Jg10.1a). Unlike most dark rays, which have reddish colors, this dark-ray system is distinctly less red (bluish) in color, and may have been formed from a comet of different composition. Another unusual crater is the bright-rayed crater Nanshe Catena (Plate Jg10.2). During the G28 orbit, *Galileo* passed quite close over this region and obtained several high-resolution mosaics to investigate the relationship between younger bright and older dark terrains, principally Harpagia Sulcus and Nicholson Regio. Several ancient palimpsests are apparent within Barnard Regio.

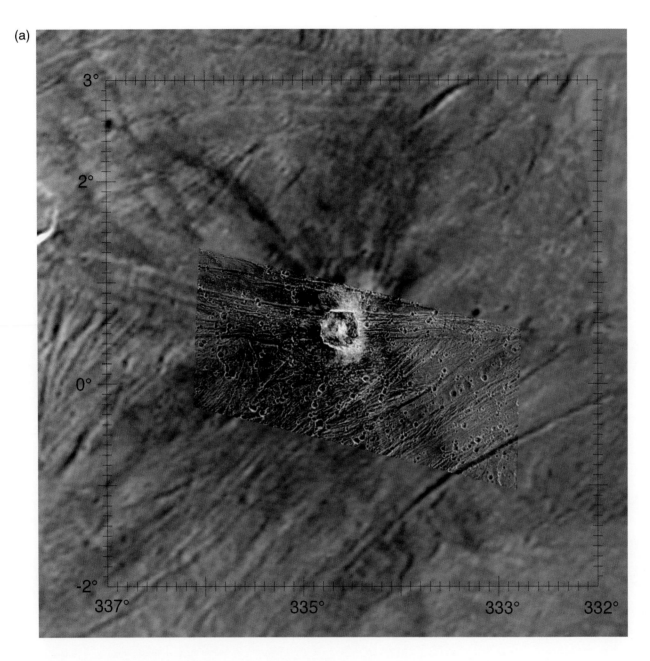

Plate Jg10.1a **Kittu**
Encounter: *Galileo* G7   Resolution: 145 meters/pixel
Orthographic Projection   Map Scale: 1 cm = 17.1 km

This prominent 17-kilometer-wide dark-ray crater is relatively young. Although part of the rim is bright, most of the ejecta rays of this crater are dark. The dark bluish color (see Plate 10.1b) is due to contamination from the projectile, most likely a primitive carbonaceous asteroid or comet. A pancake ejecta deposit (see Plate Jg2.2) can be seen extending about one crater radius from the rim. The abundant small craters are secondary craters from the nearby Misharu impact crater, located due south of Kittu.

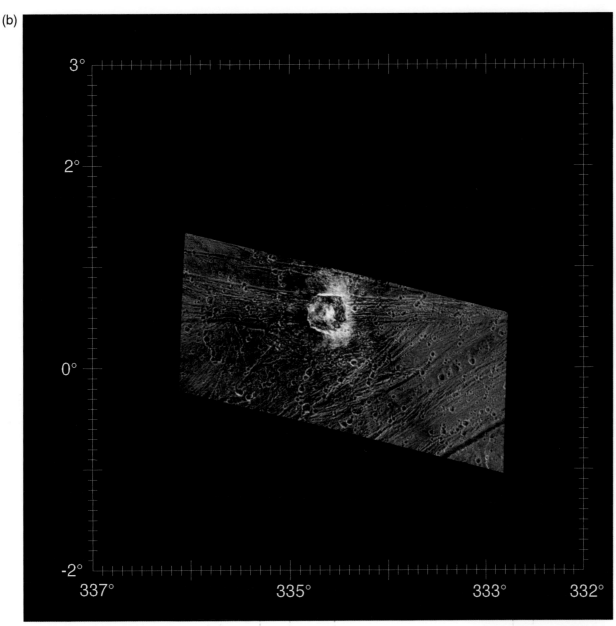

Plate Jg10.1b Kittu: Color
Encounter: *Galileo* G7   Resolution: 290 meters/pixel
Orthographic Projection   Map Scale: 1 cm = 17.1 km

Like Kenshu (Plate Jg8.7), Kittu is dark rayed, but this partially returned simultaneous color observation shows that the rays have an unusual dark bluish color, at least when compared to other dark materials. This points to a different composition for the body that struck Ganymede to make this crater.

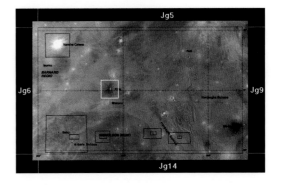

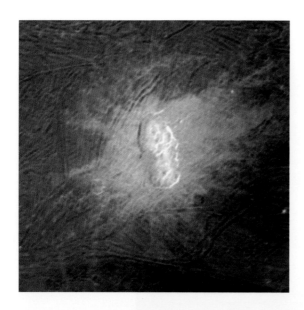
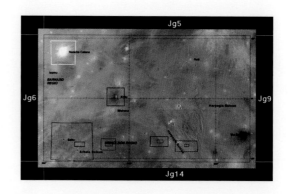

Plate Jg10.2 Nanshe Catena

Encounter: *Voyager* 1   Resolution: 1000 meters/pixel   Orthographic Projection   Map Scale: 1 cm = 52.8 km

Although no high-resolution images were taken of this most unusual impact crater, it is shown here in close-up to highlight its oblong shape and seven (!) central pits. Measuring 50 by 104 km, it formed when a chain of seven (or more) discrete closely spaced bodies (most probably a comet tidally disrupted by Jupiter) struck Ganymede simultaneously. The bright rays indicate it formed recently on the geologic timescale, probably within the past few hundred million years or so.

Plate Jg10.3.1 Arbela Sulcus and Nicholson Regio

Encounter: *Galileo* G7   Resolution: 180 meters/pixel   Encounter: *Galileo* G28   Resolution: 132 meters/pixel
Orthographic Projection   Map Scale: 1 cm = 30.7 km

This is our best regional view of dark terrains of Ganymede, and fully illustrates the complex geologic history of these ancient terrains. Like parts of Marius Regio (see Plate Jg8.5), Nicholson Regio is composed of dispersed brighter and darker patches. Pervasive narrow fractures can be seen coursing through dark terrain, especially towards the south and east. Most large craters are degraded, either by fracturing and disruption, landform degradation, viscous relaxation, or a combination of these. Examples include the crater split apart by extension at 14°S, 353°W, the fractured crater at 18°S, 347.5°W, and the relaxed crater at 16°S, 345°W. The relatively bright smooth patch near 13°S, 351.5°W, is probably an ancient palimpsest scar. Most of these craters date to a relatively old age, and formed when the crust of Ganymede was relatively warm and "soft." A pair of adjoining elliptical craters at 14°S, 351°W were probably formed by the impact of a binary object at a very low angle.

Two bands of bright terrain cross Nicholson Regio here, a band of intensely grooved material at 15°S, 348°W, and the smooth bright band Arbela Sulcus. One interpretation is that the dark terrain either side of Arbela pulled apart laterally and the bright band is entirely new crustal material that upwelled from Ganymede's mantle. This is how ocean floor on Earth and dilational bands on Europa (Plate Je8) form. The test is whether we can uniquely match split craters and fractures in dark terrain on either side of Arbela Sulcus. Alas, the two dark blocks can be put together in almost any arrangement and no fit is particularly convincing. (See also Plates Jg4 and Jg8.5.) You can try this too. Make a photocopy of this plate, cut out Arbela Sulcus, and realign the edges of dark terrain. Most lanes of bright terrain formed by fracturing and flooding by icy lavas. Shown at half scale to fit page.

Plate Jg10.3.1 **Arbela Sulcus and Nicholson Regio**

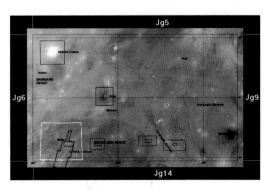

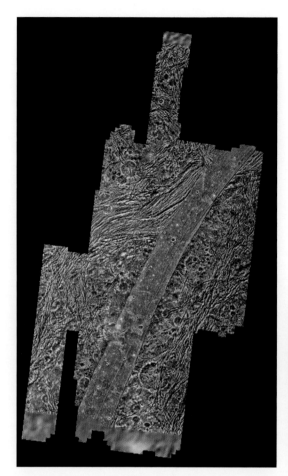

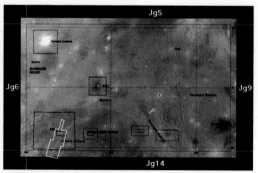

Plate Jg10.3.2-N Arbela Sulcus and Nicholson Regio: NIMS Infrared Map

Encounter: *Galileo* G8   Resolution: 3.3 kilometers/pixel
Orthographic Projection   Map Scale: 1 cm = 30.3 km

This combined NIMS–SSI infrared spectral map of the Arbela Sulcus region confirms that bright terrains are ice-rich (shown in blues) compared to the "dirty" dark terrains (shown in reds). The composition of dark terrains here is not homogenous, however. Areas of higher water-ice content are apparent as purple in this rendering. These areas correspond with regions that have been broken up by faulting and scarp formation (see also Plate Jg10.3.3). Apparently scarp formation exposes material that has higher ice content than undisturbed dark terrains. Map is centered at 17°S, 349°W.

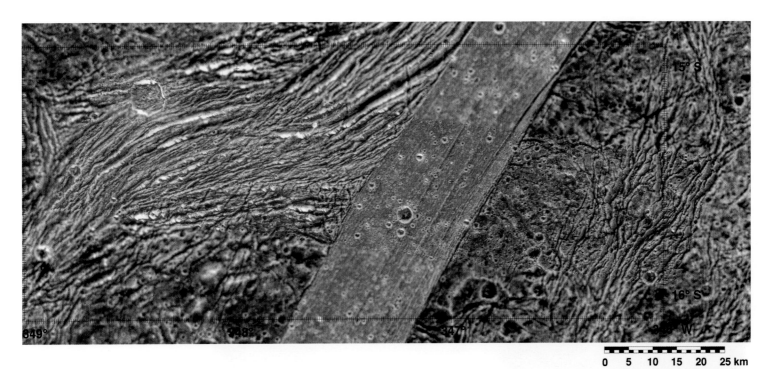

Plate Jg10.3.3 **Arbela Sulcus: High Resolution**
Encounter: *Galileo* G28   Resolution: 35 meters/pixel
Orthographic Projection   Map Scale: 1 cm = 7.6 km

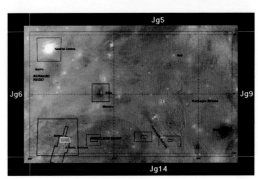

This high-resolution mosaic of Arbela Sulcus includes dark terrain, smooth bright terrain and grooved bright terrain. Arbela Sulcus is very sharp edged and mostly smooth at this high resolution. This implies that it was emplaced as very soft ice or as liquid-water volcanic flows within a fault-bounded trough, although some minor deformation evidently occurred afterwards as revealed by faint parallel lineations. The intensely fractured zone of grooved terrain to the west predates the swath of smooth terrain.

Areas of dark terrain to the east are also intensely fractured, part of a pattern that extends up from the south as seen in Plate Jg10.3.1. The scarp faces visible here indicate that dark terrain is broken into small extensional fault blocks, some of which have rotated from horizontal. Dark material has moved down slope to cover the valley floors between these blocks.

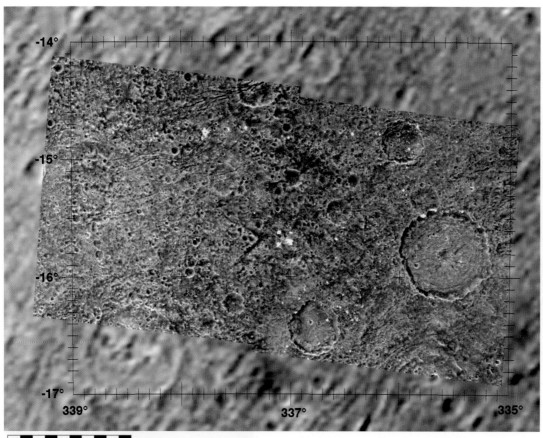

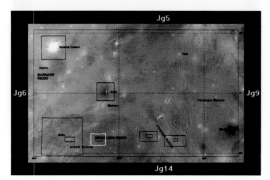

Plate Jg10.4.1 **Nicholson Regio: Context View**
Encounter: *Galileo* G28   Resolution: 127 meters/pixel
Orthographic Projection   Map Scale: 1 cm = 14.8 km

This portion of Nicholson Regio is relatively intact. Although several sets of narrow linear fractures are apparent in the upper left quadrant, there is no evidence for the extensive breakup seen in high-resolution views of Marius Regio (Plates Jg4.1 and Jg4.2) and to the west within Nicholson Regio itself (Plate Jg10.3). This lack of deformation may be due to the fact that there are no bright terrains nearby. A well-preserved central pit crater 36 kilometers across is centered at 16.0° S, 335.5°W. The ancient impact feature at 15.5°S, 338.8°W appears to be a highly degraded penedome or central dome crater. Parts of another older degraded impact structure are evident in the southeast corner at 17.0°S, 335.7°W.

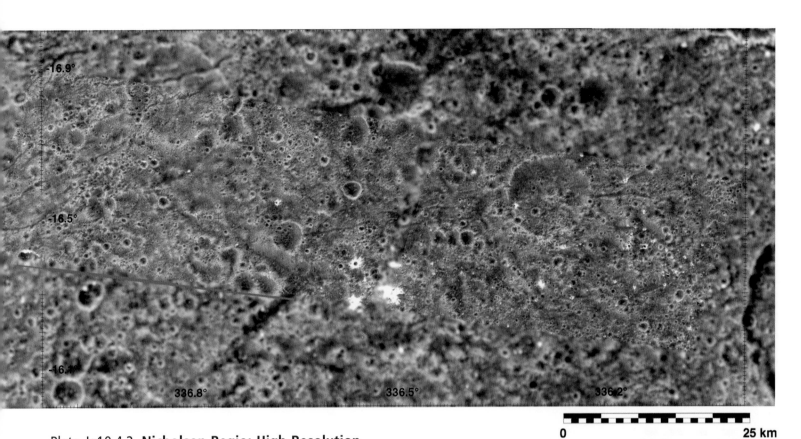

Plate Jg10.4.2 Nicholson Regio: High Resolution
Encounter: *Galileo* G28   Resolution: 28 meters/pixel
Orthographic Projection   Map Scale: 1 cm = 4.6 km

One of our highest-resolution views of dark terrain, this scene lies between the two large impact features shown in Plate Jg10.4.1. A few narrow linear fractures cross the scene, but mostly we see lots of small impact craters and several larger ones in various states of degradation. The small bright craters could be secondaries from an unknown bright-ray crater.

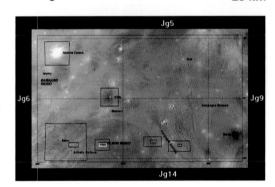

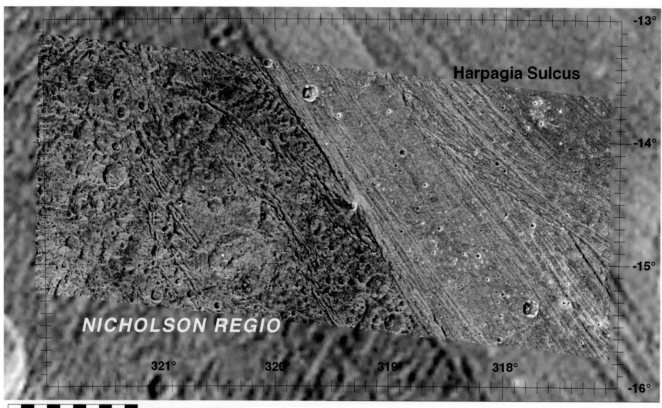

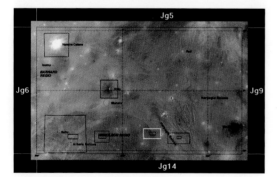

Plate Jg10.5.1 Bright Terrain–Dark Terrain Boundary: Context View
Encounter: *Galileo* G28   Resolution: 122 meters/pixel
Orthographic Projection   Map Scale: 1 cm = 14.2 km

This mosaic was targeted to investigate the nature of the boundary between older dark and younger bright terrain. Dark terrain here features several larger impact craters in different states of degradation, including two relaxed and deformed craters at 15°S, 320.3°W and 14.6°S, 320.5°W. The terrain boundary itself is a single prominent high-standing scarp, which appears to confine bright material to low-lying terrains. Several prominent sets of fractures also formed within dark terrain, parallel to the terrain boundary. Evidently, stretching and fracturing of dark terrain was extensive, but only particular sections were resurfaced and faulted to form smooth and grooved terrains. The bright-terrain material of Harpagia Sulcus has experienced varying degrees of faulting and deformation, from almost none to pervasive.

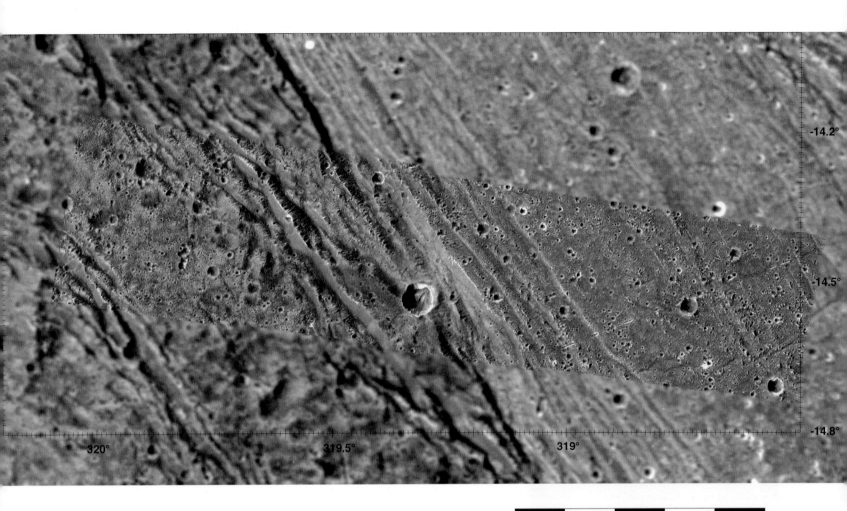

Plate Jg10.5.2 Bright Terrain–Dark Terrain Boundary: High Resolution

Encounter: *Galileo* G28   Resolution: 20 meters/pixel
Orthographic Projection   Map Scale: 1 cm = 3.5 km

The high-resolution close-up of the terrain boundary is in an area where the bounding fault scarp breaks up into a set of smaller scarps occurring within both bright and dark terrain on either side of the boundary. Most of these large fault blocks in both bright and dark terrain may have formed after bright material was emplaced. The smooth surfaces of several of these scarps are evidence of mass wasting. Like Arbela Sulcus (Plate Jg10.3.1), parts of Harpagia Sulcus are essentially undeformed. No individual volcanic flows are seen, but resolution may not be sufficient to resolve such features.

One small set of fractures cuts across the dominant trend in the southeast part of the scene. A narrow chain of pits crossing the grooves is of unknown origin. Outcrops of bedrock can be seen along the inner rim of the well-preserved crater at 14.5°S, 319.3°W. Shown at 65% scale to fit page.

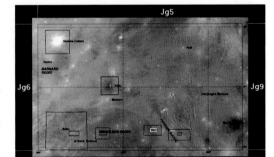

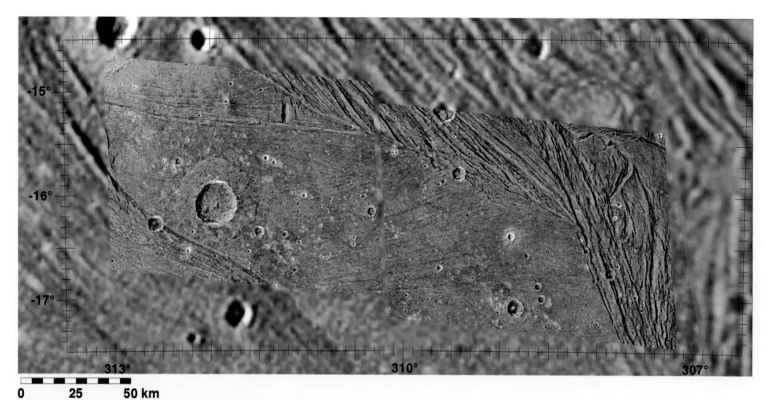

Plate Jg10.6.1 Harpagia Sulcus Smooth Bright Terrain: Context View
Encounter: *Galileo* G28   Resolution: 120 meters/pixel
Orthographic Projection   Map Scale: 1 cm = 16.7 km

This observation was targeted to investigate the origins of smooth bright terrain and search for evidence of volcanism. The irregular arcuate enclosure at far right (15.5°S, 308°W) is a caldera-like feature like those in Sippar Sulcus and scattered across Ganymede (Plates Jg5.1, Jg13.1 and Jg14.1, for example). The floor of this feature is covered by a ridged unit, which could be interpreted as a viscous lava extrusion. The smooth bright-terrain unit itself has a subtle linear fabric, suggesting that few units on Ganymede completely escape some form of deformation. A typical pancake ejecta deposit surrounds the large 8-kilometer-wide central peak crater (see Plate Jg2.2).

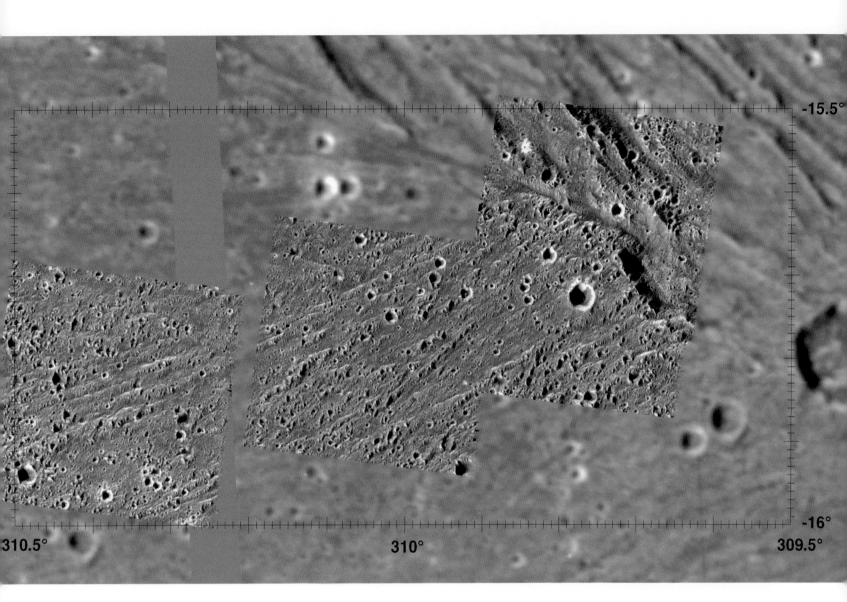

Plate Jg10.6.2 Harpagia Sulcus Smooth Bright Terrain: High Resolution

Encounter: *Galileo* G28   Resolution: 16 meters/pixel
Orthographic Projection   Map Scale: 1 cm = 2.8 km

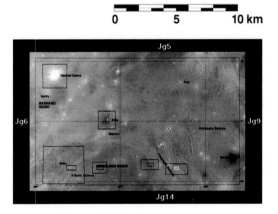

This view, the highest-resolution view of smooth terrain we have, shows a variety of surface textures. Here, smooth terrain is not completely smooth. A more subtle northerly trending fabric crosses the dominant ridge sets. These are probably not volcanic textures, but indicate the influence of several minor tectonic episodes. Whether smooth material filled depressions between ridges or the ridges deformed an original smooth deposit is not known. The only direct evidence for volcanism is the large nearby caldera (Plate Jg10.6.1) and the general flatness of these terrains. The arcuate scarp at 15.7°S, 309.65°W may be a truncated impact crater or a partial volcanic caldera like the much larger one in the context mosaic.

Lambert Conformal Conic Projection
Map Scale: 1 cm = 185 km

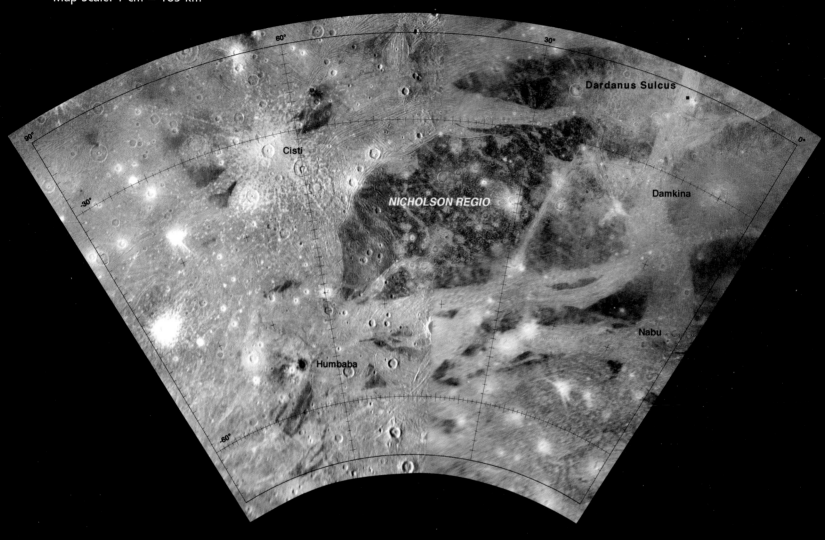

Plate Jg11 Nabu Quadrangle

The large block of dark terrain near center is the southwestern portion of Nicholson Regio. The bright linear features within it are a set of furrows similar to those in Galileo Regio (Plate Jg3) and eastern Marius Regio (Plate Jg8). Several appear to be bright, suggesting that bright terrain may have exploited some older furrows during formation. Numerous impact features of different age are seen throughout the map.

The large splotch at upper right is the floor and ejecta deposit of Damkina, a 175-km-wide central dome crater and the largest regular crater on Ganymede. Part of the southeastern rim of Damkina seems to be missing or only partly formed, similar to what was observed at Bran on Callisto (Plate Jc13.1). Dark-floored crater Humbaba is probably related to contamination of ejected ice by dark projectile material.

Lambert Conformal Conic Projection
Map Scale: 1 cm = 185 km

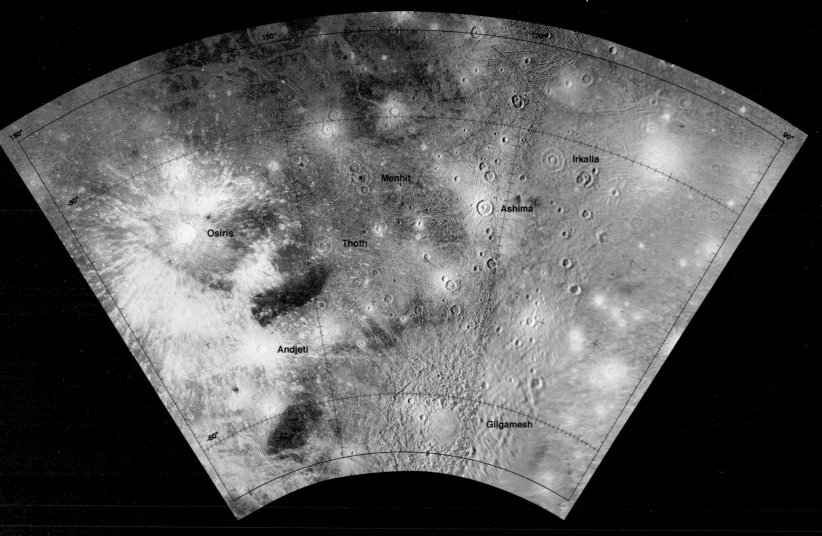

Plate Jg12 Osiris Quadrangle

Several large impact features of different ages dominate this region. The large bright crater to the west, Osiris (Plate Jg12.1), is 105 kilometers across and is probably the youngest large impact crater on Ganymede. Like Pwyll on Europa (Plate Je14), these rays are unusually bright and extensive. Some can be traced for as long as 1500 kilometers. Note that the bright rays darken whenever they cross blocks of older dark terrain (especially the triangular-shaped block that Osiris formed on), due to the higher non-ice composition of these terrains.

The second large impact feature is Gilgamesh (Plate Jg12.2), a 600-kilometer-wide impact basin that is the largest fully preserved impact feature on Ganymede. This basin is only 1.5 kilometers deep (compared to large basins on the Moon that are 8 to 12 kilometers deep) and postdates bright terrain formation, but is clearly older than Osiris and has lost most of its rays.

The third crater, Menhit (Plate Jg12.3), is almost as old as bright terrains. This feature is dominated by its central dome and is actually high standing. The fourth large impact is very ancient (predating bright terrain) and difficult to recognize at this resolution (Plate Jg12.4). This multi-ring structure, centered at 29°S, 155°W, is partly broken up by younger bright terrain, but some of the concentric ridges and graben can still be seen. Numerous secondary craters from all four impacts populate the region.

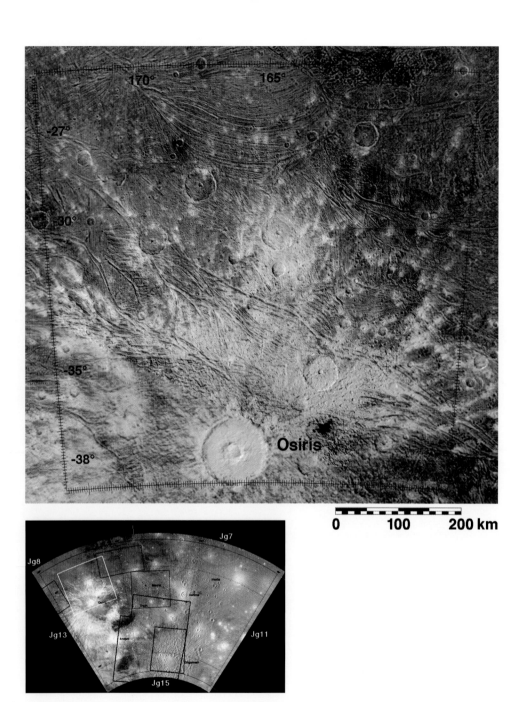

Plate Jg12.1 Osiris
Encounter: *Voyager* 2 Resolution: 495 meters/pixel
Orthographic Projection Map Scale: 1 cm = 57.9 km

This high-resolution *Voyager* view features the ejecta pattern surrounding the 105-kilometer-wide bright-rayed central dome crater Osiris, which is probably the youngest crater on Ganymede. Most of these splotches that make up the bright rays are centered on small secondary craters, which form their own small ejecta deposits. Although the image missed Osiris directly, we can see that the bright impact frost deposit covers the entire floor of the crater. Osiris itself formed on a block of dark terrain. The rays that extend from Osiris become noticeably brighter as they cross from dark to bright terrain.

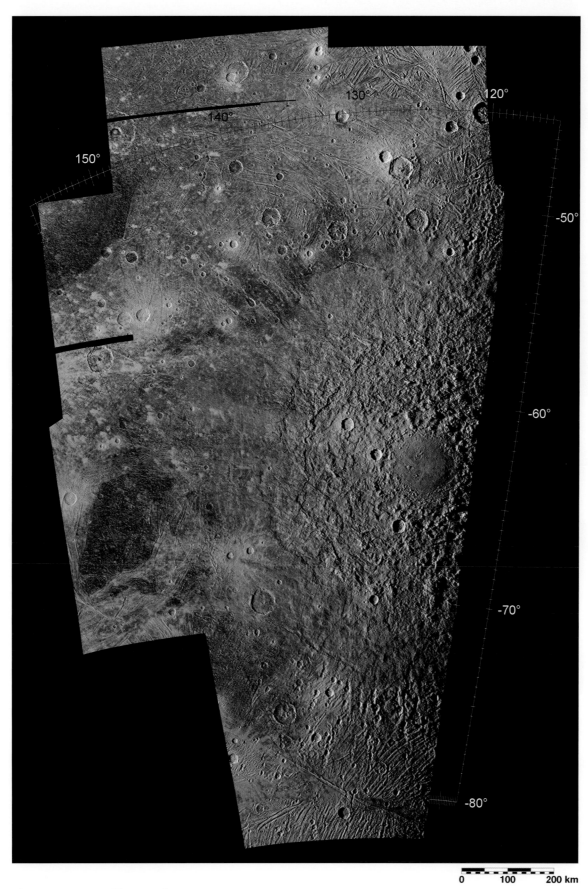

Plate Jg12.2.1 **Gilgamesh**

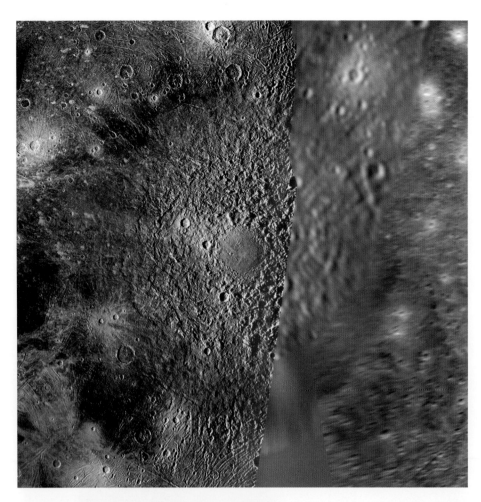

Plate Jg12.2.2 **Gilgamesh: regional view**
Encounter: *Voyager* 2
Resolution: 1000 meters/pixel
Orthographic Projection
Map Scale: 1 cm = 118 km

This basin-centered view shows the entire Gilgamesh structure. *Voyager* and *Galileo* imaging to the east (at 2 kilometers resolution) is sufficient to show additional secondary craters to the south and east but few structural details.

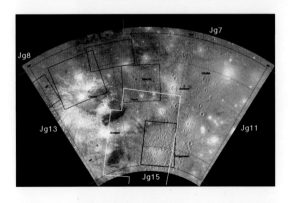

Plate Jg12.2.1 **Gilgamesh**
Encounter: *Voyager* 2   Resolution: 550 meters/pixel
Orthographic Projection   Map Scale: 1 cm = 86.6 km

The 600-kilometer-wide Gilgamesh impact basin is the largest fully preserved impact feature on Ganymede. Although similar in size to Asgard, Gilgamesh more closely resembles classical large impact basins on the Moon and Mercury. Rather than numerous closely spaced rings, Gilgamesh consists of only three to five prominent ring scarps or graben. The smooth central depression is not the crater floor but rather part of the central uplift complex and may be related to central domes. A wide zone of ridges and scarps surrounds the depression. The true rim corresponds to the most prominent of the inward-facing scarps, best expressed at 60°S, 137°W, just west of a fresh 30-kilometer flat-floored crater. Extensive fields of large secondary craters lie to the north and south of the basin. Many of these craters are oddly shaped or elongated radial to the basin center. The crater-free zone between the rim and secondaries has been scoured and buried by an ejecta deposit best expressed near 73°S, 130°W. The relatively simple structure of Gilgamesh contrasts sharply with the complex multi-ring systems that typify Callisto and ancient dark terrain on Ganymede (see Plates Jc3 and Jg12.4).

Plate Jg12.2.3
Gilgamesh from *Galileo*
Encounter: *Galileo* E12
Resolution: 160 meters/pixel
Orthographic Projection
Map Scale: 1 cm = 37.8 km

Originally intended to provide improved resolution for crater counting statistics, these low-phase-angle images proved to be too washed out by the noontime Sun to make out details such as small craters.

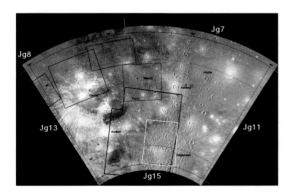

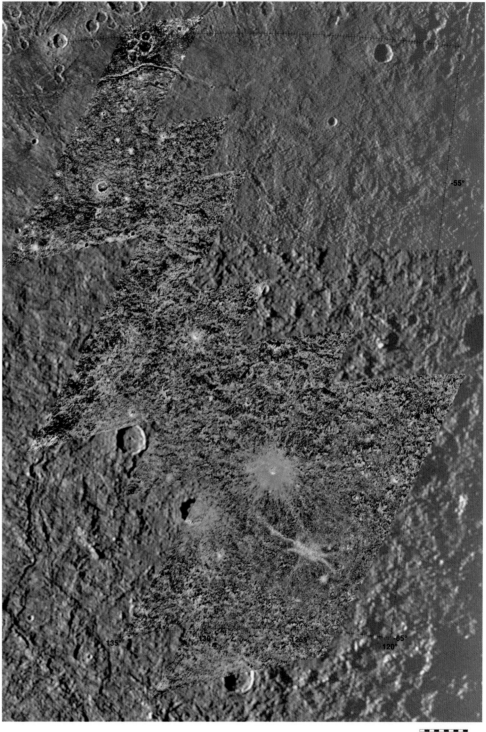

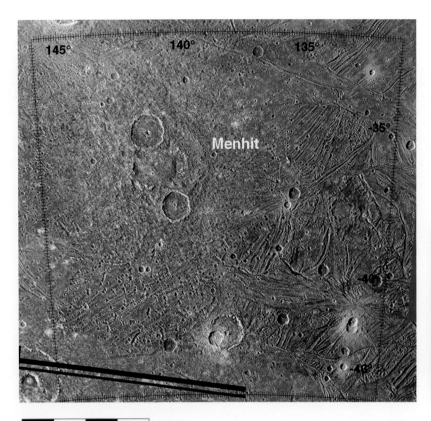
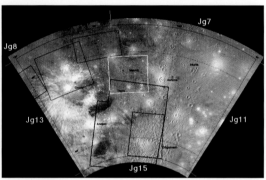

Plate Jg12.3 **Menhit**
Encounter: *Voyager* 2   Resolution: 490 meters/pixel
Orthographic Projection   Map Scale: 1 cm = 57.3 km

Partly hidden beneath the two large central pit craters in the northwest corner is the large relatively old 145-kilometer-wide penedome impact feature, Menhit. The prominent central dome at 36.2°S, 141°W lies directly between the two central pit craters, 40 and 53 kilometers across. Surrounding this dome are concentric zones of massifs and ridges. The two superposed central pit craters lie directly on top the outermost ridge, remnants of a vestigial crater rim best seen to the northeast. The concentrated fields of small craters are secondary craters from Menhit, as well as some contributions from nearby large craters. Menhit is provocative because stereo mapping shows that both the floor and the central dome of this crater are elevated 500 meters or so above surrounding terrains. Impact crater floors are normally below or near the elevation of the original surface. The origin of this floor uplift is unknown. See also Anzu (Plate Jg2).

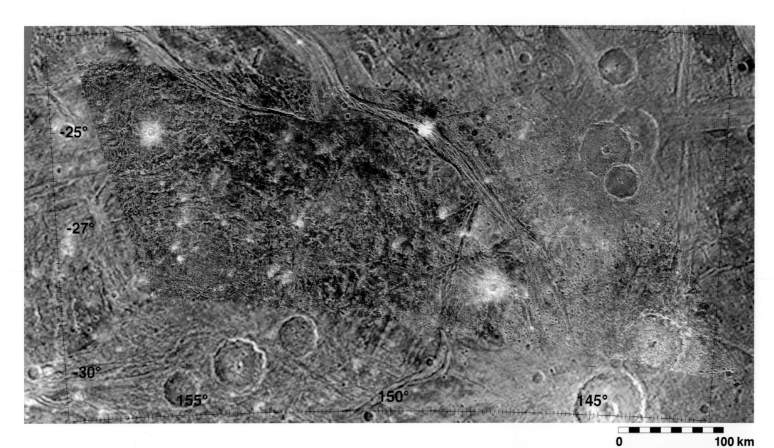

Plate Jg12.4 **Ancient Multi-ring Basin**

Encounter: *Galileo* G8   Resolution: 240 meters/pixel
Orthographic Projection   Map Scale: 1 cm = 36 km

Although only 33% of ancient dark terrain remains on Ganymede, several prominent ancient multi-ring structures are preserved. Despite the relatively high-Sun view, this *Galileo* view is one of our best views of a smaller (and almost intact) ring system of this type (see also Plates Jg2 and Jc3). This structure, centered at 27.8°S, 154.5°W, has been partly broken up by younger bright terrain to the south and north and is likely degraded by slope erosion and viscous relaxation. Several concentric bright-crested ridges up to 1 kilometer high are evident, forming a wreath of ridges almost to the center. The field of older dark craters at longitudes 147° to 150°W are secondary craters associated with this basin, which allow us to estimate the original rim diameter to be ~225 kilometers. A cluster of large central pit craters, together with their ejecta and secondaries, dominates the eastern portion of the scene.

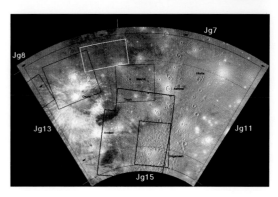

Lambert Conformal Conic Projection
Map Scale: 1 cm = 185 km

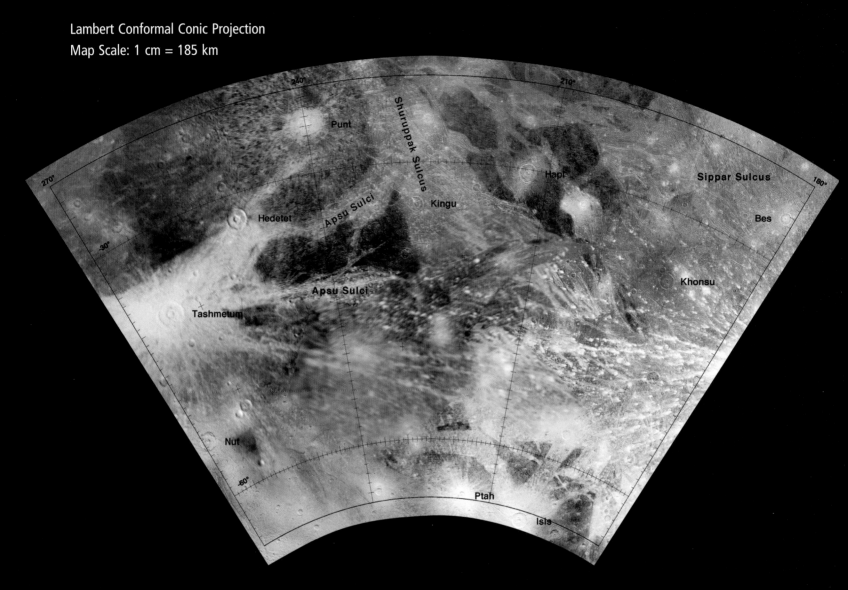

Plate Jg13 Apsu Sulci Quadrangle

Three large bright-ray systems cross this quadrangle, originating from Punt, Tashmetum, and Osiris (see Plate Jg12). Each ray system has a different brightness level, showing how ray systems fade over time. The dark terrain to the northwest is dominated by the large relatively young 155-kilometer-wide central dome crater Punt and its bright rays. Bright rays from Osiris intrude from the east. The extensive bright terrain in the northeast corner is geologically complex and is also home to an unusually high concentration of caldera-like features (Plate Jg13.1). These features are distinctly non-circular and have been attributed to water volcanism, and were observed by *Galileo* here and at numerous other locations.

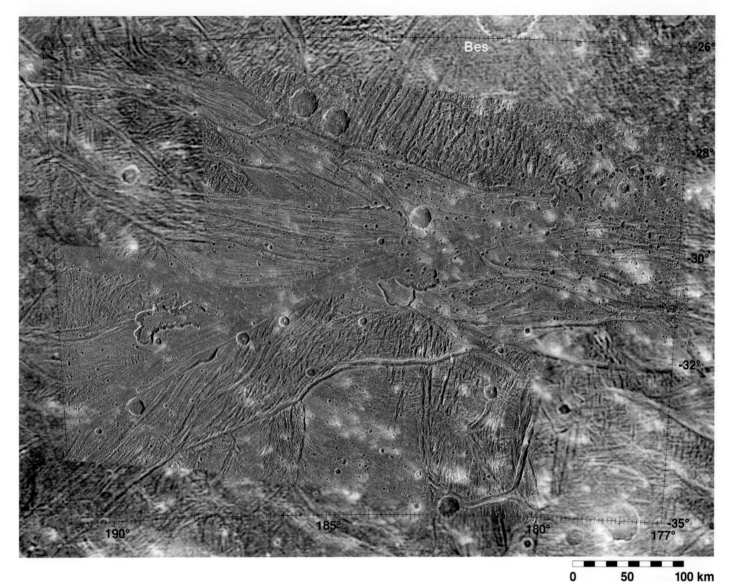

Plate Jg13.1 Calderas

Encounter: *Galileo* G8   Resolution: 180 meters/pixel
Orthographic Projection   Map Scale: 1 cm = 32.7 km

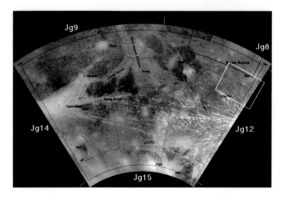

This 6-frame mosaic features no less than 6 caldera-like features (2 more lie further to the east). Most of these are adjacent to the band of smooth bright terrain crossing the scene east to west. These features resemble some volcanic calderas on Earth, which are often non-circular and can have steep-sided rims. In each case, the rim is open toward the adjoining bright band, suggesting that material flowed out into the bright band or was cut by it. The floor of the largest caldera has a rugged ridged surface. On Earth, this is characteristic of viscous, sticky lavas. The smooth surface of the band, in contrast, is indicative of runny, watery lavas. The band and the calderas may represent different stages in the volcanic evolution of bright terrain. The frequency of these features (at least 40 have been identified globally) indicates they are of some importance and that volcanism plays at least some role in the formation of smooth bright terrain. Note also the different degree of deformation within the various crosscutting bands of bright terrain. Secondaries from Osiris form the numerous bright splotches across the scene. The numerous small craters to the upper right are secondaries from Bes and a large penedome crater in the uppermost right corner of the scene.

Lambert Conformal Conic Projection
Map Scale: 1 cm = 185 km

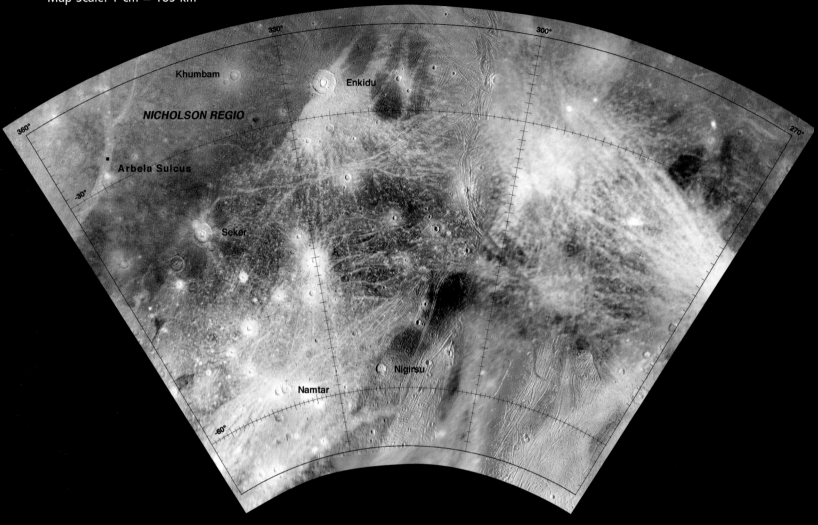

Plate Jg14 Namtar Quadrangle

This region is evenly divided between large areas of bright terrain to the east and dark terrain to the west. Between Enkidu and Namtar lies an area with characteristics of both. Although dark, this area is crossed by numerous grooves. The northern boundary is clearly delineated by bright rays of Enkidu, which are easily recognized only to the south of this line. A similar albedo pattern on bright and dark terrains was seen surrounding Osiris (Plate Jg12). This region may be characterized by tectonic resurfacing, a mechanism whereby fracturing reshapes a landscape without significant bright-terrain volcanism. A prominent north–south trending set of ridges runs along longitude 307°W. The brighter zones south of 48° are part of the south polar frost deposit.

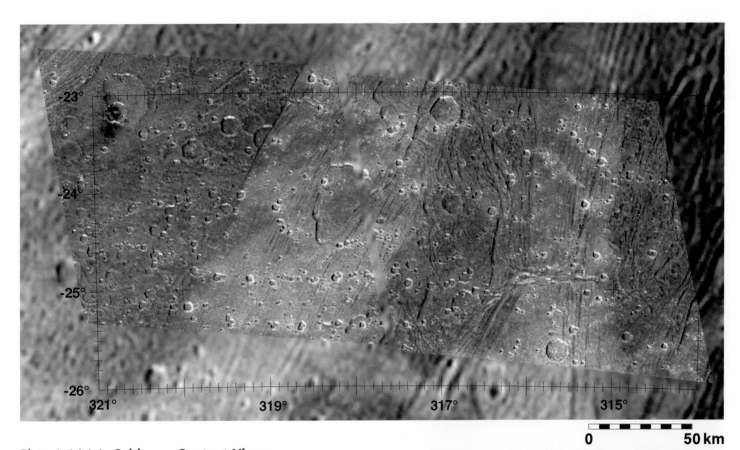

Plate Jg14.1.1 Calderas: Context View
Encounter: *Galileo* G28   Resolution: 165 meters/pixel
Orthographic Projection   Map Scale: 1 cm = 17.7 km

This mosaic features two calderas situated within a narrow band of bright terrain along longitude 318°W. Most of the small craters scattered across bright (and dark) terrains are secondary craters formed by the 125-kilometer-wide Enkidu crater located just to the west of this scene (Plate Jg14). Note the dark-floored central peak crater in the northwest corner.

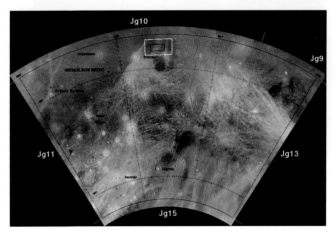

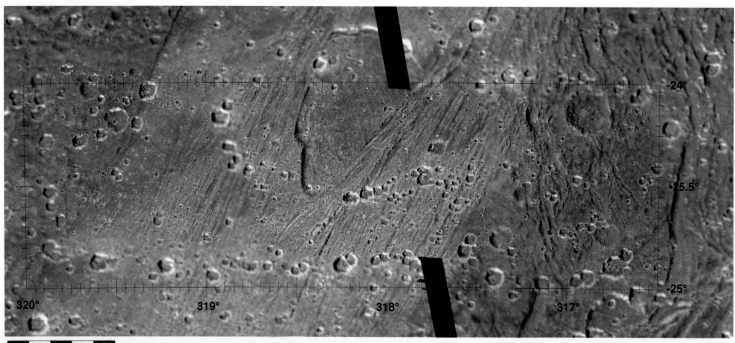

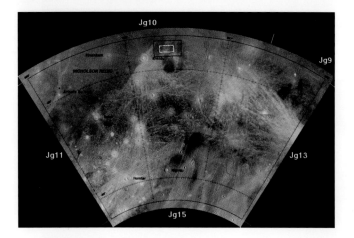

Plate Jg14.1.2 **Calderas: High Resolution**
Encounter: *Galileo* G28   Resolution: 45 meters/pixel
Orthographic Projection   Map Scale: 1 cm = 8.4 km

This view of one of the calderas in Plate Jg14.1.1 reveals a coarse surface texture, but otherwise few diagnostic surface details. The band of bright terrain is lightly grooved, and several prominent fracture zones cross the dark terrain to the east. Abundant secondary craters from Enkidu dominate the scene. High resolution reveals bright frost-covered slopes on the north walls of many of these craters. These are similar to frost-covered pole-facing slopes in the northern hemisphere (Plate Jg3.2), but these may be the most equatorial examples observed on Ganymede, indicating this process occurs beyond the limits of the obvious polar caps.

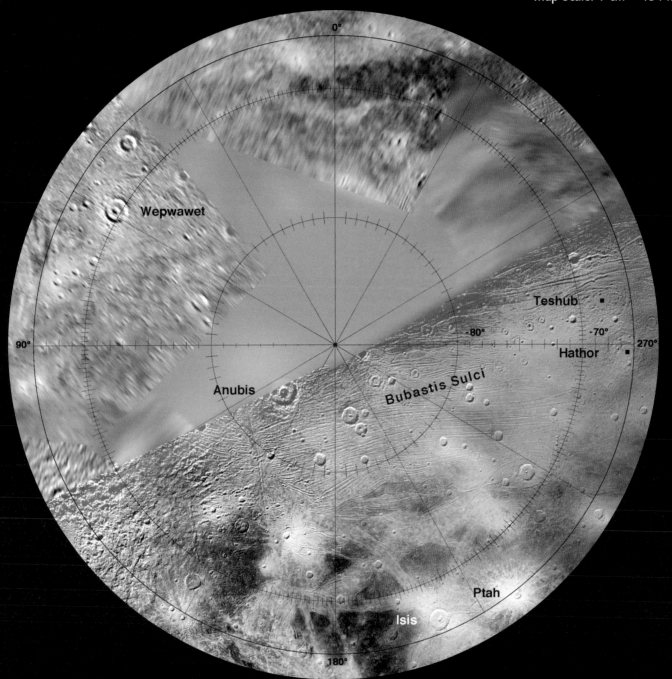

Plate Jg15 **Hathor Quadrangle**

The broad expanse of grooved bright terrain near the south pole, Bubastis Sulci, features some of the longest stretches of uninterrupted groove formation on Ganymede. Southern portions of Gilgamesh (65°S, 130°W) and its ejecta and secondary craters (see Plate Jg12) intrude from the north. Other prominent features include smooth terrain bands and several old impact scars, Hathor and Teshub. Bland areas have not yet been mapped.

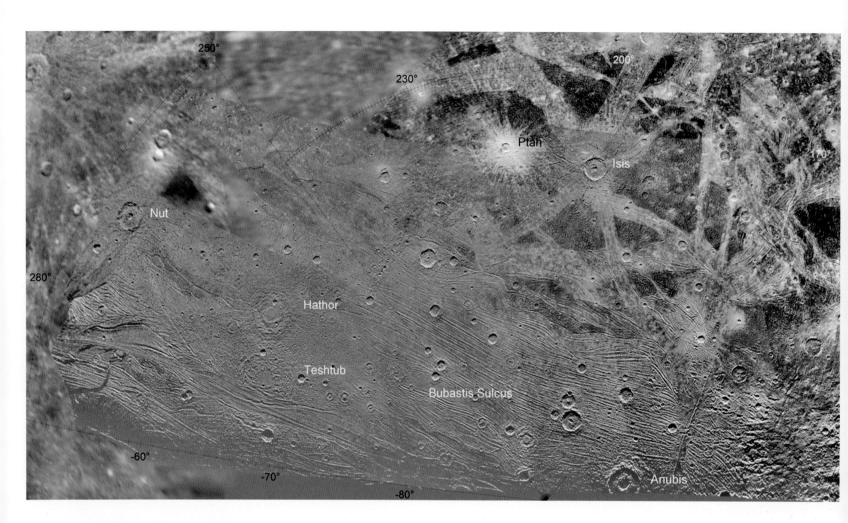

Plate Jg15.1 **Bubastis Sulci**

Encounter: *Voyager* 2   Resolution: 500 meters/pixel
Orthographic Projection   Map Scale: 1 cm = 118 km

This mosaic of the south polar region is *Voyager*'s best view of Ganymede. Groove characteristics within Bubastis Sulci are unusually variable. Many grooves die out toward the north (left in this mosaic), transitioning to smooth terrain. A narrow ridge truncates these grooves near Anubis. Hathor and Teshub are similar to Zakar (Plate Jg5.2) and other penepalimpsests. They are similar in age to bright terrain.

A late-forming band of smooth bright terrain cuts across the outer edge of Teshub. Another narrow lane of smooth terrain extends down from Isis toward Anubis. At least two caldera-like features, similar to those in Plate Jg13.1, are adjacent to this lane. One of these is only 50 kilometers south of Isis. A small patch of smooth terrain at 75°S, 165°W appears to embayolder grooved terrain. Shown at 50% scale to fit page.

Europa

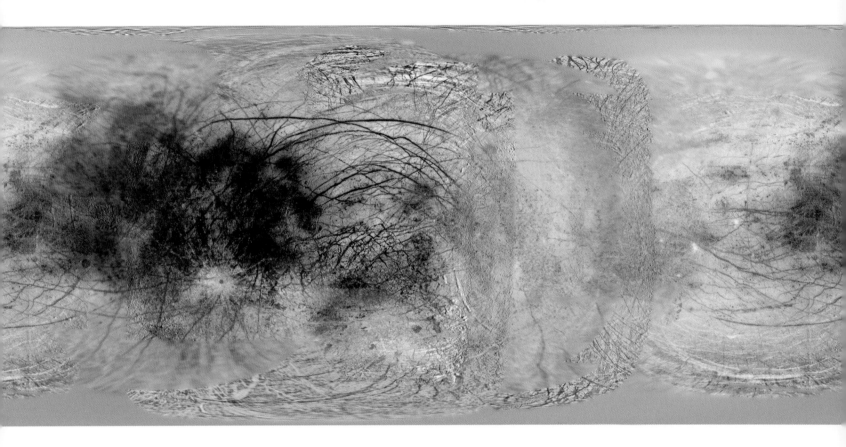

Plate Je **Global Map of Europa**

Simple Cylindrical Projection
Map Scale: 1 cm = 407 km
(valid at equator and along longitude lines)

Orthographic Projection: Center 25°S, 215°W

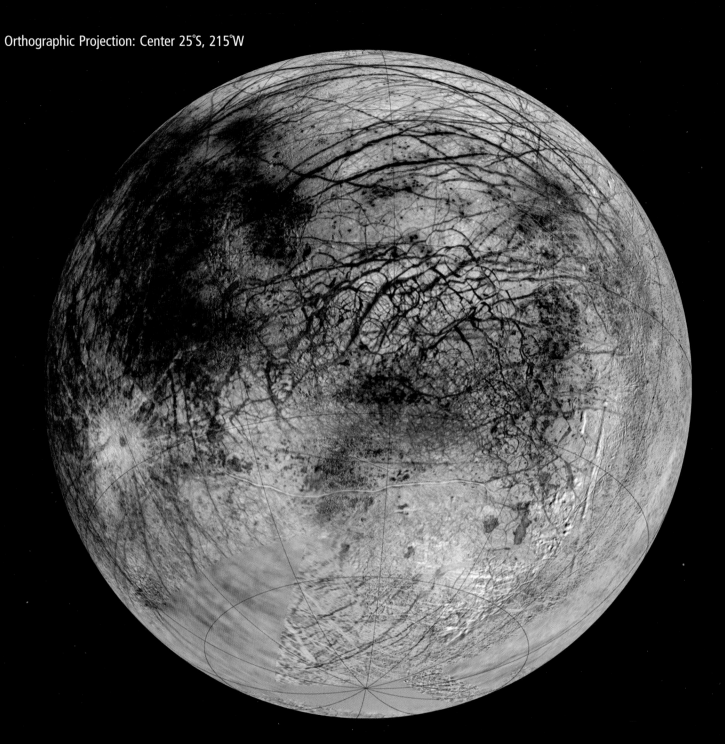

Plate Je01
Global View 1: The Dilational Bands Rift System

The zone of dark dilational bands known as Argadnel Regio sweeping across this region forms the largest, most coherent zone of rifting evident on Europa today (and also my first planetary discovery!). These short stubby bands (formerly known as wedge-shaped bands or dilational bands) are typically 25 kilometers across and up to 300 kilometers long, and formed when the icy shell cracked and was pulled apart in a north–south direction (see also Plate Je8). The bright band Agenor Linea at south center runs along the southern border of this rift zone. Additional dilational bands lie to the south of Agenor Linea, a rare arcuate trough to the east (see also Plate Je13). Long dark arcuate bands (formerly known as triple bands) dominate the regions north of the dilational bands. This symmetric global pattern may not be a coincidence. Some of these features are related and may have formed together, probably during an episode of polar wander on Europa.

Orthographic Projection: Center 35°N, 35°W

Plate Je02
Global View 2: Sub-Jovian Hemisphere

This view, roughly opposite from that of Plate Je01, shows the northern portion of the sub-Jovian hemisphere. *Galileo* mapped this hemisphere once: late in the mission during orbit I25 at 1-kilometer resolution. No dilational bands are evident, but Corick Linea, a twin to Agenor Linea (Plate Je13), snakes across the center of this view. Long arcuate dark bands and large dark splotchy areas, often associated with chaos, are concentrated in the equatorial region. Northern regions appear more diverse geologically, but there are very few higher-resolution images in this hemisphere. The small bright spots near the equator are impact craters. The odd yellowish streak near the pole may be real or an artifact of image calibration.

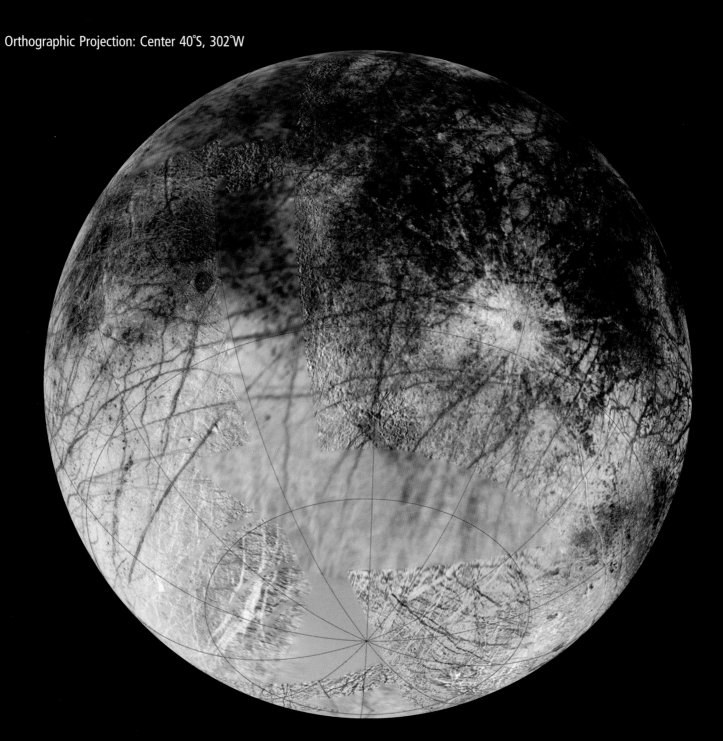

Orthographic Projection: Center 40°S, 302°W

Plate Je03
Global View 3: Large Impact Craters and Southern Regions

This view of Europa's southern hemisphere is dominated by long arcuate dark bands and by two major impact craters, the large dark spot Callanish at center left (Plate Je10.1), and the radial bright-ray system and dark spot associated with crater Pwyll at center right (Plate Je14.1). These craters measure 33 and 27 kilometers across, respectively, but they could not be more different in morphologies. Callanish is a multi-ring impact feature similar to but much smaller than Valhalla on Callisto (Plate Jc6). Pwyll, a central peak crater, is the youngest large crater on Europa. In both cases, the dark spot corresponds to the crater floor and the inner part of the ejecta deposit surrounding the crater rim. Most of the irregular dark pattern in equatorial areas corresponds to chaos or disrupted terrains, showing how dark material is strongly concentrated towards the equator. Whether bright icy frosts obscure dark material near the poles has not yet been determined.

Orthographic Projection: Center 0°N, 90°W

Plate JeHL
Global View: Leading Hemisphere

Resolution across this hemisphere is highly variable, ranging from 230 meters near center to 4 kilometers in the blurry areas west of the central meridian in this view. Very few high-resolution images were acquired here and most of this hemisphere remains largely unknown. Numerous ridges and long dark bands cross the hemisphere. The dark patterns are mostly chaos formations involving disruption of ridged plains. The small bright spots are poorly resolved impact craters (more are likely in the very poorly resolved western sectors). Despite the enhancement applied to this view, the contrast between geologic units is not as strong on this hemisphere as on the trailing side.

Orthographic Projection: Center 0°N, 270°W

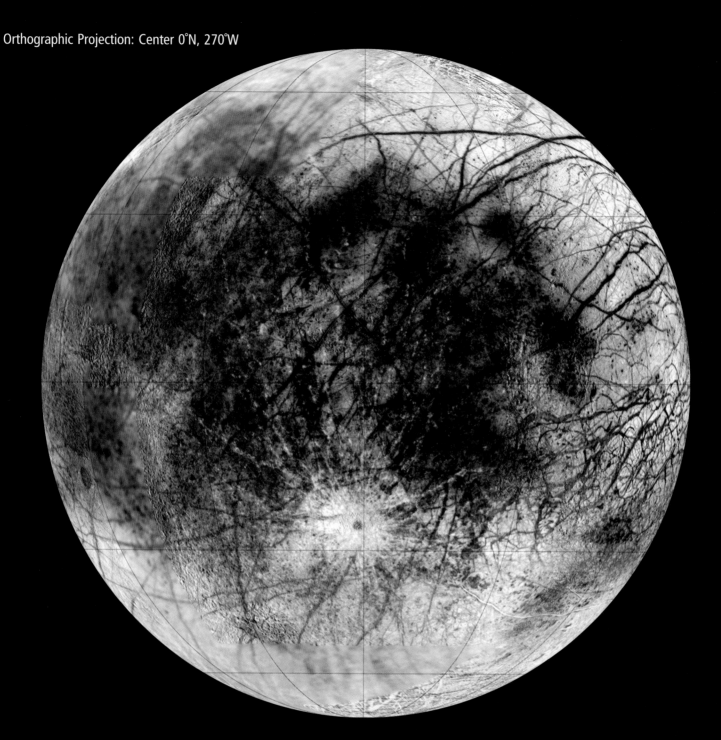

Plate JeHT
Global View: Trailing Hemisphere

Most of *Galileo*'s high-resolution mapping coverage is in this hemisphere, although most of the global mapping is restricted to 1 to 4 kilometers resolution. Sulfur contamination from Io is abundant across this hemisphere. The equatorial dark regions are mostly associated with chaos or areas of plains disruption. Near center lies well-studied Conamara Chaos, the dark spot hanging from the giant "X" formed by the intersection of Asterius and Agava Linea in the center of the hemisphere. Dilational bands can be seen in the lower right quadrant, while long dark bands dominate elsewhere. The dark spot at far left is Callanish, a multi-ring impact feature. Bright rays radiate several thousand kilometers from the crater Pwyll (south of center) and dominate the hemisphere in the same way as rays from other very young craters, such as Tycho on the Moon, or Osiris (Plate JgO2) on Ganymede. All three of these features were frequent *Galileo* imaging targets.

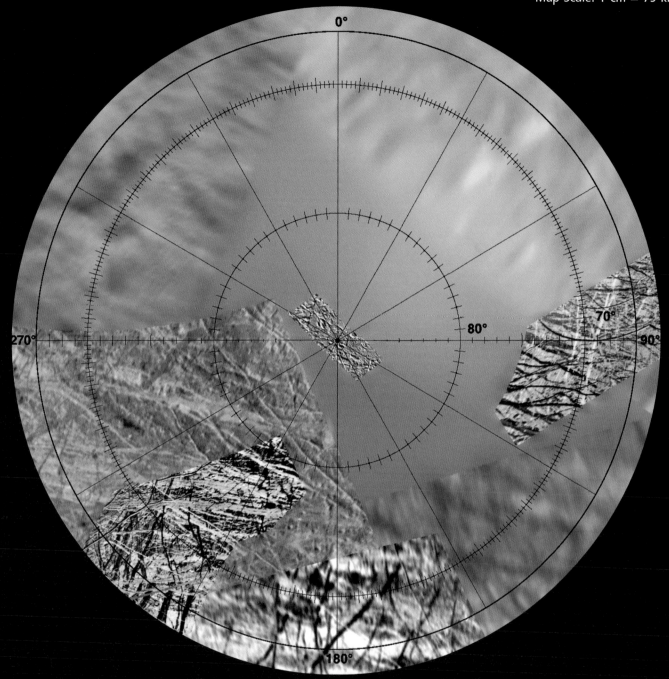

Polar Stereographic Projection
Map Scale: 1 cm = 79 km

Plate Je1 Ogma Quadrangle

The north polar region is very difficult to characterize due to the limited mapping coverage by *Voyager* and *Galileo*. What we can see of the region appears to be dominated by ridged plains and long narrow ridges, based on regional images at 1.5- to 4-kilometer resolution and two small sections of 230-meter resolution mosaics. The only high-resolution imaging is at the pole itself (Plate Je1.1), where chaos material is more abundant than ridged plains, demonstrating the potential futility of attempting to characterize Europa from low- or medium-resolution imaging.

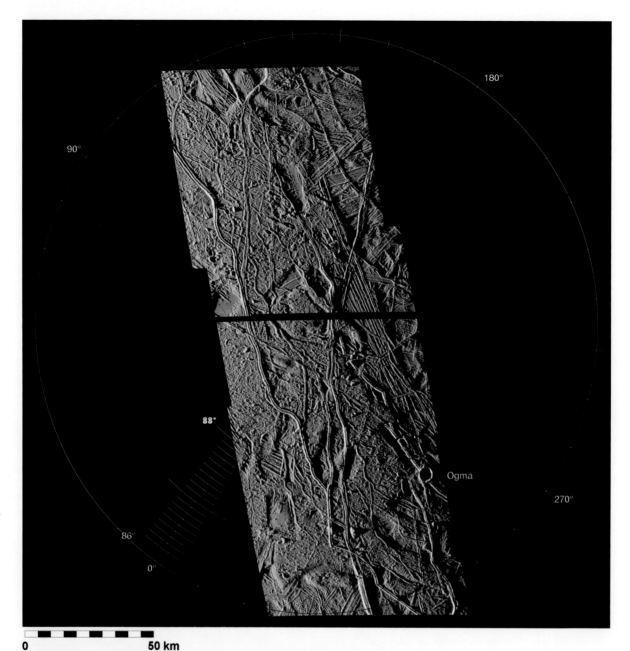

Plate Je1.1 **North Polar Region**
Encounter: *Galileo* I25   Resolution: 90 meters/pixel
Orthographic Projection   Map Scale: 1 cm = 14.2 km

The exact location of this mosaic is not known due to the lack of context imaging. The nominal (predicted) location of the original *Galileo* targeting vectors is shown here, and is likely good to within half a frame width or so. This 3-frame *Galileo* mosaic was planned to search for geologic differences between polar and more equatorial regions. The observation was not extensive enough to achieve that aim but does, however, reveal extensive areas of matrix material. These deposits rise 100 meters or so above older ridged plains and have clearly flowed across the top of them. Several linear and sinuous double ridges cross the matrix deposits and are the most recent geologic events to occur in this region. These patches of matrix material are extensive enough that they almost coalesce into a single massive deposit. A few curious patches of smooth material (e.g., 88°N, 135°W and 88.8°N, 60°W) appear to embay ridged plains, and could be evidence for limited occurrences of water volcanism. If polar wander did indeed occur on Europa, then these terrains may have formed much closer to the equator than they are now.

Europa

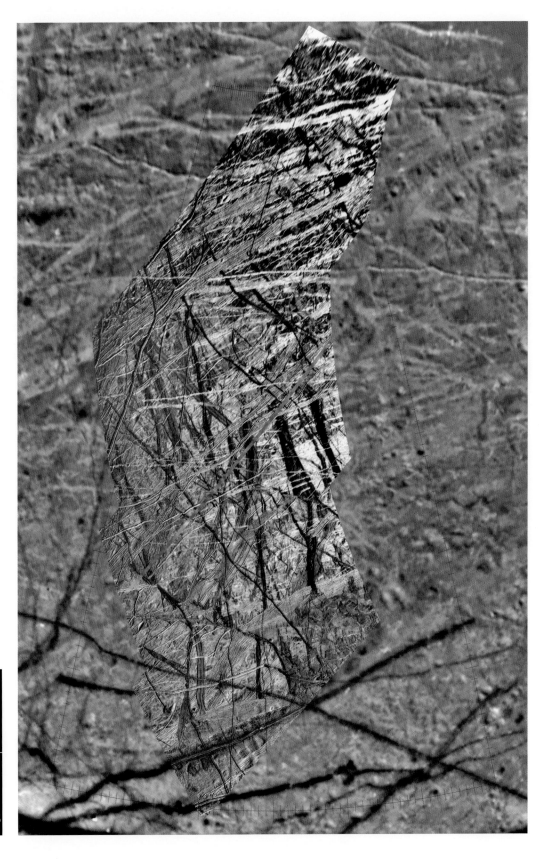

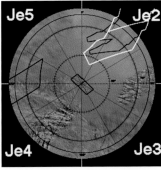

Plate Je1.2a

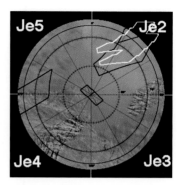

Plate Je1.2b
Northern Plains:
Regional Mapping Mosaics
Encounter: *Galileo* E19
Resolution: 170–190 meters/pixel
Orthographic Projection
Map Scale: 1 cm = 40.2 km

These two regional mapping mosaics feature endless tracts of ridged plains. Linear bands and double ridges are seen in abundance. Although a few lenticulae and surface depressions are apparent south of 55°, chaos and other regional expressions of topographic variability, such as in Moytura Regio (Plate Je14.3) and elsewhere, are noticeably absent in this region. The high contrast in some areas is due to the oblique viewing geometry. Two overlapping mosaics were acquired, ostensibly for stereo mapping. The first mosaic has been merged with the global color mosaic.

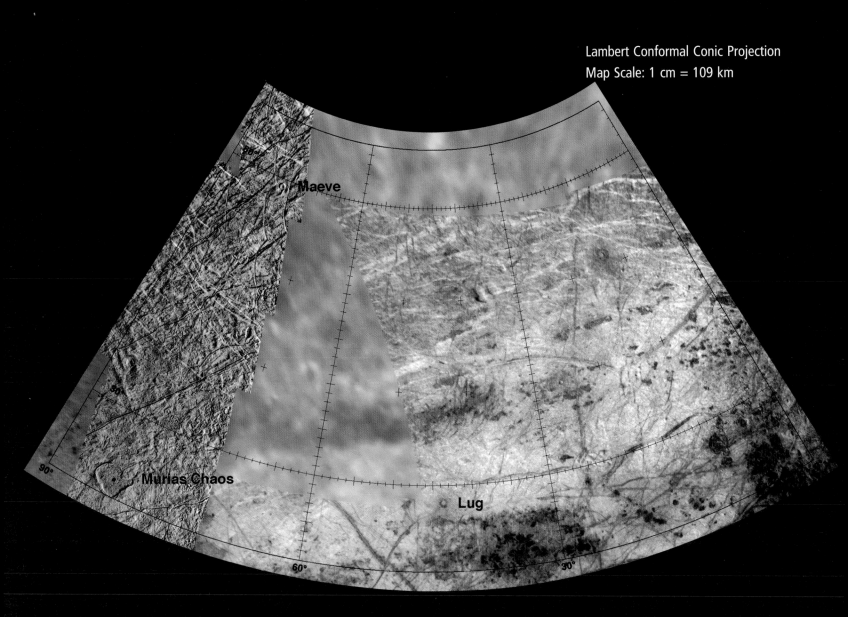

Plate Je2 Murias Chaos Quadrangle

Except for the 250-meter-resolution mosaic along the western margin, *Galileo* observed much of this quadrangle at 1-kilometer resolution during orbit I25. Most of this region is comprised of older ridged plains and scattered outliers of chaos. Sets of cycloidal ridges north of 45° (and similar to those at the south pole [Plate Je15]) form during Europa's daily tidal stressing. The most distinctive feature known in this quadrangle is the large blob-shaped feature at lower left, Murias Chaos, informally known as "The Mitten." This roughly textured material oozed onto the surface, causing the icy shell to buckle and fracture. The dark ring, Lug, is probably a 12-kilometer-wide impact crater. Dark rings like this usually represent the inner portion of a crater's ejecta deposit, as seen at Manannán, Cilix and other craters. The lens-shaped plateau at 43°N, 74°W stands 800 to 900 meters high and is one of the highest features on Europa. The short parallel troughs at 34°N, 86°W are part of a global set of concentric arcs related to polar wander (see Plate Je12). These troughs have at least 1 kilometer of relief and are among the deepest features on Europa.

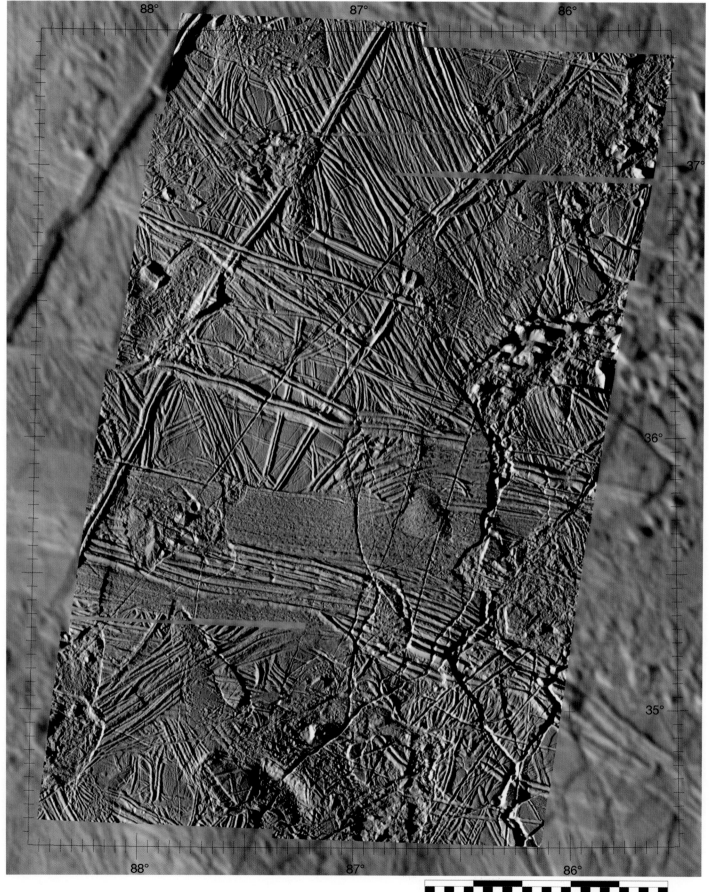

Plate Je2.1

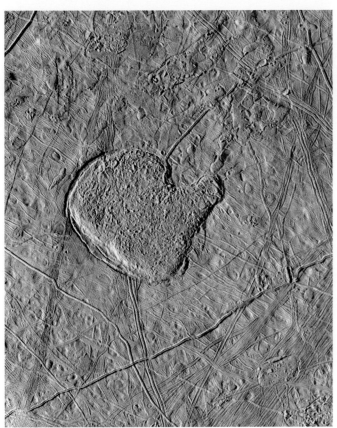

Plate Je2.2 **Murias Chaos**

Encounter: *Galileo* E15   Resolution: 230 meters/pixel
Orthographic Projection   Map Scale: 1 cm = 27.2 km

This large blob of chaos material is partly surrounded by a set of narrow fractures that formed in the surrounding older ridged plains. The fractures indicate that the icy shell fractured under the great weight of matrix material, several hundred meters thick. At one point, the ring fracture was clearly overrun by this material as it flowed outward from a smaller central area. Other examples of thick matrix deposits are located in the north polar region (Plate Je1.1). See Plate Je7.2 for location.

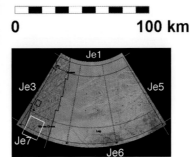

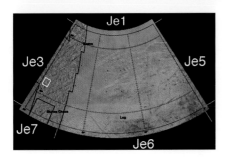

Plate Je2.1 **High-resolution Mosaic: 87°W**

Encounter: *Galileo* E11   Resolution: 32 meters/pixel
Orthographic Projection   Map Scale: 1 cm = 3.8 km

This mosaic of ridged plains includes some of most extended high-resolution *Galileo* imaging coverage of Europa, and is virtually the only such mosaic of its kind on the leading hemisphere. Stereo analysis confirms there is 1 kilometer of relief across this scene, the most of any known location on Europa. Relief of this magnitude confirms that the ice shell cannot be merely a few kilometers thick and is probably on the order of 10 kilometers thick or more. This topography is part of a set of parallel troughs that are part of a global system of symmetric circles on Europa (see Plate Je12).

The ridged plains to the east (e.g., 36°N, 86°W) have been disrupted and broken into smaller plates of various sizes. This disruption is the western section of a large zone of chaos that extends to the east of this view. Several smaller lenticulae (ovals) of chaos also cut across and obliterate older ridges and ridged plains, especially the prominent double ridge crossing from southwest to the top of the frame. Isolated blocks of ridged plains have rotated and been partly subsumed by these blobs (e.g., 36.5°N, 87.8°W), forming angular peaks that look deceptively like icebergs. These are more likely rigid plates of ice foundering into the soft ice rising from deep in the shell, as often occurs in salt diapirs on Earth. See Plate Je7.2 for location.

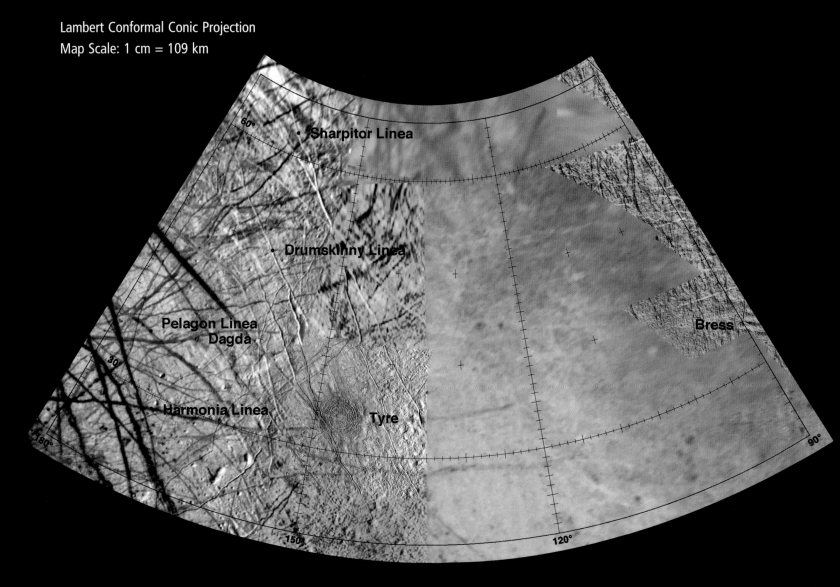

Lambert Conformal Conic Projection
Map Scale: 1 cm = 109 km

Plate Je3 Tyre Quadrangle

Long arcuate dark bands (formerly known as triple bands) dominate the western part of this region (see also Plate Je4). These bands are related to either nonsynchronous rotation or polar wander of the ice shell. A large concentric ridged dark spot in *Voyager* images, *Galileo* revealed Tyre to be a multi-ringed impact structure. Although small by Ganymede and Callisto standards, this is the largest-known impact feature on Europa, and was viewed several times by *Galileo*. The small bright spot at 37°N, 183°W is actually an impact crater, deduced from the abundance of small secondary craters in the higher-resolution mosaic adjacent to it (in the vicinity of Bress).

Plate Je3.1.1 Tyre: Medium-resolution Color

Encounter: *Galileo* G7   Resolution: 600 meters/pixel
Orthographic Projection   Map Scale: 1 cm = 70.9 km

This was *Galileo*'s first view of the large dark spot Tyre, first observed by *Voyager*. These color images confirmed that the spot was a large ringed feature and that the color was distinct from that of surrounding ridged plains. Although the numerous small dark spots surrounding Tyre resemble secondary craters, additional imaging (Plate Je3.1.2) was required to map these features reliably.

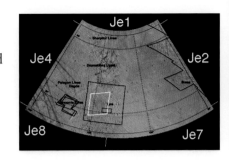
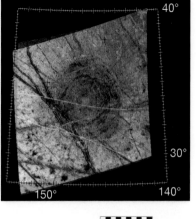

(a)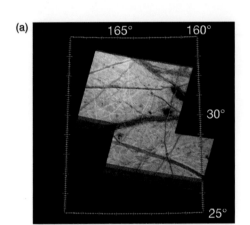
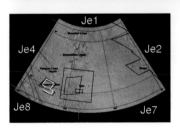

(b)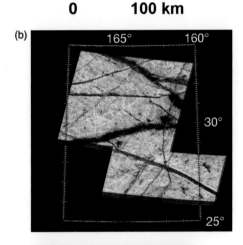
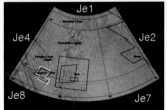

(c)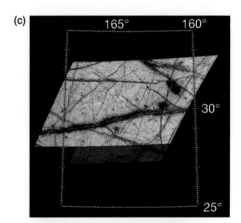
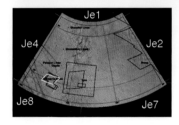

Plate Je3.3a
Encounter: *Galileo* G7   Resolution: 440 meters/pixel
Plate Je3.3b
Encounter: *Galileo* G7   Resolution: 440 meters/pixel
Plate Je3.3c
Encounter: *Galileo* E15   Resolution: 210 meters/pixel
Ridged Plains – Color and Phase-angle Mapping
Orthographic Projection   Map Scale: 1 cm = 52 km

These mosaics of ridged plains were obtained in order to characterize the phase behavior of the surface in color. The first image was obtained at nearly 0° phase angle (similar to a full Moon), the middle image at 5° and the last at 65° phase angle (similar to a gibbous Moon). Although we do not see much difference between these views, the terrains in fact darken with phase angle (these images have been stretched differently to show local features). The major difference is that the 0° phase mosaics are distinctly non-uniform in color, indicating that the albedo behavior of Europa at low phase angles is complex. Information from these images can be used to characterize the porosity and structure of the uppermost surface.

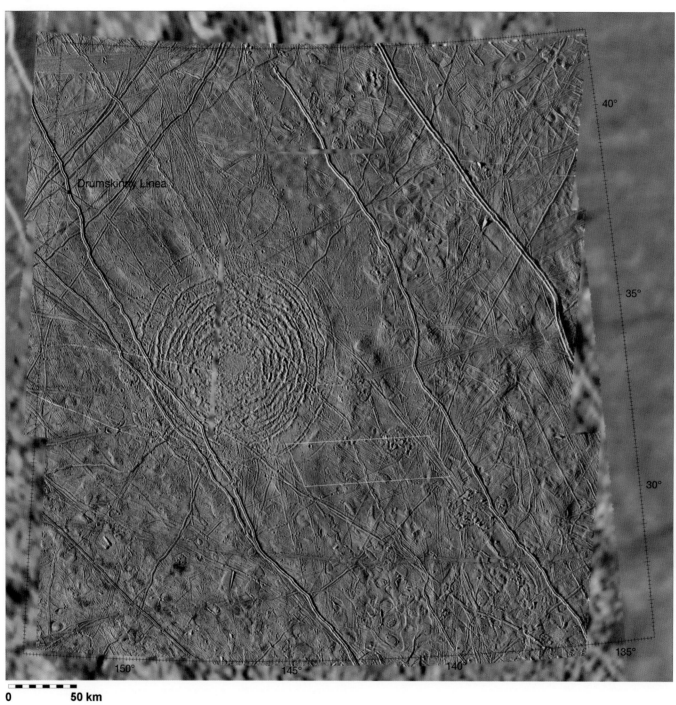

Plate Je3.1.2 Tyre: Medium Resolution
Encounter: *Galileo* E14   Resolution: 170 meters/pixel
Orthographic Projection   Map Scale: 1 cm = 26.8 km

This multi-ring impact structure resembles much larger ancient impact structures on Ganymede and Callisto (see Plates Jg3, Jc3, and Jc7). Tyre comes complete with concentric inner ridges and outer graben fractures, surrounded by a zone of thousands of small secondary craters. The secondary crater field is important because it allows us to estimate the crater size (based on the relationship between secondaries and crater rims established from study of other impact craters). The estimated 38-kilometer diameter of Tyre places the nominal rim location just within the inner zone of continuous concentric ridges.

Ringed basins of this type are found only on icy bodies. Concentric rings like this are seen at high resolution at Callanish (Plate Je10.1) and form when the brittle outer icy layers are very thin. The small sizes of Callanish and Tyre, in comparison to the larger ringed structures on Ganymede and Callisto, is a strong indicator that the icy shell of Europa is thin, on the order of 10 to 20 kilometers, and supports the consensus that Europa has subsurface ocean.

The region surrounding Tyre is typical of ridged plains, and features scattered outcrops of chaos, and isolated plateaus and basins. Although Tyre is a relatively recent event, at least two isolated tectonic features cross and thus post–date Tyre, the linear east–west fracture across the center and the two double ridges to the southwest. It is also possible that these fractures predate Tyre but continued to be active after impact. The white rectangle shows the location of Plate Je3.2, which lies deep within Tyre's secondary crater field. Color from Plate Je3.1.1 has been added. The blurry areas are gaps in the images.

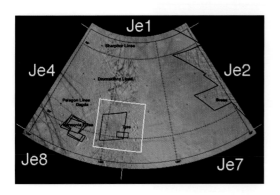

196 Atlas of the Galilean Satellites

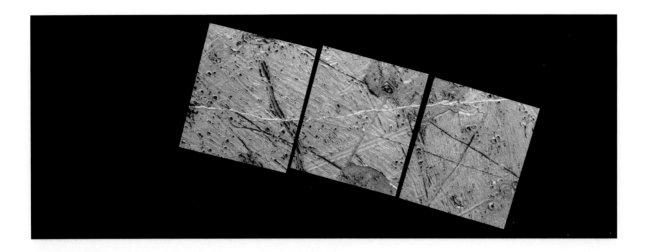

Plate Je3.2a
Encounter: *Galileo* E15   Resolution: 28 meters/pixel

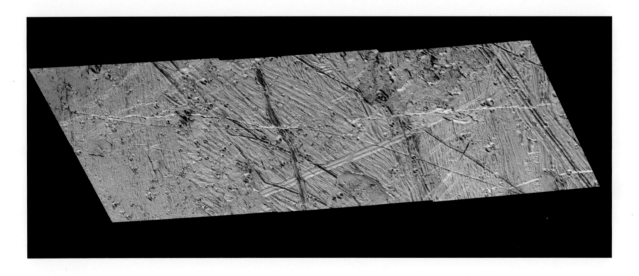

Plate Je3.2b
Encounter: *Galileo* E15   Resolution: 35 meters/pixel

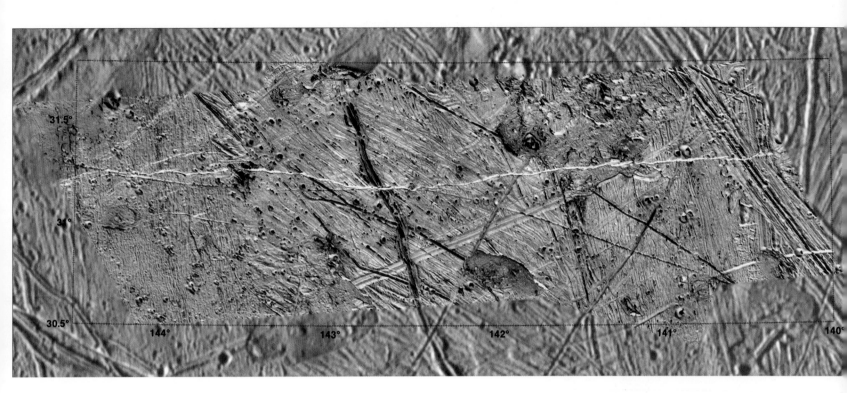

Plate Je3.2 Chaos and Tyre Secondaries

Encounter: *Galileo* E15   Resolution: 35 meters/pixel
Encounter: *Galileo* E15   Resolution: 35 meters/pixel
Orthographic Projection   Map Scale: 1 cm = 6 km

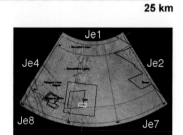

These two 3-frame mosaics form a high-resolution stereo mosaic pair featuring ridged plains and chaos material southeast of Tyre. The numerous circular depressions are Tyre secondary craters (the mosaic just missed the outermost ring of Tyre). Dark material can be seen at the bottoms of many of these craters and some of the linear troughs. The large area of chaos to the north and east (31.6°N, 141.6°W) features several blocks of icy crust that have rotated and are partly buried by matrix material. The oval lenticula at bottom center (30.7°N, 142°W) rises at least 100 meters above the adjacent plains. The interior of the chaos includes matrix material and several large tilted blocks, and formed when soft ice upwelled from the lower icy shell and broke through the surface. The most recent event appears to be a deep fissure crossing east to west across the scene. Mosaics have been reduced to fit the page.

Lambert Conformal Conic Projection
Map Scale: 1 cm = 109 km

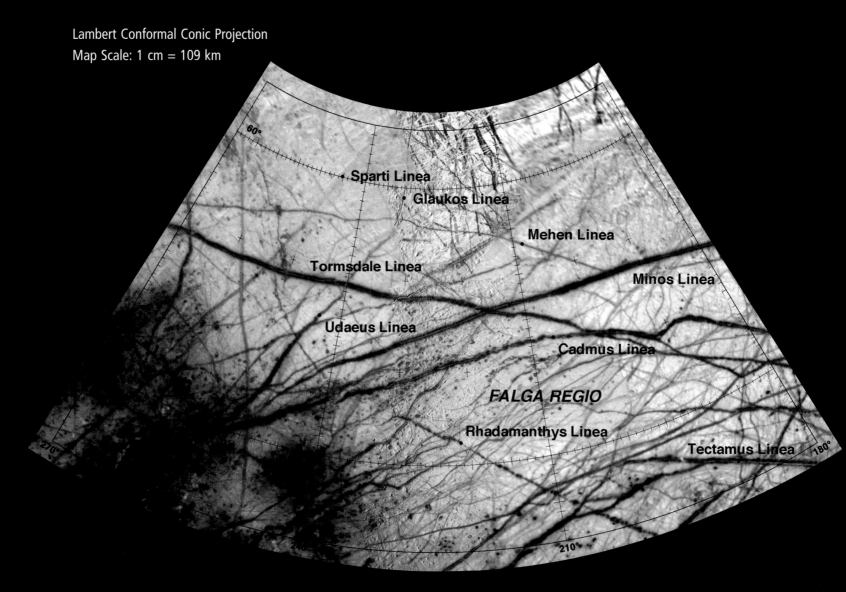

Plate Je4 Rhadamanthys Linea Quadrangle

The dark linear features crossing this region in abundance are dark bands, first seen by *Voyager*. The original "triple band" terminology, dating to *Voyager* imaging, refers to these double ridges flanked by dark material. Scattered dark spots distinguish the bright ridged plains of Falga Regio from areas to the south. These dark "freckles" comprise a large field of lenticulae (see Plate Je4.1). Dark spots are most likely diapirs or similar features formed when upwelling of soft ice from below broke through the surface, much like salt domes on Earth. The extensive dark material to the southwest is usually associated with chaos or disrupted terrains. Dark ring-shaped features (e.g., 28°N, 268°W and 30°N, 256°W) within this terrain may be unusual chaos formations, but also resemble dark impact features such as Manannán and Pwyll when viewed at low resolution.

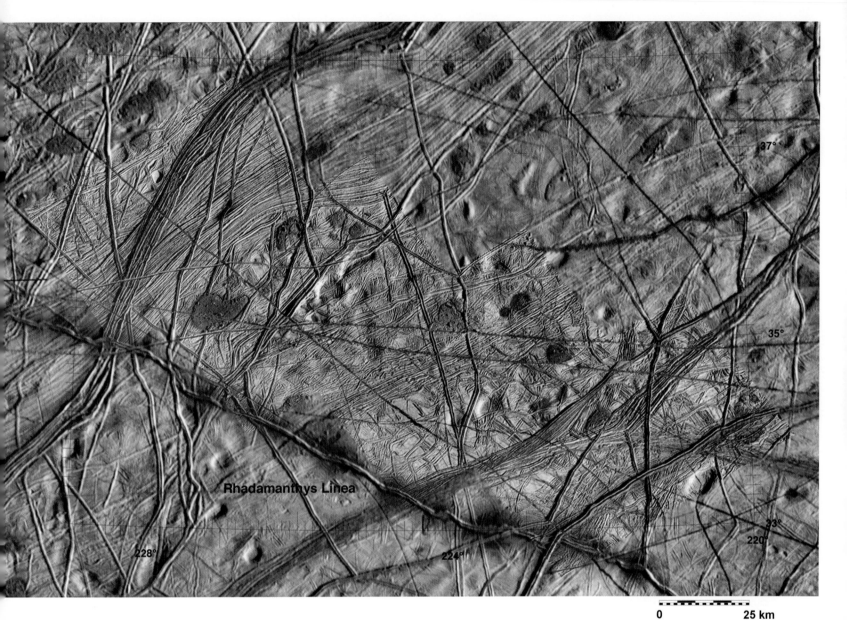

Plate Je4.1 Rhadamathys Linea

Encounter: *Galileo* E19   Resolution: 60 meters/pixel
Orthographic Projection   Map Scale: 1 cm = 10.1 km

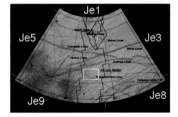

Due to a minor pointing error, this 4-frame mosaic just clipped its intended target, Rhadamanthys Linea, the prominent wiggly dark band along the bottom of the mosaic. It is interesting that the darkest areas along Rhadamanthys occur where the ridge briefly turns due east–west, implying that a particular stress orientation is required for dark material to form. Models include pyroclastic eruption of dark material and local shear heating along the ridge to remove more volatile water frosts.

The mosaic is also of interest as it is our best view of the oval to tadpole-shaped features 10 to 25 kilometers across that are common in this region of ridged plains. These ovoids include depressions, the largest of which are 350 to 450 meters deep and are among the deepest features known on Europa. Intermingled with these depressions are sets of dark ovals of similar size and shape that are filled with matrix material and sometimes termed lenticulae. These may be the expression of diapirs that pierced the surface.

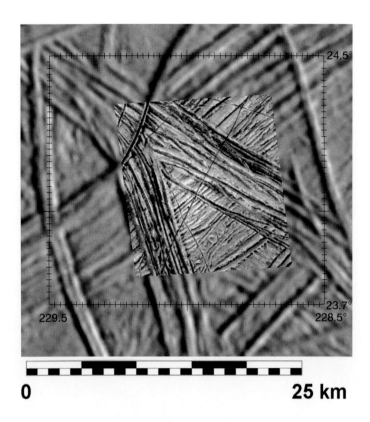
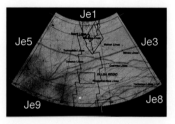

Plate Je4.2 **Band Bifurcation: High Resolution**
Encounter: *Galileo* E14   Resolution: 35 meters/pixel
Orthographic Projection   Map Scale: 1 cm = 3.3 km

This single image (one of a set of 4 identical images of the same site), highlights a ridged band where it divides into two branches, and was taken at almost 0° phase angle. Although limited in scope, the image appears to show that the southern branch actually crosscuts and postdates the branch extending to the southwest. Each branch may have formed by the growth of individual ridges within each band, rather than as coherent features. The image also shows how the appearance of Europa can change from the shadowy low-Sun background context view to the high-Sun high-resolution image at center, where it is evident that dark material prefers to lie in low areas.

Lambert Conformal Conic Projection
Map Scale: 1 cm = 109 km

Plate Je5 Annwn Regio Quadrangle

Poorly mapped, this quadrangle appears to consist of typical ridged plains, dark chaos, and dark bands. A long linear fracture near 45°N, 355°W, and several small deep linear depressions near 50°N, 305°W, are the only distinguishing features, due in part to the poor mapping coverage. The linear depressions are part of a sparse global pattern of concentric arcs (see also Plates Je2, Je10, and Je12) that formed when the icy shell rotated or flipped on its side due to polar wander, stressing and fracturing.

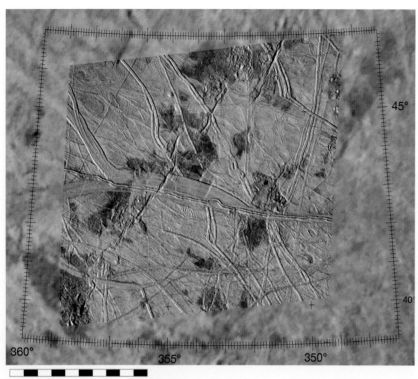
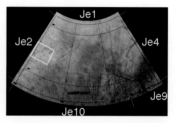

Plate Je5.1 **Fractured Plains**

Encounter: *Galileo* I25   Resolution: 225 meters/pixel
Orthographic Projection   Map Scale: 1 cm = 26.6 km

The I25 encounter provided *Galileo*'s best view of Europa's sub-Jupiter hemisphere. *Galileo* obtained a nearly complete mosaic of this hemisphere at 1-kilometer resolution, but also acquired part of a 225-meter-resolution mosaic of the same terrains. The nearly mute *Galileo* craft was able to return only a few fragments of this potential coverage, however (see also Plate Je10.2.1).

The ridged plains, double ridges, bands, and chaos are typical of Europa. The striking features here are the segmented linear fissures striking diagonally across the scene. Fissures of this type are uncommon, but form when the icy shell is stretched and ruptures. These fractures cut across most other features, but also appear to connect large outcrops of chaos to the northeast and southwest, and may be associated with their formation.

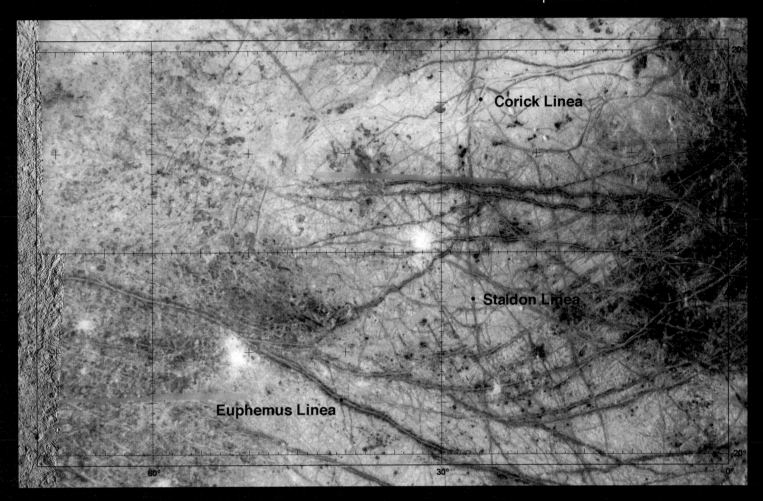

Mercator Projection
Map Scale: 1 cm = 104 km

Plate Je6 Euphemus Linea Quadrangle

This is one my favorite quadrangles in the *Atlas*. Under mid-day Sun conditions, Europa most resembles a Jackson Pollack painting. The dark lineations and dark spots could have been dripped onto the surface by the great celestial painter. The smaller dark spots seen here and elsewhere are probably diapirs 10 to 20 kilometers across, composed of water ice contaminated with salts and other compounds. Such diapirs rise slowly to the surface from the bottom of the ice shell floating over the ocean believed to lie underneath. Corick Linea, the bright lineament snaking across the top of the mosaic, is a twin of Agenor Linea, located on the opposite hemisphere (see Plate Je13).

The subtle change from brownish to orange tones (in this color exaggeration) as we track westward across the scene is part of a global pattern related to the implantation of sulfur from Io. The impact of charged ions of sulfur and other metals, carried outward from Io by Jupiter's powerful magnetic field, is at least partly responsible for Europa's general orange-reddish coloration. Other contaminants of the ice come from the lower ice shell or perhaps the ocean itself.

The few small bright spots are unresolved impact craters. The unusual circular feature at 9.5°N, 47°W may be a large impact crater. Fuzzy rectangular gray patches are gaps in the 1-kilometer-resolution mapping coverage from *Galileo* orbit I25, which makes up most of the imaging here. Portions of a 230-meter-resolution pole-to-pole mosaic can be seen on the westernmost edge (Plate Je7.2).

Mercator Projection
Map Scale: 1 cm = 104 km

Plate Je7 Brigid Quadrangle

Another poorly mapped quadrangle, this quadrangle demonstrates the nature of the losses resulting from the main antenna failure on *Galileo*. The fuzzy background images that dominate here have a resolution of only 4 kilometers. *Voyager* 2 mosaics at 1.7-kilometer resolution are visible along the western margins. *Galileo* mapping mosaics obtained on *Galileo* orbits E15 and E17 cover the eastern edge and parts of the central region at resolutions of 230 meters or better, but should have been covering the entire quadrangle, as well as those to the north and south. The next mission to Europa will complete this global mapping. Tara Regio, where imaged sufficiently, consists of severely disrupted terrains and has been almost completely resurfaced by chaos materials.

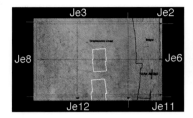
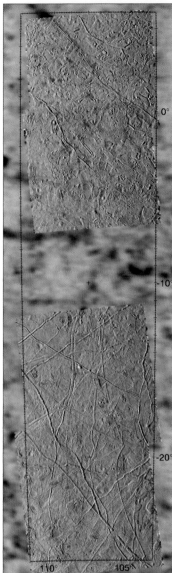

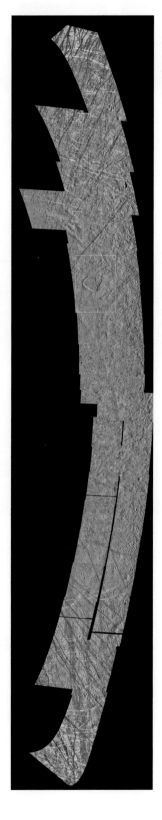
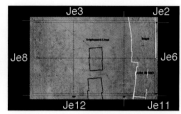

Plate Je7.2 **Regional Mapping Mosaic: 80°W**
Encounter: *Galileo* E15
Resolution: 230 meters/pixel
Encounter: *Galileo* E17
Resolution: 220 meters/pixel
Sinusoidal Projection
Map Scale: 1 cm = 246 km
Map Scale: 1 cm = 143 km
(following page)

Plate Je7.1 **Ridged Plains**
Encounter: *Galileo* G7   Resolution: 500 meters/pixel
Orthographic Projection   Map Scale: 1 cm = 59.1 km

This view is the best-available imaging of this region. Originally a more extensive mosaic, only 4 images were returned. These terrains are representative of large areas of Europa and no distinctive features stand out at this resolution.

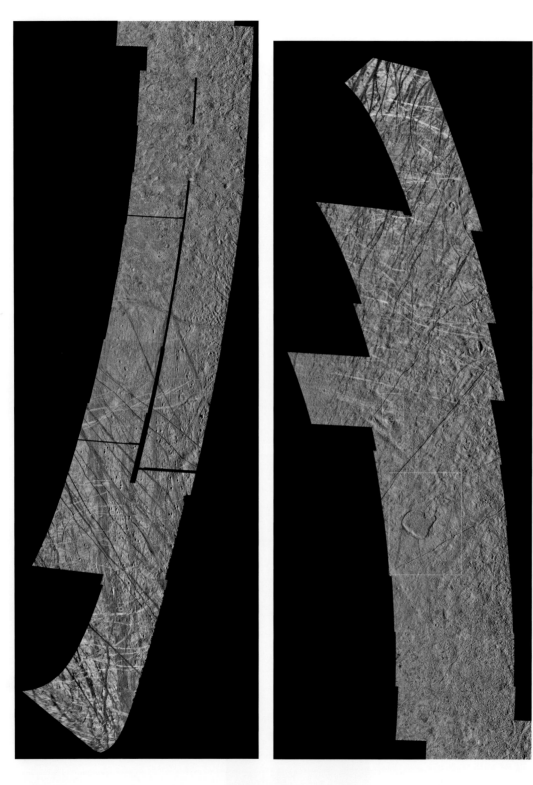

Plate Je7.2

This pole-to-pole mosaic, obtained during two orbits (E15 to the north and E17 to the south) is the only such mosaic in the leading hemisphere. If the *Galileo* mission had gone as planned, these narrow strips would have covered the entire visible hemisphere during each orbit, providing extensive stereo coverage. Chaos units dominate in the center area, grading into ridged plains to the north and south. The plains to the south also feature large numbers of oval depressions, similar to those in found in quadrangle Je4. The irregular mitten-shaped feature to the north is featured in Plate Je2.2, as is a high-resolution mosaic to the north (Plate Je2.1), indicated by the white rectangles.

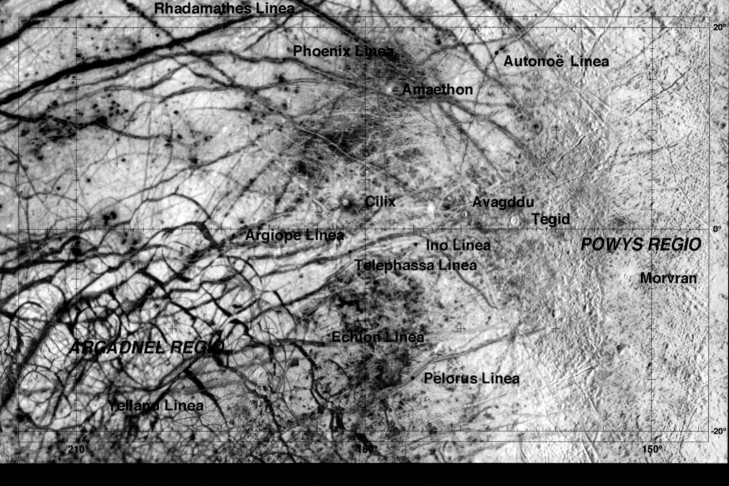

Plate Je8 Cilix Quadrangle

The long arcuate dark bands to the north give way in the south to shorter lineations called dilational (or originally wedge-shaped) bands. Argadnel Regio (see also Plates Je01, Je9, Je13, and Je14) is one of only two regions where these bands occur: the other is to the south (see also Plates Je13 and Je15). The remarkable degree to which the sides of these bands match exactly indicates the bands formed when the icy outer shell of Europa cracked and literally pulled apart. Warm dark ice from the base of the shell intruded upward to fill the gaps. At least one hundred kilometers of new crust formed when these bands opened up. These bands were a popular *Galileo* target, especially near Yelland Linea.

Several impact craters, including Cilix and the bright spot Amaethon, are located near the equator. Dark spots to the northwest are part of a pattern formed by upwelling diapirs of soft ice (see also Plates Je4 and Je9). The subtle break in the color pattern in this quadrangle is due to the use of a different filter to substitute for a missing infrared image in one of the mosaics used to construct the color components.

(a)

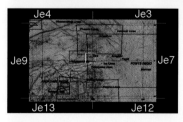

Plate Je8.1.2a Cilix: High Resolution
Encounter: *Galileo* E15   Resolution: 60–100 meters/pixel
Orthographic Projection   Map Scale: 1 cm = 9.4 km

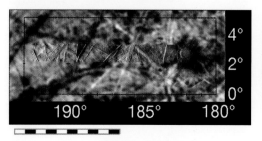

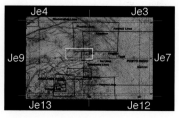

Plate Je8.1.1 Cilix
Encounter: *Galileo* C3   Resolution: 880 meters/pixel
Orthographic Projection   Map Scale: 1 cm = 70 km

This low-resolution image is a fragment of a global mosaic that was only partly returned to Earth. It confirmed that Cilix, an irregular dark spot in *Voyager* images, was in fact a circular impact crater. The characteristic raised rim of an impact crater is evident, but the interior was in shadow at the time of the observation.

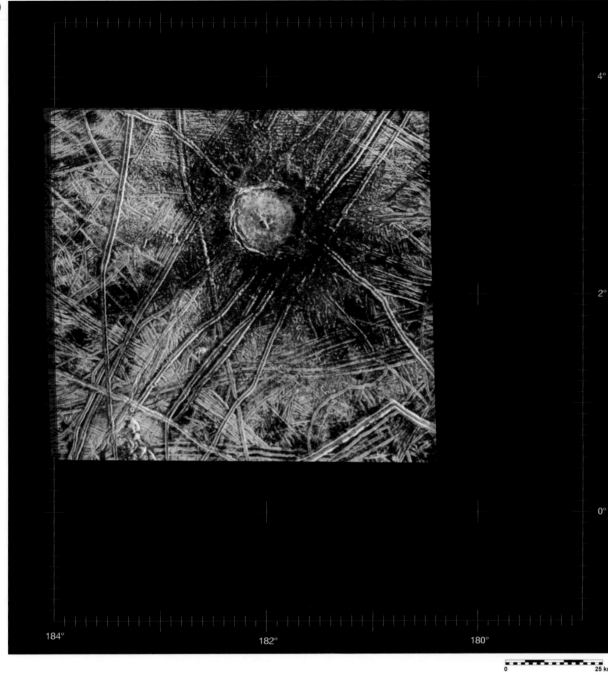

Plate Je8.1.2b Cilix: Color View
Encounter: *Galileo* E15   Resolution: 100 meters/pixel
Orthographic Projection   Map Scale: 1 cm = 9.4 km

At 19 kilometers diameter, Cilix is one of the largest "normal" craters on Europa. The dark reddish deposits surrounding Cilix are part of the ejecta deposit and were observed from *Voyager*. The inner part of this deposit clearly disrupts or blankets preexisting ridges next to the crater rim. The composition of the dark material is poorly understood, but could be salt or sulfate enriched material excavated from deeper in the icy shell during impact. The crater floor is dominated by a chaotic but relatively flat impact deposit and a small central peak complex. Most of the crater rim consists of a steep inward-facing scarp, but at least one rim terrace is visible to the southwest. Rim terraces are formed when a block or rim material slides downward and are very rare on icy satellites.

The sharp bend seen in the large double ridge at 1°N, 181°W is typical of cycloidal ridges (Plate Je15), believed formed by Europa's diurnal tidal stress cycle. The stereo version confirms there is nearly 1 kilometer of relief across these rolling plains.

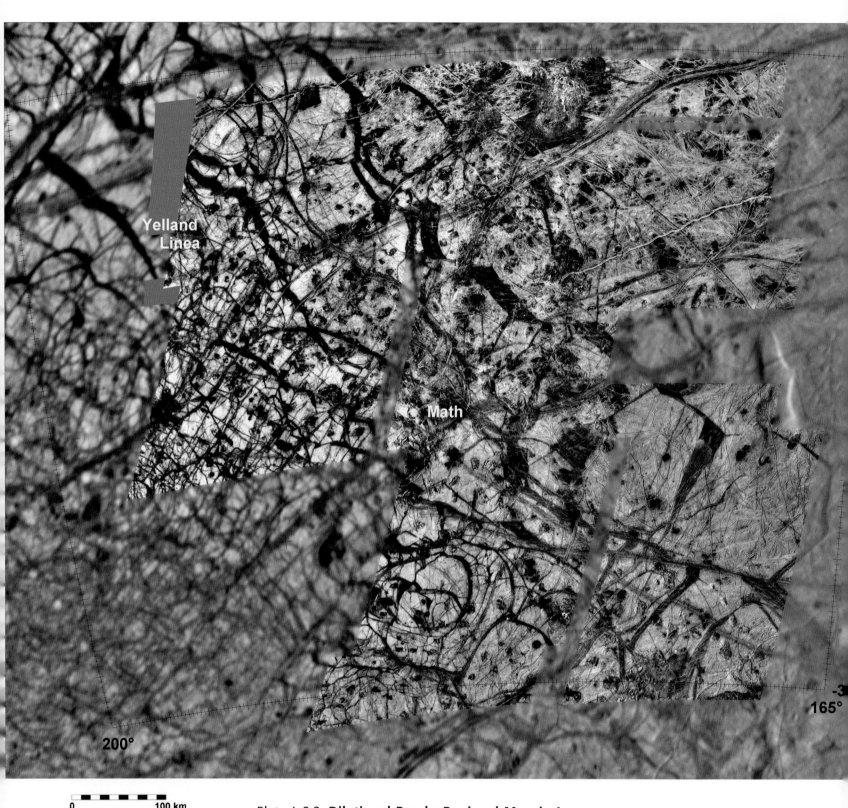

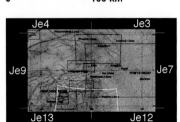

Plate Je8.2 **Dilational Bands: Regional Mosaic 1**

Encounter: *Galileo* E14   Resolution: 230 meters/pixel

Orthographic Projection   Map Scale: 1 cm = 38.8 km

Dilational bands are highly concentrated in Argadnel Regio (see Plate JeO1) and form a well-organized set of parallel bands connected by narrow orthogonal dark bands. A region roughly 2000 by 1000 kilometers across has been torn apart by these bands, forming one of the largest rift zones in the Solar System. *Galileo* obtained several

Plate Je8.3 Dilational Bands: Regional Mosaic 2 (incomplete)

Encounter: *Galileo* C3   Resolution: 420 meters/pixel
Orthographic Projection   Map Scale: 1 cm = 39.7 km

Dilational bands formed when the icy shell fractured into blocks of various sizes, some of which then spread apart. The dark lanes represent entirely new crustal material. This is similar to how continents pull apart and oceanic crust forms on Earth, but contrasts with the style of resurfacing seen on Ganymede and on most other planets in the Solar System. On Ganymede, most lanes of bright terrain form when ancient dark terrain is stretched and broken apart but new material is extruded on top of the older crust (Plates Jg8.5 and Je15.3).

This image is the largest surviving fragment of a planned mosaic that would have covered 20% or more of the surface of Europa at a resolution of ~400 meters. Only fragments of four frames (Plates Je8.1.1, Je8.7, and Je13.4.1) were returned due to the limited time to play back data off the tape recorder. This surviving frame includes several dilational bands and clearly shows the sharp matching boundaries on either side. The narrow ridges within the bands probably formed incrementally in a series of smaller steps.

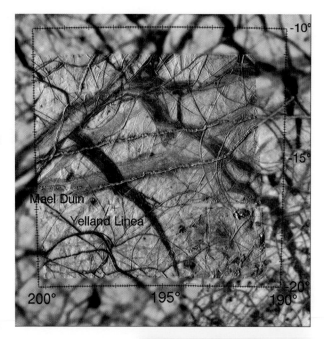

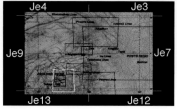

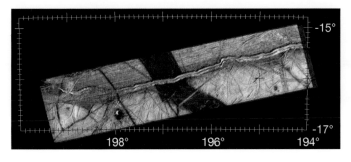

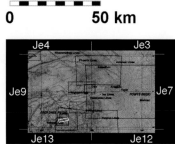

Plate Je8.4.1 Dilational Bands: Yelland Linea, Color

Encounter: *Galileo* E17   Resolution: 170 meters/pixel   Orthographic Projection   Map Scale: 1 cm = 20.1 km

This color view of Yelland Linea highlights the relatively reddish color of the dilational band and the intermediate nature of the gray-colored band trending along the top of the mosaic. This suggests that bands may age by brightening with time, perhaps by frost deposition.

mosaics of these features, especially in the region of Yelland Linea, shown in this and the following mosaics.

Argadnel Regio is geologically complex. Numerous dark spots (chaos) and ridges, some of which are younger than the numerous dilational bands, are evident. Also present are the 14.5-kilometer-wide impact crater Math at 26°S, 184°W, and a large irregular depression at 25°S, 165.25°W. This depression is roughly 1 kilometer deep and is the deepest known feature on Europa. It is part of a system of concentric depressions that formed when the pole of Europa's icy shell shifted toward the equator – "polar wander" (see Plate Je12). Shown at 70% scale to fit page.

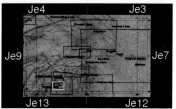

Plate Je8.4.2a Dilational Bands: Yelland Linea, High Resolution

Encounter: *Galileo* E12   Resolution: 42 meters/pixel
Orthographic Projection   Map Scale: 1 cm = 7.2 km

Galileo's view of this dilational band reveals a succession of different geologic events. The surrounding ridged plains have a complex geologic history. Trending across the top of the scene is a broad band (15.5°S, 195°W) with a brightness intermediate between ridged plains and the dark band. This "gray" band is younger than ridged plains generally but older than the Yelland Linea. Reconstructions show that ridged plains may have formed from a succession of dilational bands similar to Yelland Linea, each succeeding band cutting the previous bands before it into a patchwork of ridged blocks. The dark bands were the last to have formed. A younger double ridge cuts and displaces Yelland Linea and other features at 17.2°S, 195.3°W, demonstrating strike-slip faulting. Yelland Linea itself is a ridged band. The ridges running along Yelland Linea appear to be symmetric. This suggests that these bands opened incrementally, with each short opening allowing another thin wedge of material to intrude and expand the feature.

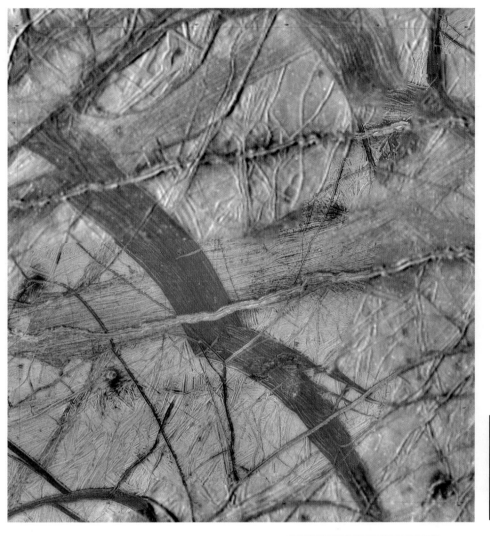

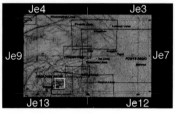

0   50 km

Plate Je8.4.2b **Dilational Bands: Yelland Linea, Very High Resolution**

Encounter: *Galileo* E12   Resolution: 15–25 meters/pixel
Orthographic Projection   Map Scale: 1 cm = 12 km

This observation consists of two crossing linear mosaics, forming a giant "X" on the surface. They are rich in detail, including packets of ridged units of various ages, stray linear fissures, and numerous double ridges, some of which transition into narrow dilational bands, and parts of Yelland Linea itself. This large mosaic has been reduced to 50% to fit the page.

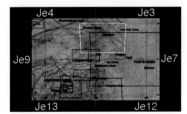

Plate Je8.5 Dark Plains: Regional View

Encounter: *Galileo* E19   Resolution: 220 meters/pixel
Orthographic Projection   Map Scale: 1 cm = 34.6 km

This 2-frame mapping mosaic was intended to provide insights into the stratigraphic relationships in dark plains units. The low resolution and high-Sun viewing make such details difficult to interpret. The scene shows that ridges can be densely concentrated in some areas. It also highlights the overlapping spaghetti-strand nature of Europa's innumerable complex ridge and band formations. The small bright spot near center is a very young bright-ray crater a few kilometers across.

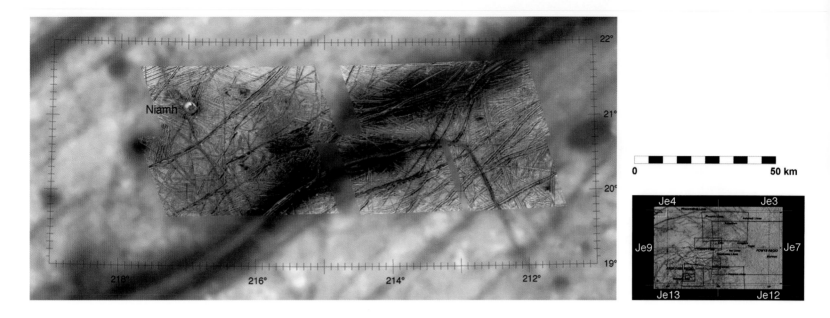

Plate Je8.6.1 **Belus Linea Band Discontinuities**
Encounter: *Galileo* E14   Resolution: 66 meters/pixel
Orthographic Projection   Map Scale: 1 cm = 13 km

This 8-frame mosaic features an odd aspect of some ridged bands (née triple bands), whereby they appear to split apart and form two parallel segments before reconnecting on their "original" trend. This view shows that while the ridges that run down the center of this band do in fact reconnect, the dark deposits on their flanks do not continue, forming the gap we see at lower resolution. This pattern occurs on Earth when two faults approach each other and are influenced by the overlapping stress fields, causing them to deflect around each other. See Plate Je9.8 for more of Belus Linea.

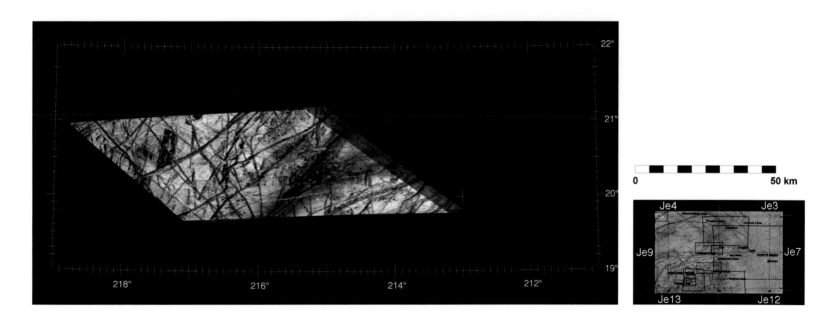

Plate Je8.6.2 **Belus Linea Band Discontinuities: Color**
Encounter: *Galileo* E19   Resolution: 85 meters/pixel
Orthographic Projection   Map Scale: 1 cm = 13 km

This oblique color mosaic of the dark ridged band in Plate Je8.6.1 provides higher-resolution color information for the western part of this structure.

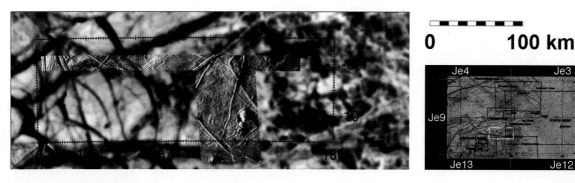

Plate Je8.7 Ridged Plains
Encounter: *Galileo* C3   Resolution: 420 meters/pixel
Orthographic Projection   Map Scale: 1 cm = 39.7 km

Although only this small fragment of one image (part of a planned global mosaic: see also Plates Je8.3, Je13.4.1) was returned, it provides a good opportunity to compare Europa's morphology and albedo characteristics. Ridges appear to correlate with both bright and dark linea. Topographic undulations are apparent even in this small image fragment. The small mound at 10°S, 183.5°W does not correspond to any specific feature, except a few small dark spots, illustrating that not all topography is related to albedo features.

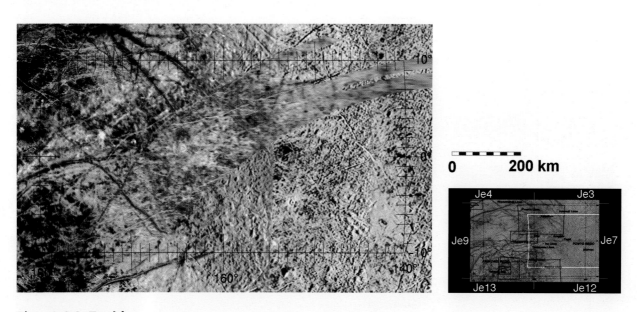

Plate Je8.8 Tegid
Encounter: *Galileo* E19   Resolution: 900 meters/pixel
Orthographic Projection   Map Scale: 1 cm = 106 km

The crater Tegid, located at 0.5°N, 165°W, has an unusual concentric morphology but was not observed at high resolution. Tegid may represent a crater form on Europa transitional between central peak craters (see Plate Je8.1) and multi-ring basins (see Plate Je10.1). This low-resolution observation was intended to help define the characteristics of this crater, and replace an image lost during orbit E16, but little can be learned from it except to remind us of the many unanswered questions in the Jupiter family.

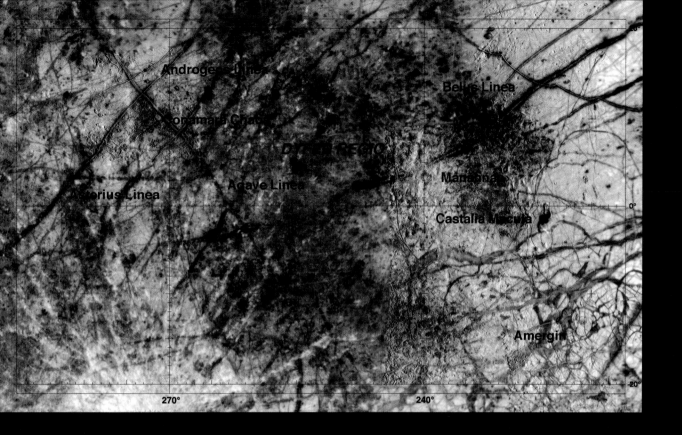

Plate Je9 Castalia Macula Quadrangle

This quadrangle also shows a region in transition. From the short dark dilational bands in the southeast and bright ridged plains in the northeast, we move into a region consisting of dark chaotic disrupted material, Dyfed Regio, crossed by occasional arcuate dark bands coming in from the northeast. The bright splatter-marks across the south and west are bright-ray ejecta from crater Pwyll (see Plate Je14).

Three features were targets of special interest for *Galileo* in this quadrangle. The large dark patch just below the large "X" is the well-mapped Conamara Chaos. The unusual crater Manannán is visible as an irregular dark ring in this view, and was also observed at high resolution. Brighter ejecta from Manannán appear to partially obscure darker portions of Belus Linea just north of the crater. The third target was the dark spot Castalia Macula. It is one of the darkest and reddest features on Europa. Immediately north and south of Castalia Macula lie two large plateaus, one of which is 850 meters high and among the tallest known features on Europa.

The subtle break in the color pattern in this quadrangle is due to the use of a different filter to substitute for a missing infrared image in one of the mosaics used to construct the color components.

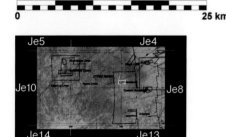

Plate Je9.1.1a **Manannán**

Encounter: *Galileo* E14   Resolution: 20–82 meters/pixel
Orthographic Projection   Map Scale: 1 cm = 4.7 km

These high-resolution views are the best we have for an impact crater on Europa. Although high-Sun illumination washes out some topographic details, mapping of this 23-kilometer-wide crater reveals many interesting details. Very little of a classic inward-facing rim scarp (as seen at craters such as Melkart on Ganymede [Plate Jg8.11] or Cilix on Europa [Plate Je 8.1]) remains to mark the rim of Manannán. A small rim section can be identified in the northeast quadrant as a dark arcuate scarp (3.5°N, 239.35°W). The crater floor consists of broken hills and a coarsely textured deposit that is inferred to be an impact deposit of fragmented and refrozen melted ice. This deposit filled topographic lows, but also was deposited on the rim and in some areas outside the rim. A small depression at 3.3°N, 239.6°W marks the center of the crater and coincides with a small radiating web of dark lines. A small mound lies just east of this feature and is a remnant of a central peak. Outside the rim lies a zone of mostly dark ejecta excavated from the crater during impact. Numerous small depressions to the east of 238.5°W and west of 240.7°W are secondary craters from Manannán.

The unusual morphology seen here and similarly at Pwyll (Plate Je14.1) are due to the very warm temperature deep inside the relatively thin icy shell. As a crater forms, the icy shell responds to the enormous stresses by fracturing, melting and moving inward, and warm ice deforms much more easily than cold hard ice. The larger the crater on Europa, the deeper it excavates into the warm ice below, causing the crater cavity to be more unstable as it forms and leading to the unusual landforms we see today. Mosaic shown at 50% scale to fit page.

 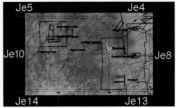

Plate Je9.1.1b Manannán: Color View
Encounter: *Galileo* E14   Resolution: 82 meters/pixel
Orthographic Projection   Map Scale: 1 cm = 9.7 km

The ridged plains on which Manannán formed are characterized by mottled white and brown patterns in this enhanced-color view. Just outside the crater rim lie dark brown deposits to the northeast, northwest and south, corresponding to the inner, or pancake, ejecta deposit. Within the rim, however, we see pale brown and bluish deposits intermingled in an irregular pattern. Crater floor deposits typically consist of mixed fragmented and melted crustal rock, in this case water ice (the deposit has since refrozen as the crater cooled). The color and brightness of water ice we see here are related to both grain size and the amount of non-ice contaminants.

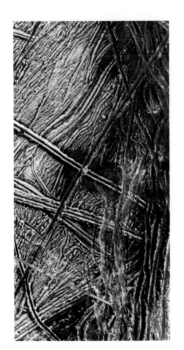

Plate Je9.2.1a-X1 Castalia Macula: Close-up I
Encounter: *Galileo* E14   Resolution: 20 meters/pixel
Orthographic Projection   Map Scale: 1 cm = 2.4 km

This view of southwestern Castalia Macula and the northern tip of South Castalia Mons shows how dark material clearly embays and fills inter-ridge valleys, especially at center and top, evidence of flow. These materials are high up on the flank of South Castalia Mons, indicating that the dark material (and perhaps Castalia Macula itself) formed before the uplift of the plateau. Several deep fissures run north–south through the center of the plateau, cutting several east–west trending double ridges.

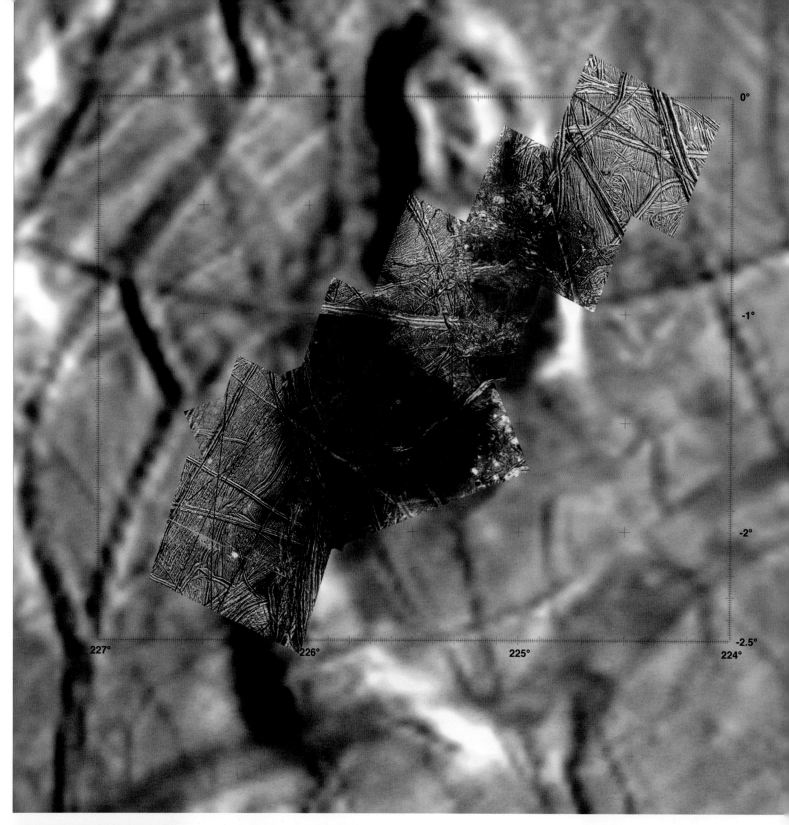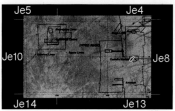

Plate Je9.2.1a Castalia Macula: Very High Resolution
Encounter: *Galileo* E14   Resolution: 20 meters/pixel
Encounter: *Galileo* E14   Resolution: 55 meters/pixel
Orthographic Projection   Map Scale: 1 cm = 4.7 km

Castalia Macula, at center, is one of the darkest and reddest features on Europa. This view combines a high-Sun high-resolution 6-frame mosaic with low-resolution low-Sun context images. The dark heart-shaped spot is a 350-meter-deep depression covered by dark red material. Ridged plains cross into the dark spot but are also covered, except for the tops of the largest ridges. Apparently, dark material once filled this depression but has since been removed, leaving this dark "stain", or high heat flow drove off water frosts. This material may be related to Europa's ocean beneath in some way, and is an object of interest as a potential exposure of deeper ice shell or perhaps ocean-rich material.

Castalia Macula is flanked by two unusual plateaus. To the north is a peanut-shaped plateau, 900 meters high. This is one of the highest known features on Europa (see also Plate Je2). The southern half of the plateau is disrupted and has a dark brownish color. The eastern side of this feature, nicknamed here North Castalia Mons, is a very steep sheer wall and likely an exposed fault scarp. To the south, boomerang-shaped South Castalia Mons rises ~500 meters. A deep fracture runs along the center of this plateau, indicating it has been pushed up, cracking the top. Mountains such as these are probably more common than realized and make it very unlikely that the ice shell is only a few kilometers thick. The ice shell requires a significant thickness to support such steep topography here, as well as the deep depressions found in other locations (e.g., Plate Je4.1).

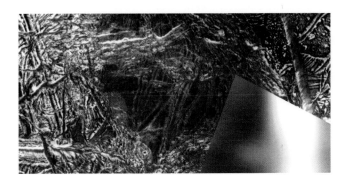

Plate Je9.2.1a-X2 Castalia Macula: Close-up II
Encounter: *Galileo* E14   Resolution: 20 meters/pixel
Orthographic Projection   Map Scale: 1 cm = 2.4 km

This view across the center section of North Castalia Mons shows the steep eastern fault scarp at upper right. This scarp has an elevation of ~900 meters! The dark interior of North Castalia Mons is disrupted by several small flow-like lobes extending onto the dark smooth material at frame center. The double ridge that enters the scene at bottom left becomes progressively more destroyed as it is traced eastward.

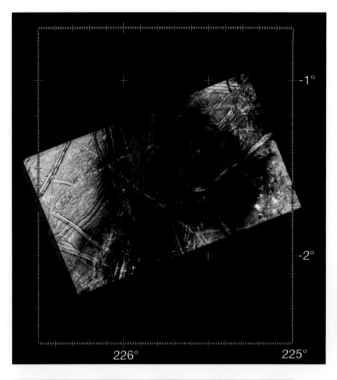

Plate Je9.2.1b **Castalia Macula: High-resolution Color**
Encounter: *Galileo* E14   Resolution: 55 meters/pixel
Orthographic Projection   Map Scale: 1 cm = 6.5 km

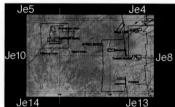

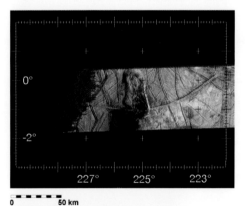

Plate Je9.2.3 **Castalia Macula: Regional Color**
Encounter: *Galileo* E11   Resolution: 300 meters/pixel
Orthographic Projection   Map Scale: 1 cm = 35.4 km

This moderate-resolution color mosaic shows that the disrupted northern half of North Castalia Mons is not dark or red, unlike the southern half.

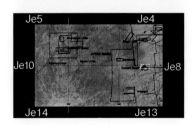

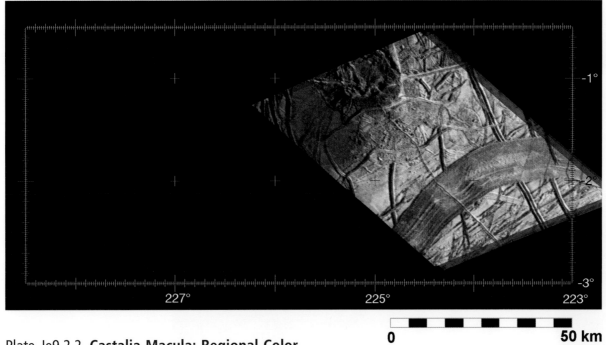

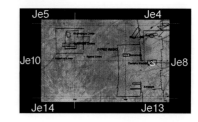

Plate Je9.2.2 **Castalia Macula: Regional Color**
Encounter: *Galileo* E19   Resolution: 85 meters/pixel
Orthographic Projection   Map Scale: 1 cm = 10 km

These higher-resolution color images confirm that the southern half of Castalia Mons has a color and albedo similar to the dark spot, Castalia Macula. Whether flows from the plateau were a source for the dark material that covers Castalia Macula is not known. The dilational band to the southeast is part of the Argadnel Regio rift zone and has an albedo and color intermediate between the dark spots and surrounding ridged plains. It may be intermediate in age between those two features.

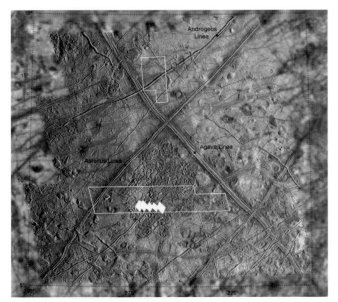

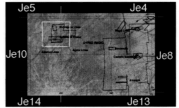

Plate Je9.3.1

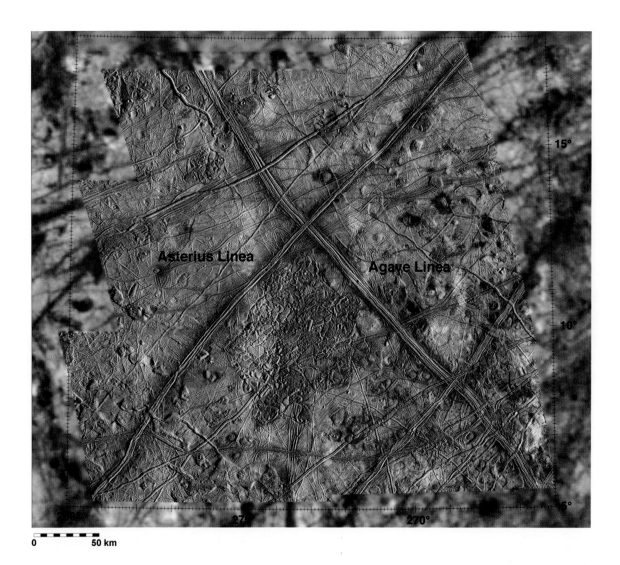

Plate Je9.3.1 **Conamara Chaos: Regional Mosaic**
Encounter: *Galileo* E6   Resolution: 175 meters/pixel
Orthographic Projection   Map Scale: 1 cm = 27.5 km

Galileo acquired a series of progressively higher resolution mosaics of Conamara Chaos, the disrupted zone nestled under the crossing of the two large ridged bands. Originally targeted for the intersection of the two large dark-flanked bands, my friend Jeffrey Moore suggested that it be retargeted for the dark blob to the south, now known as Conamara Chaos. This regional mosaic, combined with the global color mosaic, shows this entire large chaos unit. Both the ridge flanks and Conamara Chaos itself are dark and reddish in color.

The exception is the western third of Conamara, which is covered by a bright whitish ray from crater Pwyll several thousand kilometers to the south (Plate Je14). This ray is seen in both high-resolution mosaics. Isolated lenticulae of matrix material 5 to 20 kilometers across are scattered throughout the region. Conamara Chaos itself consists of numerous broken blocks of ridged plains intermingled with disrupted matrix material. The white rectangles in the reduced version show the locations of Plates Je9.4, Je9.3.3 and Je9.3.4.

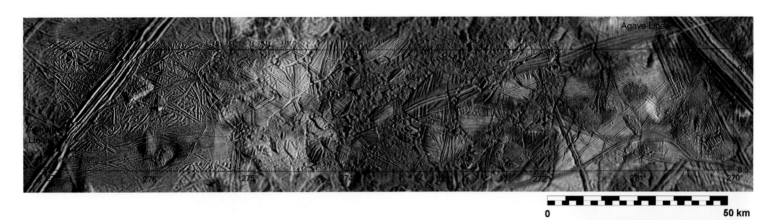

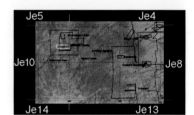

Plate Je9.3.3 **Conamara Chaos: High Resolution**
Encounter: *Galileo* E6   Resolution: 55 meters/pixel
Orthographic Projection   Map Scale: 1 cm = 10 km

Clearly evident in the western third of this mosaic are the bright splotches along longitude 275°W, part of a bright ray from Pwyll (Plate Je14.1). Some large ridges or bands can be traced across the field of broken blocks of ridged plains, indicating these blocks have not moved very far. Other blocks have shifted location. Ridge plains along the eastern margin are (or were) beginning to fracture and pull apart, especially at 9°N, 271.8°W. The very high resolution mosaic (Plate Je9.3.4) is centered at 9.8°N, 274°W. Color information in this mosaic is from Plate Je9.3.2, which unfortunately does not cover the western third.

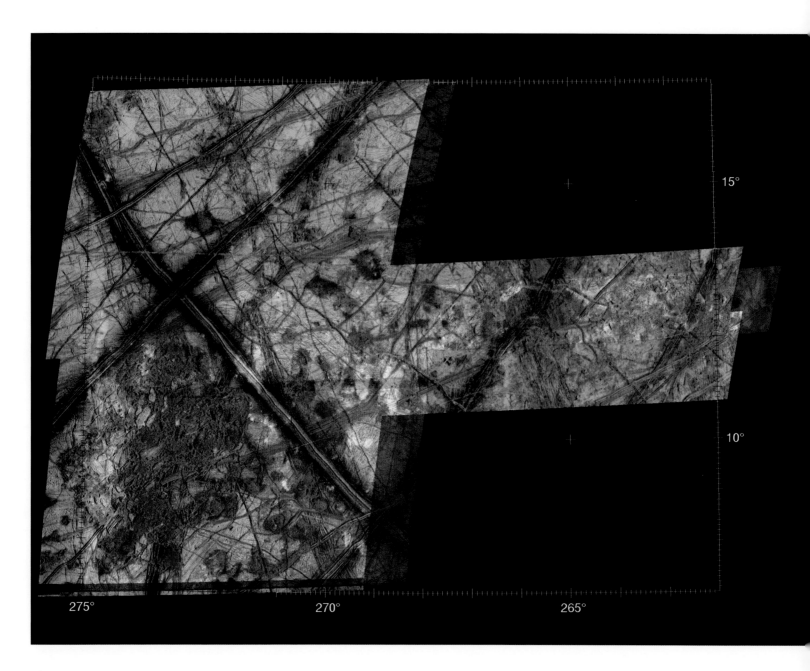

Plate Je9.3.2a

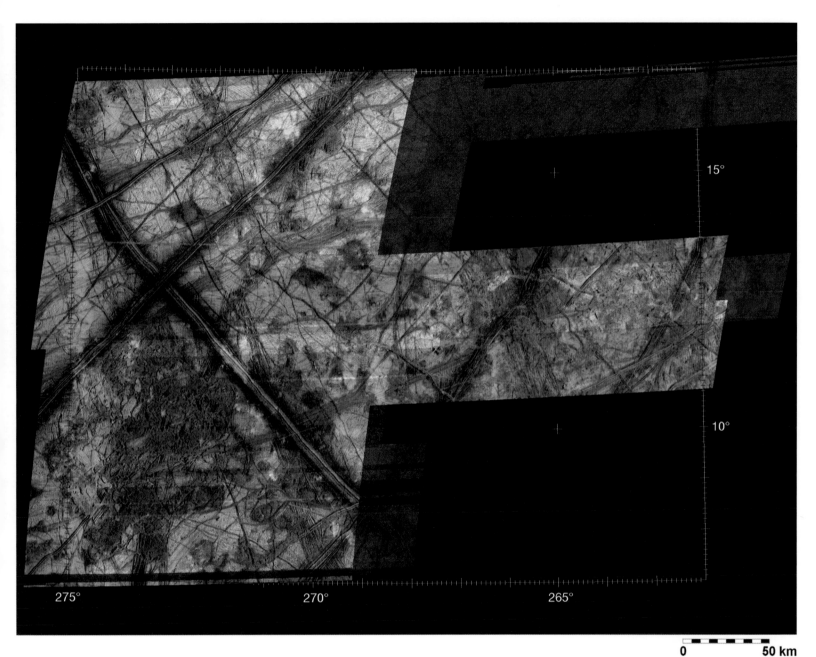

Plate Je9.3.2b Conamara Chaos: Regional Color
Encounter: *Galileo* E12   Resolution: 170 meters/pixel
Orthographic Projection   Map Scale: 1 cm = 20.1 km

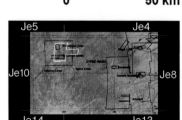

These fragmentary higher-resolution color mosaics (two different versions using two different infrared filters) show color variations within the chaos in greater detail. Some of the remnant blocks of ridged plains retain their original color and albedo. Others appear to have been covered by the dark brownish material from surrounding matrix. Dark material is abundant across the scene, especially flanking the two large ridged bands. (a) IR(0.968 nm), green and violet filters. (b) IR(0.756 nm), green and violet filters.

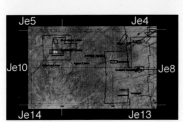

Plate Je9.3.4 **Conamara Chaos: Very High Resolution**
Encounter: *Galileo* E12   Resolution: 9 meters/pixel
Orthographic Projection   Map Scale: 1 cm = 2.4 km

These are the highest-resolution non-oblique images of Europa for which we have context imaging. Using color data from Plate Je9.3.2, these images reveal details of chaos formation. Matrix material is very rugged and in some areas can be resolved as an aggregate of severely fragmented remnants of older plains material. The largest of these fragments poke above the rugged terrain. Talus, unconsolidated erosional debris, has accumulated at the base of the sheer cliffs on the edges of the larger intact blocks, which stand up to 100 meters high. Numerous small secondary craters can be resolved within the bright splotches in the western part of a Pwyll bright ray. The most recent event was formation of the narrow fissure trending diagonally across the scene. See also Figure 5.3.5 for an alternative view of these images.

Plate Je9.4
Androgeos Linea and Ridged Plains
Encounter: *Galileo* E6
Resolution: 20 meters/pixel
Orthographic Projection
Map Scale: 1 cm = 3.2 km

The prominent double ridge shown here, Androgeos Linea, rises 350 meters above the plains. It also sits in a shallow 100-meter-deep depression along its entire length, formed by the weight of the ridge itself. Small cracks flank the double ridges, evidence of fracture due to bending of the ice shell as this ridge formed the depression. Shear heating along faults is currently thought to be a major component, but the origin of the topographic ridge is unclear. Possibilities include crushing and crumpling of the ice shell, accumulation of debris being squeezed out along the fault line, intrusion of ice from below, or a combination of these or other forces.

Ridged plains generally consist of complex crossing sets of parallel ridges. Close examination shows that some ridges and ridge sets are laterally displaced. Examples include the fractures at 15.4°N, 273.6°W and 15.2°N, 274.1°W. Strike-slip faulting of this type is common on Europa, indicating that the ice shell is broken into numerous plates. Tidal deformation and internal convection move these plates around. This is among our highest-resolution images of ridged plains.

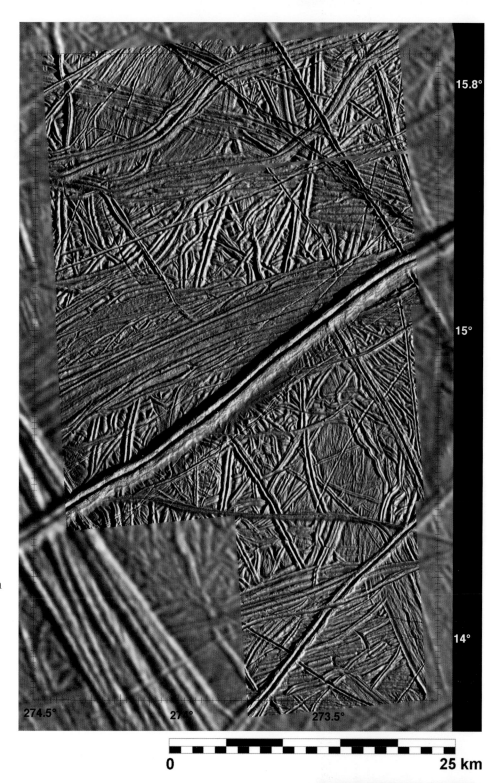

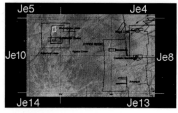

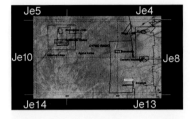

Plate Je9.5.1 **High-resolution Oblique**

Encounter: *Galileo* E12   Resolution: 12 meters/pixel
Orthographic Projection   Map Scale: 1 cm = 4.7 km

Except for the first image, this very oblique set of images was acquired at 12-meter resolution. The mosaic tracks across ridged plains that feature a variety of landforms, including small chaos units at 12.95°S, 235.25°W, at 12.9°S, 231.6°W, and elsewhere. Aside from the high-resolution views of dark and bright material distribution, the most interesting feature is the side view provided of a fault scarp on the eastern side of a polygonal plateau at 12.9°S, 233.45°W. The top part of the scarp is bright but the bottom half is dark and appears to be cluttered with talus debris. The fault scarp also cuts across a double ridge. This area is shown in its original perspective in Figure 5.3.3. and Plate Je9.5.2.

Plate Je9.5.2 **High-resolution Oblique: Highest Europa Resolution**

Encounter: *Galileo* E12   Resolution: 6 meters/pixel
Orthographic Projection   Map Scale: 1 cm = 0.7 km

This image is the first and westernmost in the sequence comprising Plate Je9.5.1 and was acquired at full resolution. It is the highest-resolution view we have of Europa. Ridged plains dominate in the foreground and background. A patch of darker chaos cuts across the center third of the image. Also clearly visible is the tendency for dark material to aggregate in topographic lows between the frost-covered ridges, which would imply that dark material migrates down-slope over time. This view is shown in its original highly oblique perspective, and suggests the view from a low-flying craft.

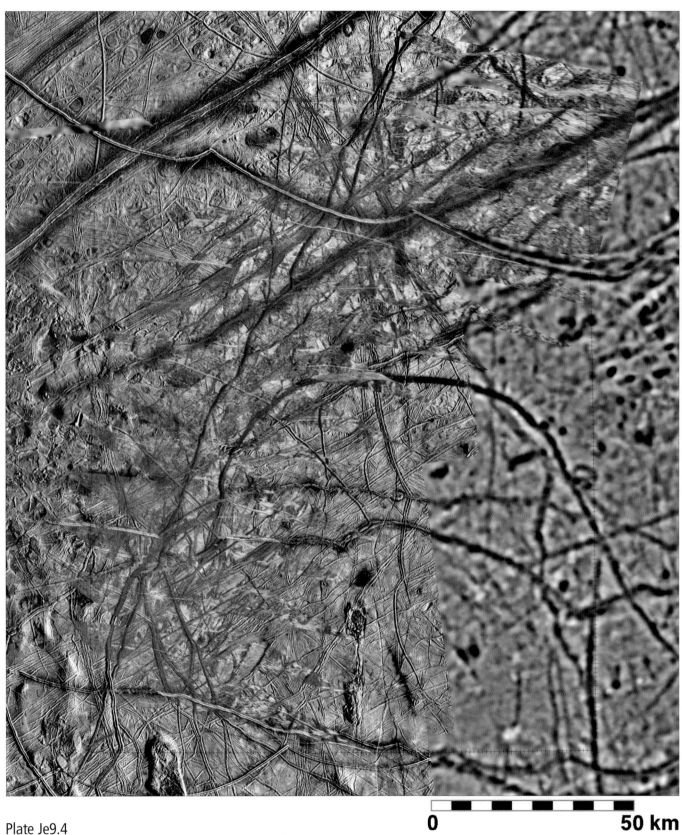

Plate Je9.4

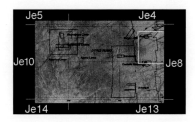

Plate Je9.6 **Plumes Search**
Encounter: *Galileo* E19   Resolution: 75 meters/pixel
Orthographic Projection   Map Scale: 1 cm = 8.9 km

These images are part of a planned search for plumes on Europa. *Galileo* found Europa's surface to be very young and potentially geologically active. Plumes of water ice and particulates have since been discovered on Saturn's geologically active icy moon, Enceladus, and logically similar (if smaller) plumes could have been or are active on Europa. These frames are the first set of images planned to image the limb of Europa. They landed on the surface and missed the limb. These and associated frames to the north are extremely oblique and cover less than one percent of Europa's surface. The lack of evidence for plumes on Europa is due only to *Galileo*'s inability to conduct a realistic global search at the phase angles and viewing conditions necessary to detect such plumes. A new mapping mission to Europa is required to determine if Europa is currently active. This version is shown at 150-meter resolution to fit on the page. Also shown in inset above is a smaller version without background context images.

Europa

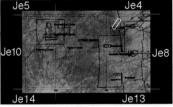

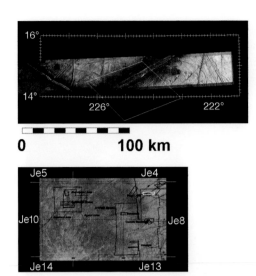

Plate Je9.7 Regional Mapping Mosaic: Fragment
Encounter: *Galileo* E19   Resolution: 200 meters/pixel
Orthographic Projection   Map Scale: 1 cm = 23.6 km

This fragmentary image illustrates the topographic complexity of Europa. The odd circular pits and depressions are related to convection within the icy shell and are part of a pattern of similar features seen near Rhadamanthys Linea (Plate Je4.1). The gentle undulations across the scene are part of this general pattern of instability.

Plate Je9.8.1 Belus Linea: Color
Encounter: *Galileo* E11   Resolution: 285 meters/pixel
Orthographic Projection   Map Scale: 1 cm = 33.7 km

Two *Galileo* color mosaics were targeted near the intersection of two dark bands. The larger dark-ridged band, Belus Linea, is crosscut by a younger cycloidal double ridge. An origin of the darker reddish material flanking the bands remains unclear. The mosaics illustrate the difference in morphology in Belus Linea and the adjacent cycloidal ridge. The white rectangle shows the location of the high-resolution view.

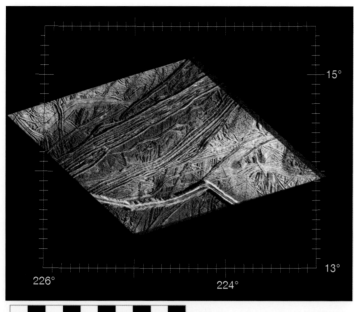

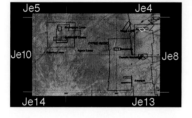

Plate Je9.8.2 **Belus Linea: Color**
Encounter: *Galileo* E19   Resolution: 90 meters/pixel
Orthographic Projection   Map Scale: 1 cm = 10.6 km

This view clearly shows that both sides of the cycloidal double ridge continue intact around the cusp of the angle at 13.8°N, 224°W. Belus Linea is broader and lower in relief, more typical of arcuate dark bands globally. These are now referred to as ridge complexes or ridged bands. The double ridge is 200 to 300 meters high, whereas Belus Linea is on the order of 50 meters high at most. The bright core of Belus Linea is more highly evolved and has experienced more deformation, perhaps in a series of small incremental widenings over time. See Plate Je8.6 for more of Belus Linea.

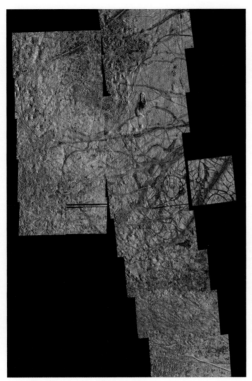
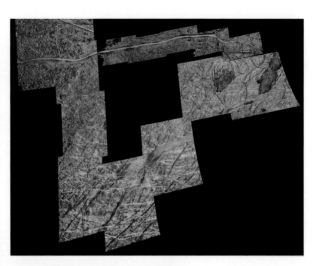
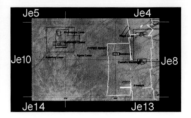

Plate Je9.9 **Regional Mapping Mosaic: 225°W**

Encounter: *Galileo* E11   Resolution: 220 meters/pixel
Encounter: *Galileo* E15   Resolution: 230 meters/pixel
Encounter: *Galileo* E17   Resolution: 225 meters/pixel
Encounter: *Galileo* E19   Resolution: 200 meters/pixel
Sinusoidal Projection   Map Scale: 1 cm = 148 km

This pole-to-pole mosaic (similar in concept to Plate Je7.2), was obtained during four orbits: E11 to the west, E15 to the north, E19 to the far north, and E17 to the south. Several high-resolution images are located within this site, including Rhadamanthys Linea (Plate Je4.1), Castalia Macula (Plate Je9.2), Dissected Terrain (Je13.1), and Manannán (Plate Je9.1), among others. Here we see a similar transition from ridged plains up north to disrupted terrains near the equator, as well as a number of dark dilational bands (the western part of the dilational band rift zone). Additional ridged plains lie to the south, including the large dark spots, Thrace and Thera Macula (Plates Je12.2, Je13.4). This Plate has been divided into three sections to fit the page.

Mercator Projection
Map Scale: 1 cm = 104 km

Plate Je10 Callanish Quadrangle

The large dark multi-ring impact crater Callanish stands out amid the dark chaotic material that covers most of this region. Similar to Tyre (Plate Je3), Callanish was viewed twice at high resolution. Multi-ring impacts of this type strongly imply that the icy shell is relatively warm and easily fractured during impact. The bright splotchy material at lower right is a ray from the young crater, Pwyll (Plate Je12) (Plate Je14). The bright terrain at lower left is likely the bright rays of a recent impact. The origin of the odd circular crater-like feature Midir is unknown. The two prominent north–south trending arcuate depressions (at longitudes 340° and 349°W) are part of a large set of concentric features related to polar wander (see also Plate Je5 to the north). A second set occurs on the leading hemisphere (see Plate Je12).

Europa | 237

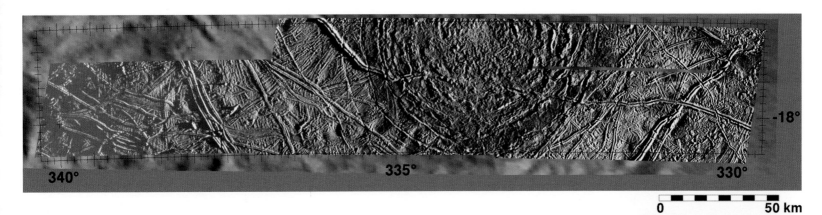

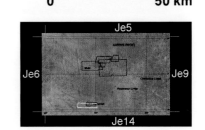

Plate Je10.1.1 **Callanish: High Resolution**
Encounter: *Galileo* E4   Resolution: 120 meters/pixel
Orthographic Projection   Map Scale: 1 cm = 14.2 km

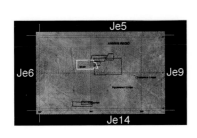

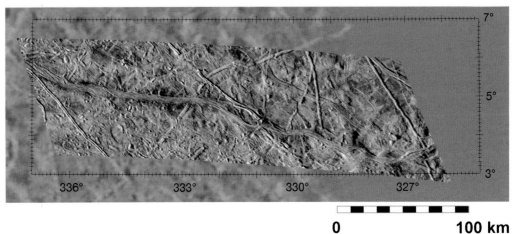

Plate Je10.2.1
Encounter: *Galileo* I25   Resolution: 235 meters/pixel
Orthographic Projection   Map Scale: 1 cm = 27.6 km

See Plate Je10.2.2 for description.

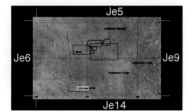

Plate Je10.1.2 Callanish: High Resolution II

Encounter: *Galileo* E26   Resolution: 47 meters/pixel
Orthographic Projection   Map Scale: 1 cm = 7.8 km

Galileo acquired two high-resolution mosaics across the central and southern sections of the multi-ring impact basin Callanish, a close cousin of Tyre (Plate Je3). The inner rings take the form of disconnected steep-sided ridges, grading outward into graben-like ring depressions, abundant secondaries. Although the entire structure is nearly 100 kilometers wide, mapping of secondary craters indicates that the original rim diameter was only 33 kilometers, approximately the location of the innermost identifiable ring.

Multi-ring basins are the result of concentric fracturing of the icy shell of Europa surrounding the crater. Multi-ring impacts form when the cold surface layers of an icy satellite are warm and hence thin enough to fracture extensively during impact. Below this brittle outer shell is warm soft, or ductile, ice (and perhaps liquid water deeper down), which moves inward to fill the crater, fracturing the surface. Both Pwyll and Callanish probably penetrated relatively deeply into the ice shell, but definitely not to the ocean itself.

The prominent double ridge crossing the scene becomes increasingly degraded toward the center of the ring system, most likely due to burial by Callanish ejecta, until it finally disappears just where it grazes the nominal crater-rim location. The center of the system is free of rings but has a coarse jumbled texture not unlike textures seen in the floors of large fresh craters on the Moon and Ganymede or in some chaos units on Europa. This is most likely a deposit of melted and highly fragmented debris from the impact. Dark reddish smooth deposits extend beyond the ring system in some locations. Note that Plate Je10.1.2 was acquired while the Sun was on the opposite side of the sky as Plate Je10.1.1, reversing local shadow contrasts. Plate Je10.1.2 is shown at 65% scale to fit page.

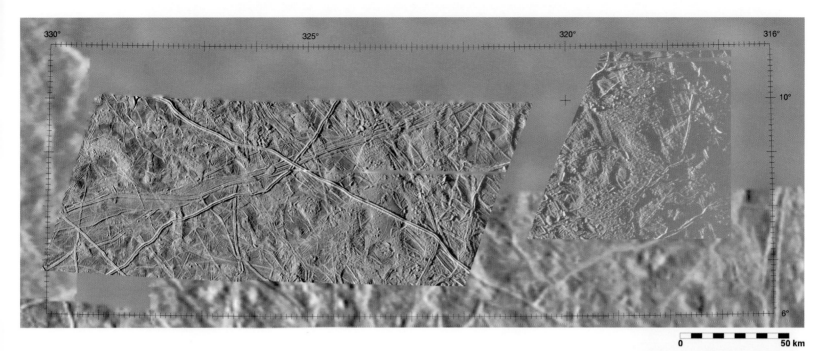

Plate Je10.2.2 **Ridged Plains**

Encounter: *Galileo* E26   Resolution: 100 meters/pixel
Orthographic Projection   Map Scale: 1 cm = 16.8 km

These mosaics, acquired in different orbits, are regional studies of ridged-plains formation. At both locations, ridged plains are partially or severely disrupted by irregular units of chaos. The area in Plate Je10.2.2 around 8°N, 319°W is almost completely replaced by high standing matrix material.

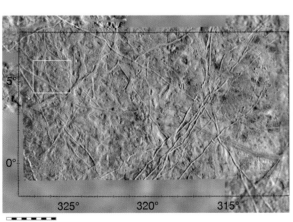

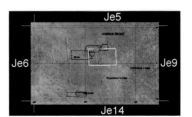

Plate Je10.3.1 **Ridged Plains: Context View**

Encounter: *Galileo* E4   Resolution: 630 meters/pixel
Orthographic Projection   Map Scale: 1 cm = 74.4 km

This partial frame was intended to provide regional context for the high-resolution mosaic located in the small white rectangle (Plate Je10.3.2). Ridged plains here are highly disrupted by what appears to be chaos and matrix material, although the low resolution makes mapping difficult.

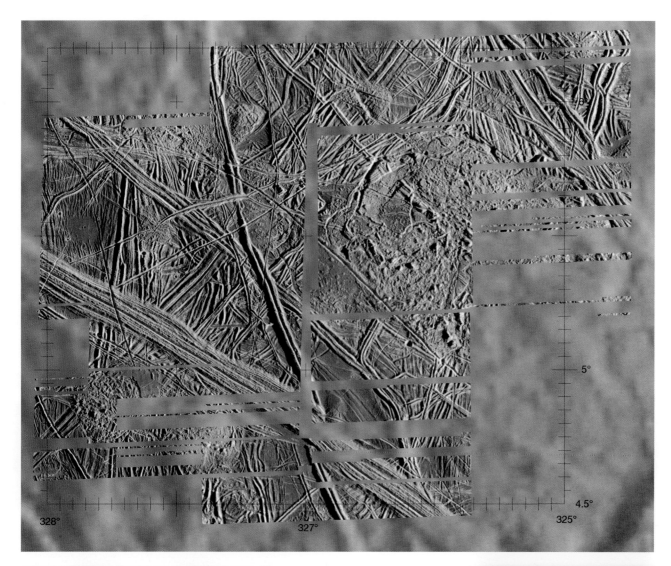

0   50 km

Plate Je10.3.2 **Ridged Plains: High Resolution**
Encounter: *Galileo* E4   Resolution: 25–35 meters/pixel
Orthographic Projection   Map Scale: 1 cm = 3 km

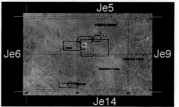

Roughly half of the two planned 4-frame *Galileo* mosaics that cover this site were returned, accounting for the numerous gaps. Nonetheless, the mosaics show the diversity and complexity of ridged plains in some areas. Typical double ridges as well as broader ridge complexes of varying complexity (5.6°N, 327.5°W and 5°N, 327.5°W) are present. Ridges on either side of the large complex ridged band to the southwest are realigned precisely if the sides of this band are closed together indicating that most of the band is new material intruded from below as the ice shell pulled apart.

A large zone of chaos is visible at right center (5.5°N, 326°N); several smaller occurrences are scattered across the terrain, identified by their jumbled morphology. Several patches of ridged plains are relatively smooth and may simply have escaped deformation. The small smooth feature at left center (5.5°N, 327.8°W) is unusual, however. The smooth topography and filling of local valleys between ridges along its edge indicates that liquid water once ponded here and has now frozen over. The small crater at its center is probably just a random impact crater and not a volcanic vent. How liquid water got to this location is uncertain, and very few similar features have been seen anywhere else on Europa.

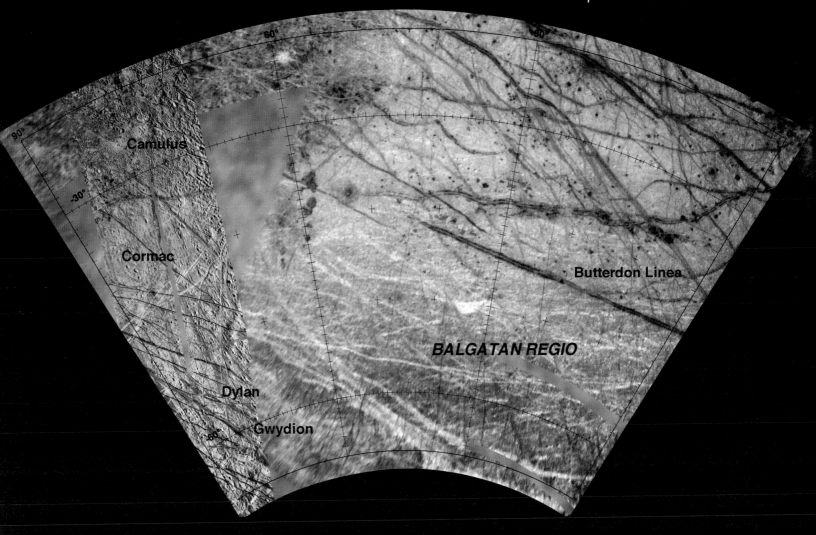

Lambert Conformal Conic Projection
Map Scale: 1 cm = 109 km

Plate Je11 Butterdon Linea Quadrangle

This quadrangle, seen mostly in high-Sun 1-kilometer *Galileo* mapping coverage, consists of ridged plains to the south and east and disrupted terrains to the northwest. Several long dark bands cross the plains. The widely scattered dark spots are similar to dark spots in quadrangles to the north (see Plates Je3 and Je4) and are probably the expression of ice diapirism. The bright spot at 23°S, 59°W is a young bright rayed 5- to 6-kilometer-wide impact crater. The dark ring at 38°S, 53°W may be an older impact crater similar to Lug (Plate Je2). The albedo patterns are similar to those of Euphemus Linea quadrangle (Plate Je6), but appear to die out south of latitude 45°.

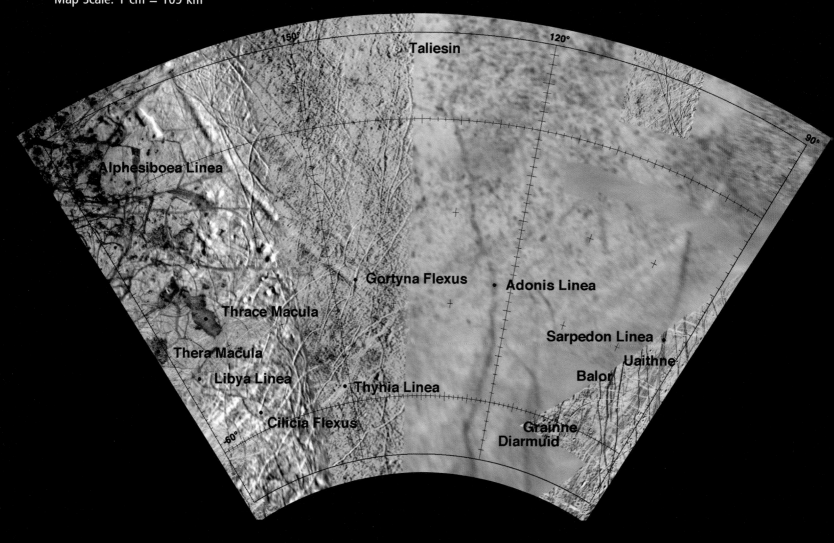

Lambert Conformal Conic Projection
Map Scale: 1 cm = 109 km

Plate Je12 Taliesin Quadrangle

Both *Voyager* 2 and *Galileo* viewed the western half of Je12 at 1.5 to 2 kilometers resolution, the eastern half at only 4-kilometer resolution. Several dilational bands are seen to the west (see Plate Je8), as well as Thynia Linea to the south. Numerous cycloidal ridges (see Plate Je15) are also apparent. Thrace and Thera Maculae are unusually large areas of chaos disruption or resurfacing.

The long curved shaded feature that arcs across the western portion is a few hundred meters deep. The smaller depression centered at 26°S, 168°W is at least 1 kilometer deep, and may be the deepest-known feature on Europa. These features are part of a global system of concentric arcuate troughs (see Plates Je2, Je5, and Je10 for additional examples). These are informally referred to as "crop circles," due to their startling global symmetry and circularity (Figure 5.3.4), which fit predicted stress patterns for polar wander on Europa, indicating that the entire ice shell has flipped over on its side.

Plate Je12.1 **Libya Linea**
Encounter: *Galileo* E17
Resolution: 42 meters/pixel
Orthographic Projection   Map Scale: 1 cm = 5 km

This (née gray) band is one of several first observed by *Voyager*. These dilational bands resemble their darker cousins like Yelland Linea (Plate Je8.4) except for the smoother surfaces. These bands might be older. Why they lack the parallel ridges (see Plate Je8.4) is unknown. Perhaps they formed more quickly or continuously and not incrementally. This 2-frame mosaic shows only a small part of Libya Linea, but highlights the subtle striations that mark parts of its surface. The real curiosity here is the relatively recent enormous jagged fissure crossing Libya Linea. The band is also folded (or buckled) along longitude 178°W, which can be seen by the subtle shading contrasts. The large crack and the smaller cracks and scarps parallel to it formed when the upper surface fractured in response to this buckling. Similar buckling was observed at Astypalaea Linea (Plate Je15.2). The small clusters of impact craters scattered across the mosaic are secondary craters ejected from a distant impact crater, possibly Tyre. Part of Libya Linea to the east can be seen in Plate Je12.2.

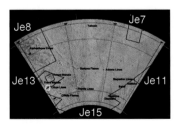

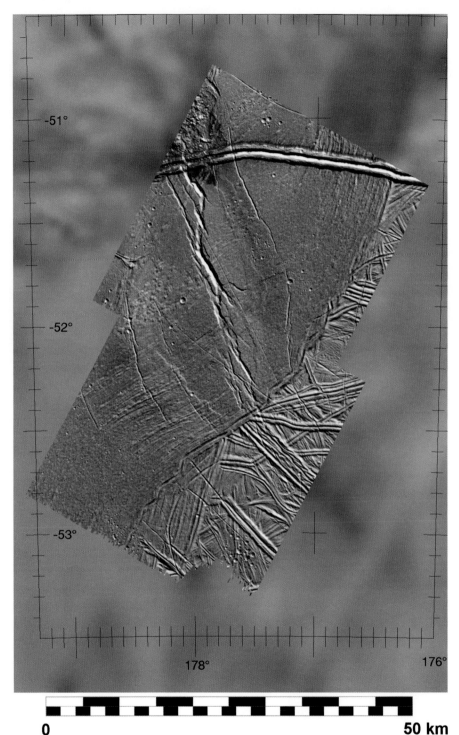

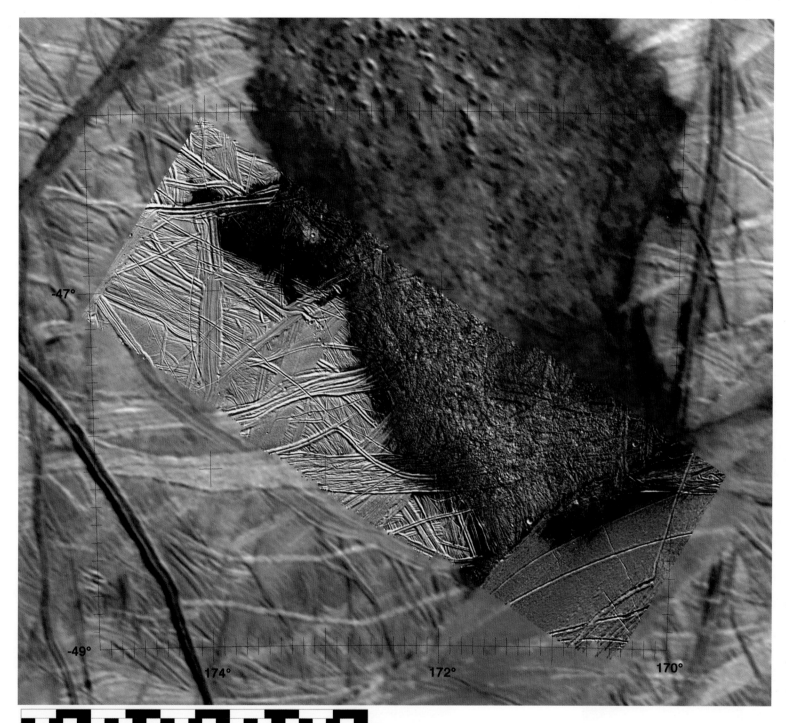

0 50 km

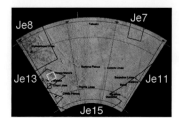

Plate Je12.2 **Thrace Macula and Libya Linea**
Encounter: *Galileo* E17   Resolution: 40 meters/pixel
Orthographic Projection   Map Scale: 1 cm = 5.25 km

This 3-frame mosaic of the southern end of Thrace Macula was all that could be spared to cover this large dark spot first seen by *Voyager* 2. The goal was to determine the nature of the dark material and how it was emplaced. Dark material can be clearly seen in some locations embaying older ridges, indicating that dark material flowed into low spots. The broad smooth band (Libya Linea) at the southern margin of Thrace also blocked some dark material. Surprisingly then, disrupted remnants of some bands and ridges can be easily traced deep within Thrace Macula, indicating that not very much new material was involved in resurfacing Thrace Macula, at least in this area.

Lambert Conformal Conic Projection
Map Scale: 1 cm = 109 km

Plate Je13 Agenor Linea Quadrangle

Agenor Linea dominates the scene and is one of only two known large bright bands on opposite sides of Europa (see Plate Je6). In this hemisphere, Agenor Linea (and a small companion, Katreus Linea) may be associated with the zones of dilational bands and ice shell rifting, which lie to the north (see also Plates JeO1, Je8, and Je9) and to the south (e.g., Libya Linea and Plate Je15). Dark dilational bands have strongly disrupted the areas north and northeast of Agenor. Global mapping suggests a common origin for these features, perhaps related to polar wander of the icy shell. Bright icy ejecta from the crater Pwyll (see Plate Je14) are visible to the west (the strong red color here is an artifact of the way in which the color mosaic was assembled). The "crater" Luchtar looks not like a crater but rather an area of deformed terrain. Eochaid is a distinctive 16-kilometer-wide crater with an outer dark ring, small central "peak" and bluish interior similar to Manannán (Plate Je9.1).

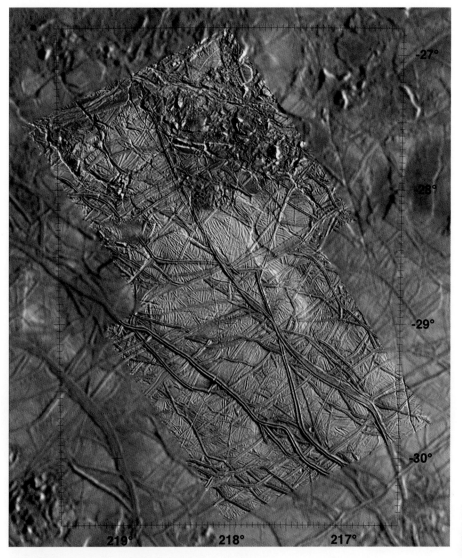

0   50 km

Plate Je13.1 **Dissected Terrain**
Encounter: *Galileo* E17
Resolution: 55 meters/pixel
Orthographic Projection
Map Scale: 1 cm = 7.6 km

The southern half of this view features relatively undisturbed ridged plains. The northern half has been disrupted and dissected by a variety of processes. To the west, ridged plains have been broken into several blocks or plates, some of which have subsided downward and are partly obscured by smooth material, possibly refrozen water flows. To the east, ridged plains have been "ground up" into blocks and fragments of various sizes, forming a large zone of chaos, most of which is elevated. This mosaic also demonstrates the importance of stereo imaging on a surface like Europa. When viewed in stereo with the lower-resolution context images, a prominent plateau becomes apparent. A fault scarp 400 meters high marks the western edge (the thin dark lineament at 19.1°N, 218.8°W).

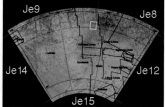

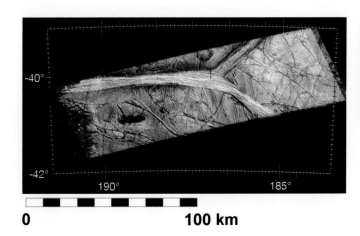

0   100 km

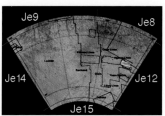

Plate Je13.3.2 **Eastern Agenor Linea: Color**
Encounter: *Galileo* E17   Resolution: 180 meters/pixel
Orthographic Projection   Map Scale: 1 cm = 21.3 km

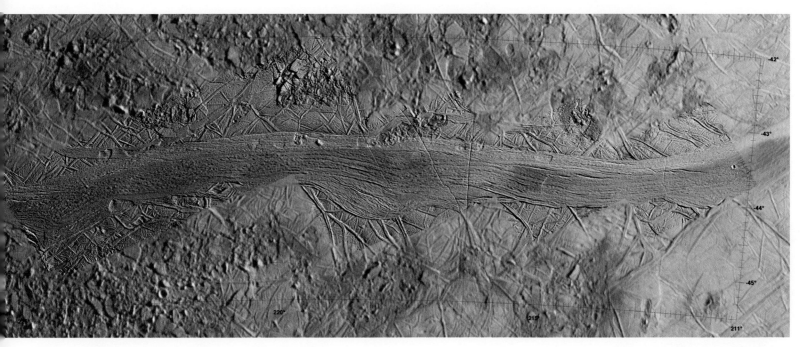

Plate Je13.2 Agenor Linea: High Resolution
Encounter: *Galileo* E17 Resolution: 55 meters/pixel
Orthographic Projection Map Scale: 1 cm = 17.1 km

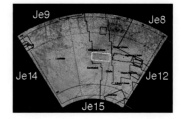

Galileo obtained two high-resolution mosaics in an effort to determine the origin of this unique feature. Agenor Linea is distinguished from dark bands (see Plate Je4) by its prominent white band, and the lack of symmetry between opposite sides of the band. At high resolution the surface of the band is intensely deformed, giving the surface a ropey texture. The band has two major components: a darker southern lane of variable width and a brighter northern component of more uniform width. Extensional cracks appear to be associated with this white northern portion. Detailed mapping by Louise Prockter and others suggests that the deformation is complex here, involving both compression and lateral shear, not unlike some strike-slip fault zones on Earth. All attempts to correlate features on the north and south side of the band (and determine the degree of lateral motion) have thus far failed.

Agenor Linea appears relatively young but the area has experienced some post-formation deformation. Several small walled depressions mark the surface. Examples are at 43.5°S, 215°W and 43.5°S, 223°W. These are down-dropped fault blocks, which could be incipient chaos. Several narrow north–south trending fissures cross the band, especially at longitude 217°W. Chaos has replaced the surface at 43.4°S, 268°W.

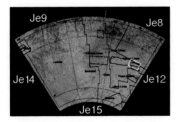

Plate Je13.3.1 **Eastern Agenor Linea: High Resolution**
Encounter: *Galileo* E17 Resolution: 55 meters/pixel
Orthographic Projection   Map Scale: 1 cm = 8.7 km

The brighter northern component in Plate Je13.2 dominates the eastern section of Agenor Linea. Here again we see deep extensional-style fissures parallel to the trend of Agenor, but also the same inability to match up features either side of Agenor, suggesting that any extensional deformation likely occurred after the band formed. The edges of the band are exceptionally sharp, however, indicating there was little or no "overflow" of bright material onto the ridged plains. In some areas, the darker flanking material appears to embay inter-ridge depressions, in others it appears to be a light dusting over all features, and in other areas appears to be completely absent. There is surprisingly little topographic expression across Agenor Linea, but the dark oval south of Agenor Linea is ~400 meters deep.

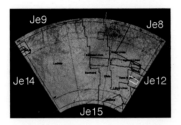

Plate Je13.4.1 **Thera Macula 1**
Encounter: *Galileo* C3   Resolution: 440 meters/pixel
Orthographic Projection   Map Scale: 1 cm = 41.6 km

Thera Macula (at upper right) is one of two large dark spots observed by *Voyager*, the other being Thrace Macula (Plate Je12.2). Although higher-resolution imaging was acquired (Plate Je13.4.2), this image highlights the relief along the margin of the dark spot. Shadows and highlights along the edge show that the dark spot is depressed in some areas and elevated in others. Ridged plains also have a rolling topographic nature.

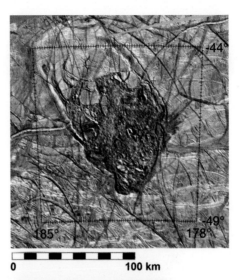

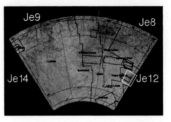

Plate Je13.4.2 **Thera Macula 2**
Encounter: *Galileo* E17   Resolution: 225 meters/pixel
Orthographic Projection   Map Scale: 1 cm = 29.5 km

This extract from the regional mapping mosaic (Plate Je9.9) shows that Thera Macula looks much like other chaos, including Conamara Chaos (Plate Je9.3). The most interesting feature is not the dark chaos but the disturbed areas to the north. Two subtle arcuate depressions extend northeast from the dark spot, and together these form a distinctive oval feature. This depression is also fractured and blocks of ridged plains have broken off from the northern edge, indicating that this oval extension is or was in the process of breaking apart. The oval shape and nature of the breakup is one of the strongest pieces of evidence in support of a diapiric origin for chaos.

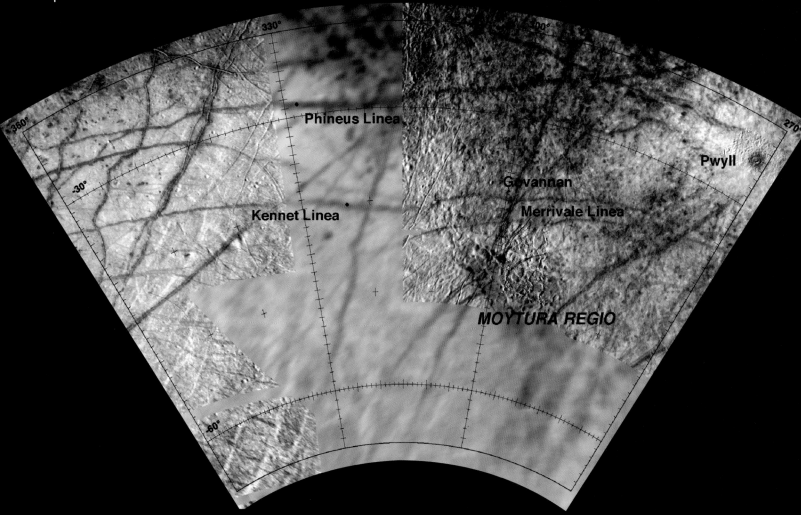

Plate Je14 Pwyll Quadrangle

At first glance, this quadrangle is divided between relatively mundane dark disrupted terrains to the northeast and bright ridged plains elsewhere, but as we have seen in other quadrangles, the limited resolution of available images probably masks many interesting features. Several prominent dark bands cross these plains. The central ridges of these features become visible as the lighting changes between 345° and 330°W longitude. Rugged relief characterizes Moytura Regio and dark disrupted terrain to the north generally. Bright ejecta from the 27-kilometer-wide impact crater Pwyll dominate the northeastern quadrant of the quadrangle. Pwyll is the youngest-known feature on Europa.

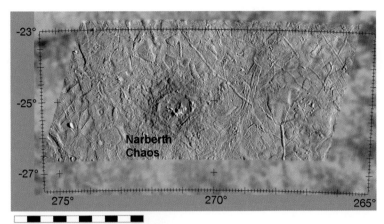
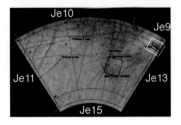

Plate Je14.1.1 Pwyll: Regional Mosaic

Encounter: *Galileo* E6   Resolution: 240 meters/pixel
Orthographic Projection   Map Scale: 1 cm = 28.3 km

At 27 kilometers, Pwyll is the largest-known, coherent, non-ringed impact crater on Europa. Yet its irregular and 500-meter-high central peak complex, irregular knobby rim, and very shallow depth are all distinctly different from similar-sized craters on Ganymede (Achelous: Plate Jg2.2) or on Europa (Cilix: Plate Je8.1). Pwyll is very similar in morphology to the crater Manannán (Plate Je9.1). Although neither Pwyll nor Manannán penetrated through the icy shell, as demonstrated by the high rugged central uplift, their formation and final morphology were strongly influenced by the warm soft lower ice shell beneath the impact sites, which allowed the floors of the craters to be uplifted more than usual. These and similar craters are direct evidence that the icy shell of Europa is much thinner than on Ganymede or Callisto. Parts of the crater floor appear flat, but whether this is impact melt is unclear. The dark annulus outside the rim also corresponds closely with a thick high-standing unit of impact ejecta, equivalent to pancake ejecta seen on Ganymede (Plate Jg2.2).

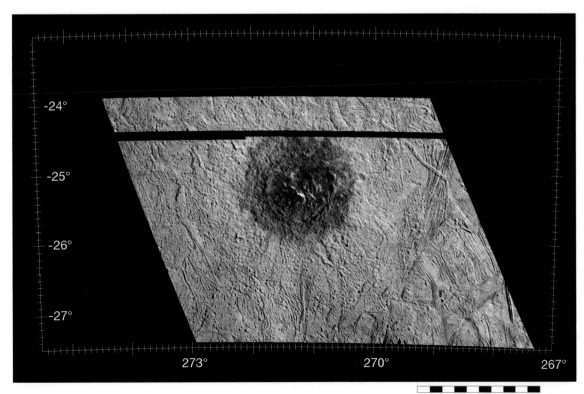

Plate Je14.1.2

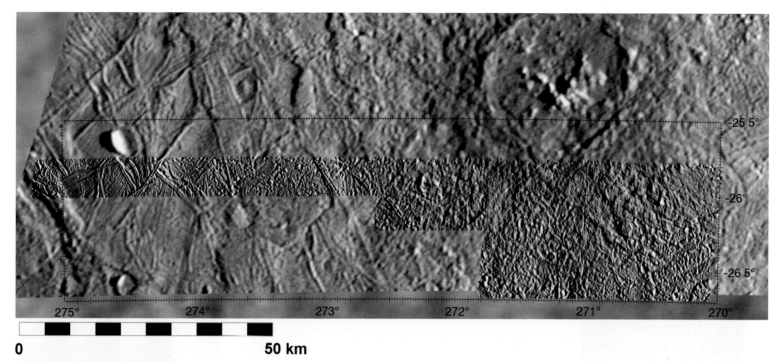

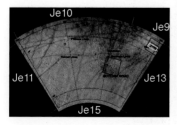

Plate Je14.2 Pwyll Ejecta: High Resolution
Encounter: *Galileo* E6   Resolution: 58 meters/pixel
Orthographic Projection   Map Scale: 1 cm = 6.9 km

This mosaic was originally intended to cross through the center of Pwyll but a minor pointing error displaced it to the south and it just missed the rim of the crater. Nonetheless, the eastern half of the mosaic provides a close view of the rugged ejecta deposit of a large Europan crater. Much of this terrain has a scoured appearance consistent with high-velocity deposition of ejecta. Hints of buried preexisting topography can be seen within the ejecta deposit. Small secondary craters are recognized west of longitude 272.6°W.

Plate Je14.1.2 Pwyll: Regional Color
Encounter: *Galileo* E12   Resolution: 125 meters/pixel
Orthographic Projection   Map Scale: 1 cm = 14.8 km

The fragmentary color mosaic of Pwyll highlights the ejecta deposits. About one-crater diameter from the rim, fields of small secondary craters are evident. The inner pancake ejecta deposit (see also Plates Je9.1.1b, Je15.4, Jg2.1 and Jg10.6.1) is evident as a gentle topographic plateau and corresponds very closely to the brownish deposit outside the crater rim. Only fragments of the 0.7-micrometer and violet images were return. These are shown merged with the green image, accounting for the horizontal color striping.

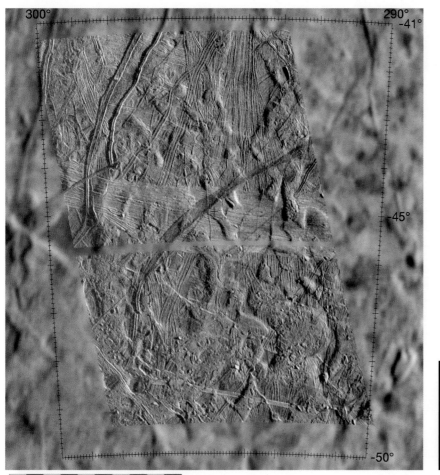

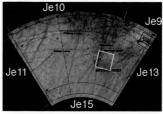

Plate Je14.3 **Moytura Regio**
Encounter: *Galileo* E12   Resolution: 180 meters/pixel
Orthographic Projection   Map Scale: 1 cm = 21.3 km

This single image highlights the topographic complexity of some areas on Europa. The undulating character of the surface is due to several events. Much of the topography is warped. Some undulations are due to fracturing and uplift of blocks of icy crust or to buckling of the surface due to compression. In some areas, material has also broken through to the surface in the form of knobby textured matrix and/or chaos, especially to the southeast. Ridged plains, including a relatively bright band that crosses the middle of the scene, dominate the original surface.

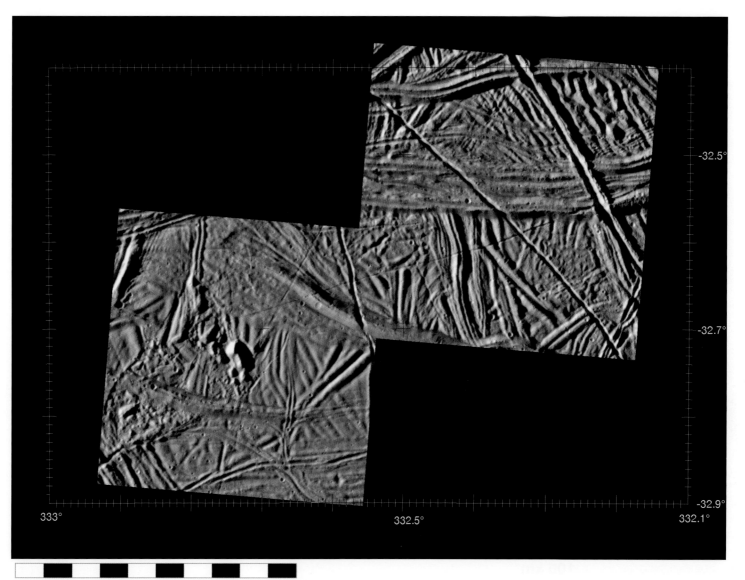

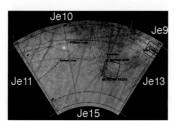

Plate Je14.4 **Ridged Plains: Very High Resolution**
Encounter: *Galileo* E26   Resolution: 11 meters/pixel
Orthographic Projection   Map Scale: 1 cm = 1.3 km

The precise location and regional context of these two images are unknown due to lack of context images. (The mosaic is less than one pixel wide in the context image mosaic, hence the lack of background information. The provisional coordinates used here are probably within half an image frame of the true location.) Nonetheless, they provide our highest-resolution undistorted views of ridged plains on Europa. Many of the ridges in the western image have a softened or rounded appearance compared to other ridges to the east or in other areas, suggesting that some ridges may be very old. The small mounds are part of an area of disrupted terrain to the west. The small craters are probably secondary craters from a large unseen impact crater, possibly Callanish to the north.

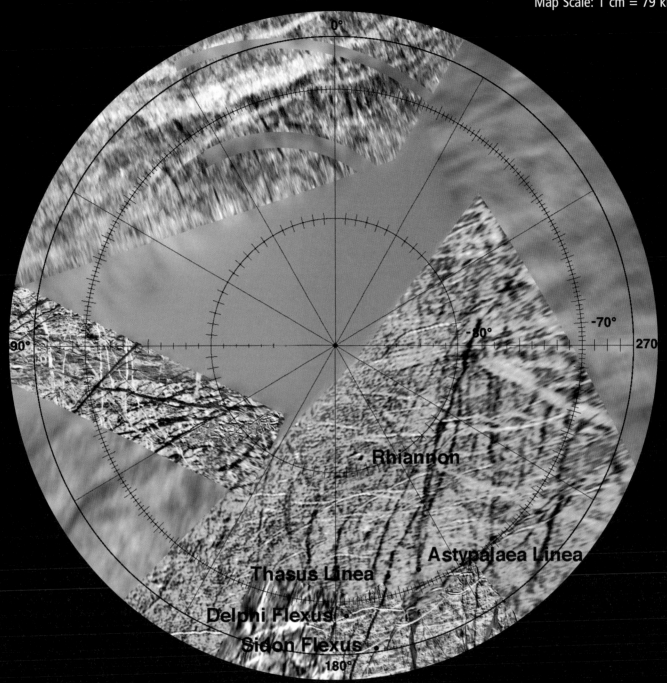

Polar Stereographic Projection
Map Scale: 1 cm = 79 km

Plate Je15 Sidon Flexus Quadrangle

The south polar region of Europa is famous for its curved cycloid ridges, first observed obliquely by *Voyager* 2 and supplemented by additional oblique *Galileo* views. These narrow concatenating arcuate structures form broad sweeping arcs across the surface. *Galileo* observations confirmed their presence elsewhere on Europa (see also Plates Je8.1.2b, Je9.8.2 and Je15.2) and led to models demonstrating that each arc segment forms during a Europan day as the diurnal tidal stress field sweeps across the surface. The topographic ridges are several hundred meters high and form some time after the initial crack, however. Whether cycloids formed in ancient times or are forming now is not known. Also seen here are additional dilational bands, including the very long prominent Astypalaea Linea. Unfortunately, a planned *Galileo* mosaic across most of this area during the very first orbit was cancelled due to a spacecraft anomaly.

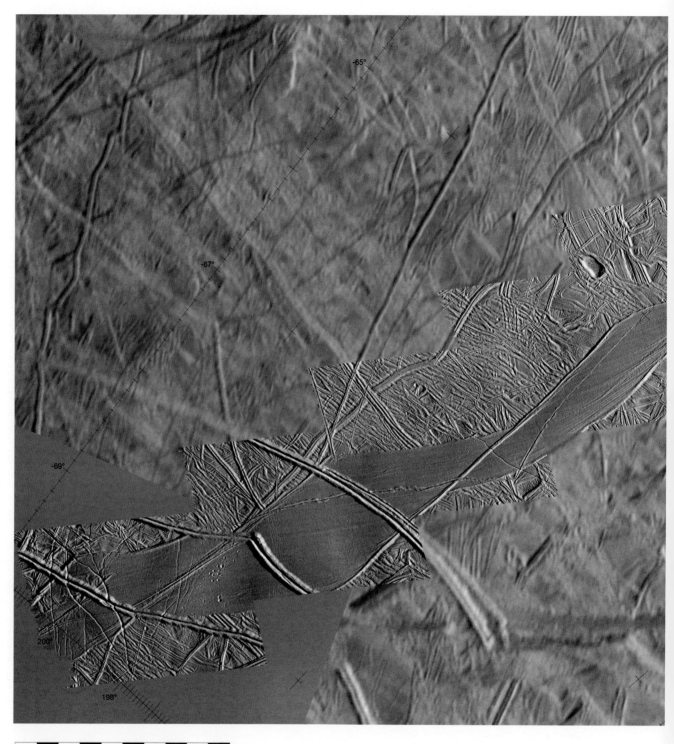
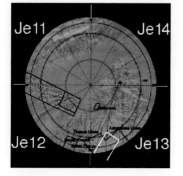

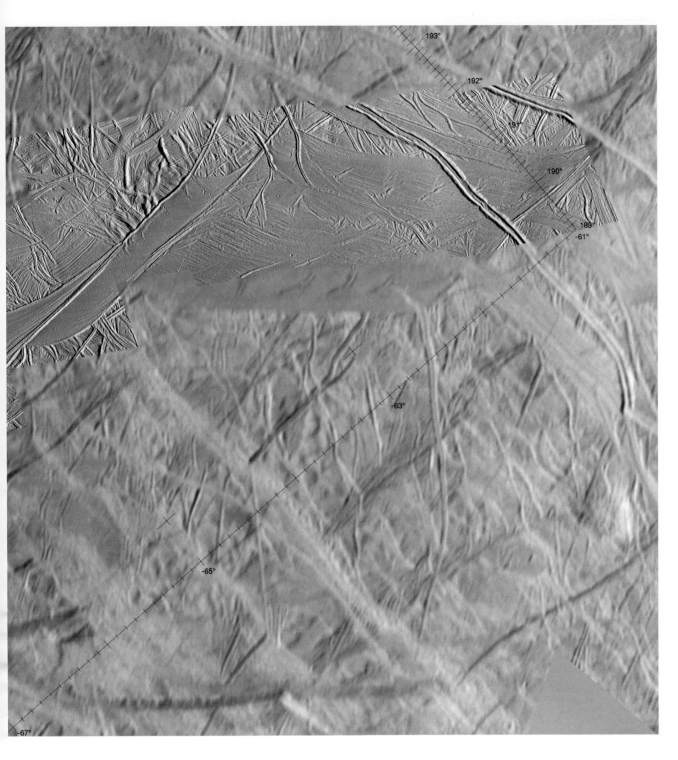

Plate Je15.2 **Astypalaea Linea**
Encounter: *Galileo* E17   Resolution: 43 meters/pixel
Orthographic Projection   Map Scale: 1 cm = 8 km

This extended 9-frame mosaic targets one of a number of dilational bands similar to the dilational bands in quadrangles Je8 and Je9. Astypalaea Linea is recognized here by its smoother surfaces (relative to surrounding ridged plains) and polygonal outlines. Like dilational bands Libya and Thynia Linea (Plate Je15.3), this band is subtly striated and the sides of this band can be fit together rather closely, indicating the smooth material is new crustal material formed from deep in the ice shell. The sense of dilation can be inferred by noting that the angles at 63°S, 194°W and 64.9°S, 194.1°W were originally together. (Try it yourself for this and other sites on Europa and Ganymede, but be sure to use a photocopy!)

Other features include clusters of secondary craters, especially at 69°S. At least three younger cycloidal double ridges cross this band. The double ridge at 61.4°S features an unusually wide central band and may be transitional between classical double edges and more evolved features such as Belus Linea (Plate Je9.8).

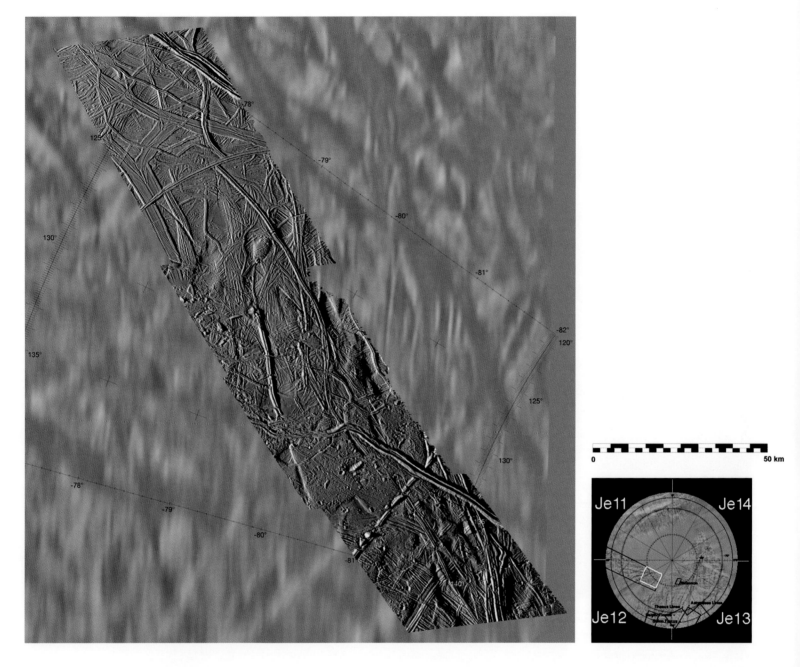

Plate Je15.1 **South Polar Terrain**
Encounter: *Galileo* E17   Resolution: 40 meters/pixel
Orthographic Projection   Map Scale: 1 cm = 10.5 km

Originally targeted to search for polar deposits, this mosaic reveals an interesting array of geologic processes. Towards the north (near 77.5°S, 122–125°W) is an area dominated by bands of relatively smooth material. Several of these bands form "triple-junctions," where three bands converge. These have been interpreted as zones where three icy plates have pulled apart simultaneously, allowing material to ooze up from below. In several cases, later strike-slip faults have cut across bands and displaced blocks of terrain laterally (Figure 4.2).

The arcuate ridges at 78°S, 124°W and 77.8°S, 122°W are classic examples that cut and displace numerous preexisting structures. To the south, the terrain is increasingly broken up and partly replaced by matrix material, forming a zone dominated by chaos. Small smooth patches at 79°S, 124°W and 78.5°S, 125.5°W may be related to the smooth patch seen in Plate Je10.3.2. If polar wander did indeed occur on Europa, then these terrains originally formed much closer to the equator than they are now. Shown at 45% scale to fit page.

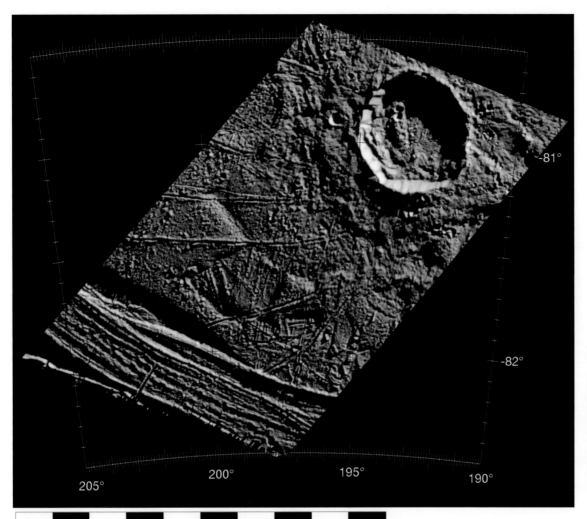
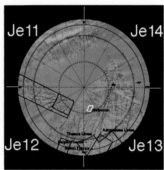

Plate Je15.4 Rhiannon
Encounter: *Galileo* E17   Resolution: 42 meters/pixel
Orthographic Projection   Map Scale: 1 cm = 5 km

This 15-kilometer-wide impact crater was one of only a few observed by *Voyager* in 1979. *Galileo*'s view reveals that it has a flat floor and no central peak of the type found in larger craters, such as Cilix (Plate Je8.1). Hummocky deposits are visible at the base of the rim scarp (81°S, 194.5°W). These formed when part of the rim wall collapsed. The topographic margin of the pancake ejecta deposit (at 81.4°S, 195°W, for example) is well expressed in this view. Roughly half as wide as the crater itself, this annular plateau outside the crater rim is similar to that seen in craters on Ganymede (see Achelous: Plate Jg2.2) and Europa (Pwyll: Plate Je14.1). Ridged bands and chaos are visible to the south and west.

260 Atlas of the Galilean Satellites

Plate Je15.3

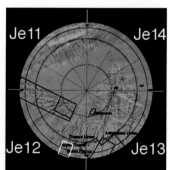

Plate Je15.2-X Astypalaea Linea: Close-up

Encounter: *Galileo* E17   Resolution: 43 meters/pixel
Orthographic Projection   Map Scale: 1 cm = 5.1 km

Another unusual feature of Astypalaea is the subtle shading variations between 68°S and 70°S. These are ripple-like undulations in the surface discovered by Louise Prockter and Bob Pappalardo and interpreted to be large-scale folds 200 to 300 meters in amplitude in the icy shell of Europa. Small fissures and wrinkle ridges can be seen on the crests and floors of several of these folds (oriented vertically in this view). They are the only clear expression of compressional deformation documented on the surface. A cluster of secondary craters is also visible to the left. The source crater is unknown but is likely one of the larger craters, such as Tyre, Callanish, Cilix, or Rhiannon.

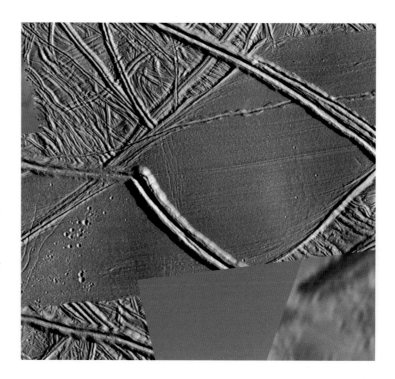

Plate Je15.3 Thynia Linea

Encounter: *Galileo* E17   Resolution: 40 meters/pixel
Orthographic Projection   Map Scale: 1 cm = 6.3 km

This smooth (née gray) band, first observed by *Voyager*, is similar to Libya Linea (Plates Je12.1 and Je12.2) and to dilational bands to the west. This partial 4-frame mosaic shows subtle parallel banding to the south and knobby hills to the north within the darker band material. Although different in detail, these subtle features appear to have a general axial symmetry across the band. This would be consistent with 28 kilometers of lateral opening of the icy shell and upwelling of material in the center of Thynia Linea and other bands. With the exception of the younger prominent cycloidal double ridge, all features on either side of Thynia line up very well if the sides of the bands are digitally closed back together, demonstrating that Thynia Linea is composed of entirely new crustal material.

Io

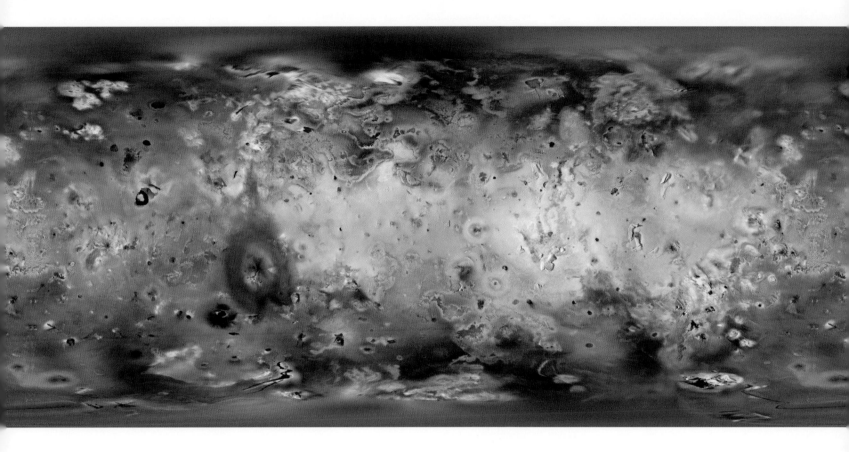

Plate Ji Global Map of Io

Cylindrical Projection   Center: 0°N, 180°W
Resolution at Equator: 1 kilometer/pixel
Map Scale: 1 cm = 480 km

Resolution at Equator: 2 kilometers/pixel

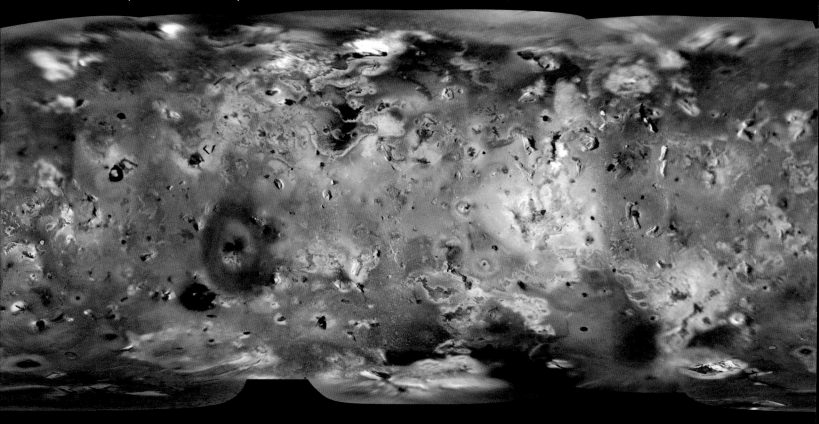

Plate JiM Global Map of Io – Mountains

This mosaic of Io was assembled from low-Sun terminator images in order to highlight Io's mountains. These images are typically lower in resolution than those used to construct the global mosaic. A few mountains are missing in this mosaic due to gaps in coverage. These missing mountains are visible in stereo or limb images.

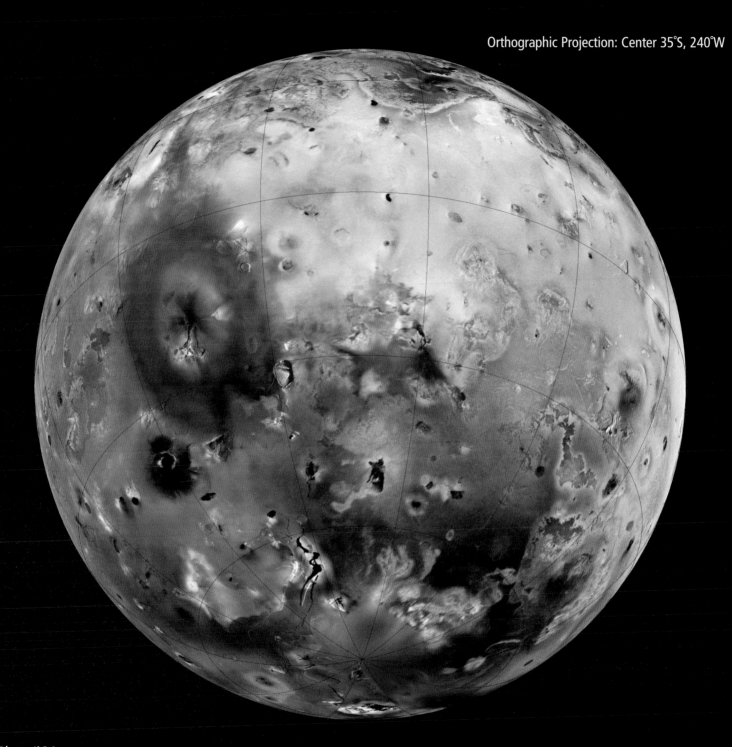

Orthographic Projection: Center 35°S, 240°W

Plate JiO1
Global View 1: Pele and South Polar Region

Of Io's 500+ volcanoes, the gigantic reddish Pele and dark Babbar Patera eruption sites dominate this hemisphere. These huge deposits formed as volcanic plumes of gas and dust fall back to the surface. Smaller reddish fan-shaped deposits betray the locations of recent or ongoing eruptions at Marduk and Culann Paterae. The different colors likely represent differences in sulfur composition in the eruption plumes. The dark deposits at Babbar probably include magnesium-rich orthopyroxenes. Whitish materials near the south pole are probably sulfur dioxide frosts outgassed from mountain slopes or at scarp failures. In the Pele plume the reddish colors may be due to radiation damage to short-chain sulfur molecules.

Orthographic Projection: Center 45°N, 170°W

Plate Ji02
Global View 2: North Polar Region and Prometheus

Looking down toward the north polar region of Io, we see the prominent differences between polar and equatorial terrains on the anti-Jovian hemisphere. The polar regions have more contrasting albedo patterns and are dominated by darker redder materials, possibly sulfur deposits darkened by radiation damage. The equatorial regions are dominated by bright sulfur and sulfur dioxide deposits. The ring feature along the equator is the Prometheus plume deposits, just west of the bright whitish Emakong region. The reddish Culann Patera eruption site lies just south of Prometheus.

Orthographic Projection: Center 30°S, 330°W

Plate Ji03
Global View 3: Sub-Jovian Hemisphere

Cruising outward from Jupiter, we might see this view, similar to what *Voyager* 1 saw as it passed the volcanic moon in March 1979. Shadows highlight relief in the southern regions. The view is dominated by four great volcanoes. The enormous reddish plume deposit ring from Pele to the east is visible from Earth-based telescopes. The whitish sulfur-dioxide-rich deposits surrounding the Masubi complex are evident to the far southwest. Loki Patera volcano, the dark horseshoe-shaped feature to the north, may be a lava lake. Finally, just above center is the Ra Patera eruption site, which last erupted in 1996–1997. The features shown here are as they were in 1979.

Orthographic Projection: Center 0°N, 90°W

Plate JiHL
Global View: Leading Hemisphere

This view dramatically shows Io's dark reddish sulfur-rich polar deposits, the color of which may be due to radiation-induced chemical changes on the surface. White and yellow deposits rich in sulfur dioxide dominate equatorial areas. Although irregular and globally variable, the boundary between the equatorial and polar zones falls roughly at 35° north and south latitudes. The two large bright patches, Bosphorus and Tarsus Regiones, are active volcanic centers and are brightened by diffuse deposits of sulfur dioxide frost. Mountains are common in this hemisphere.

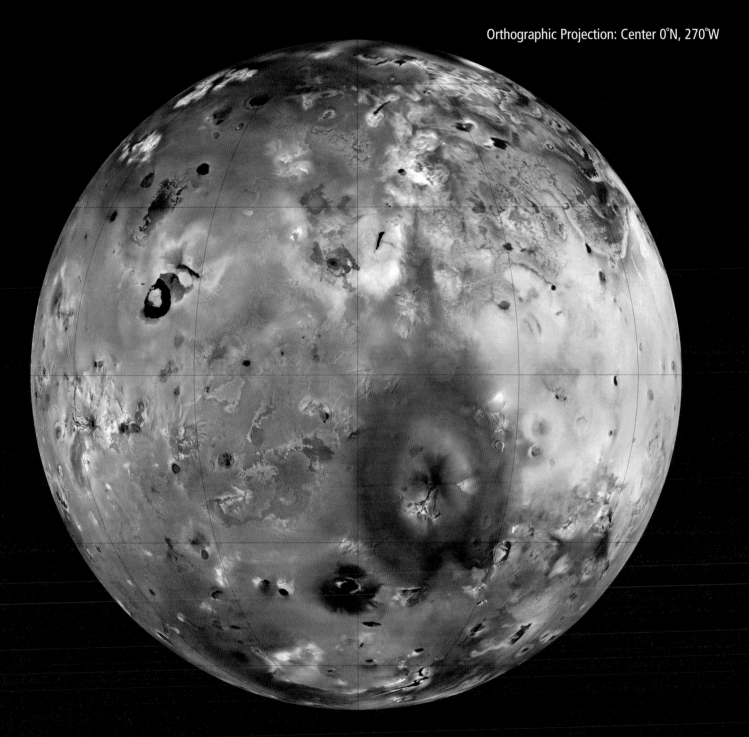

Orthographic Projection: Center 0°N, 270°W

Plate JiHT
Global View: Trailing Hemisphere

The trailing hemisphere includes two great volcanic eruptions, Pele and Loki. Both may be active lava lakes, as their appearance did not changed appreciably between 1979 and 2001. The dark red ring at Pele may be due to a much higher abundance of volatile sulfur in the lavas, compared to Loki. Although the dark reddish polar deposits continue here, the bright yellowish equatorial deposits that characterize the leading hemisphere (Plate JiHL) are missing in many areas.

The differences between polar and equatorial zones so prevalent in the leading hemisphere are less intense here. Whether these differences might be due to the Jovian radiation zones, which preferentially strike this hemisphere, is unknown. Large-scale lava flows (west of Pele, for example) are much more common in equatorial latitudes on this hemisphere, however, suggesting that geologic history may also be important.

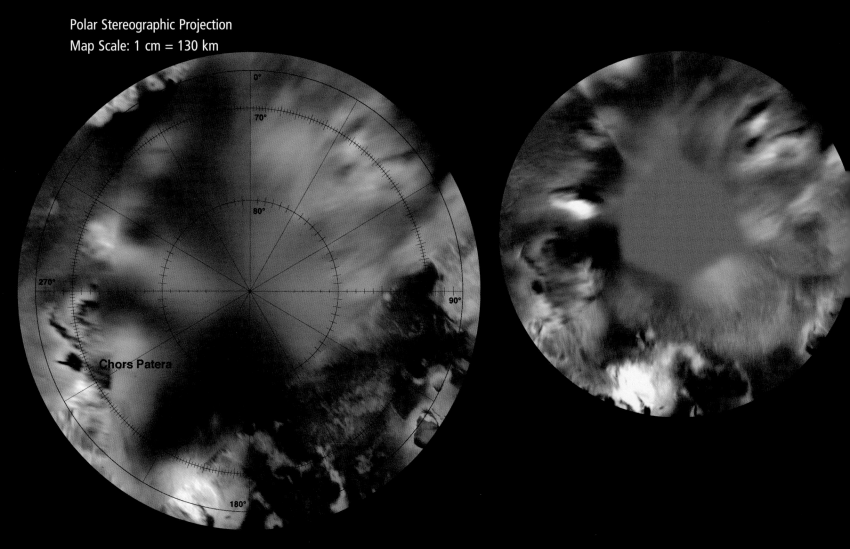

Plate Ji1/Ji1-m Chors Patera Quadrangle

Io's north polar region is the least well known territory on the satellite. The reddish color of both polar regions stands out in this view. Several mountains are apparent in the lower-resolution relief map, but were not observed in detail. They are associated with whitish deposits, probably sulfur dioxide. Part of the region near longitude 105°W was covered in 2001–2007 by the Tvashtar eruption plume deposit (see Plate Ji3.2). Each quadrangle includes a smaller version from Plate JiM.

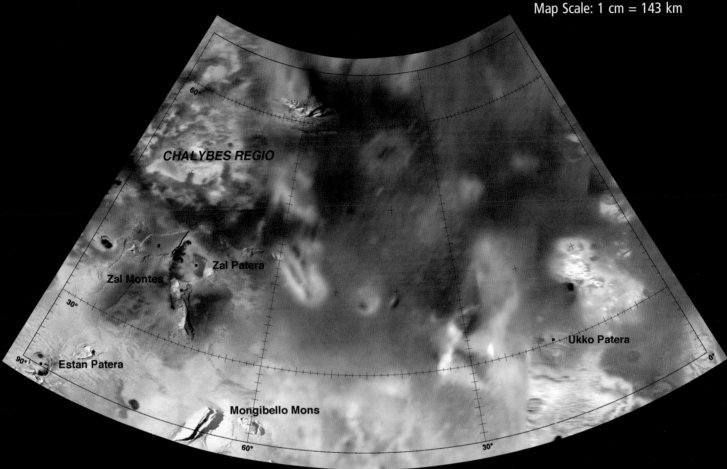

Lambert Conformal Conic Projection
Map Scale: 1 cm = 143 km

Plate Ji2/Ji2-m Zal Quadrangle

Despite the high-resolution images near Zal Montes, this area was not well covered by *Voyager* or *Galileo*. Numerous mountains break the smooth volcanic plains here, and this region has a relatively high concentration of mountains. Mongibello Mons is striking for its double-ridge morphology. Zal Patera, sandwiched between the two components of Zal Montes, is the site of a volcanic hotspot. The red and greenish colors and the dark lava flows suggest it is the site of recent or ongoing volcanism. The northward transition from yellowish to reddish coloration of the plains is again evident.

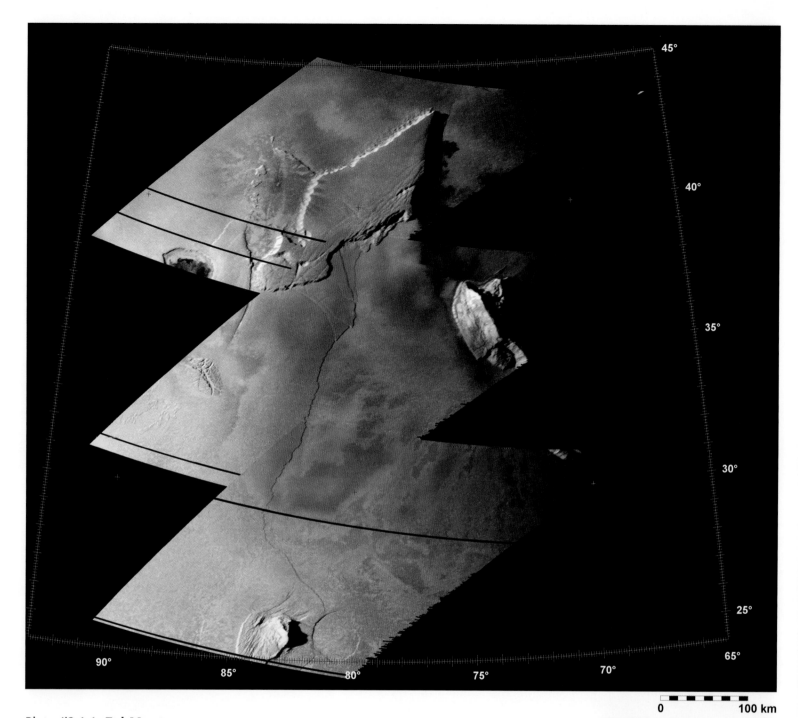

Plate Ji2.1.1 **Zal Montes**

Encounter: *Galileo* I25   Resolution: 250 meters/pixel
Orthographic Projection   Map Scale: 1 cm = 42.2 km

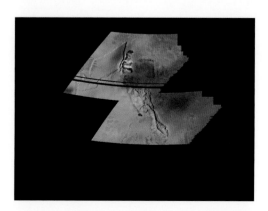

Plate Ji2.1.3-N **Zal Montes: NIMS Thermal Maps**

Encounter: *Galileo* I27   Resolution: 8.2 kilometers/pixel

Dark flows within Zal Patera correspond to a thermal hotspot seen by NIMS. Estimated temperatures are on the order of 500 K. No plume has been detected to date, however.

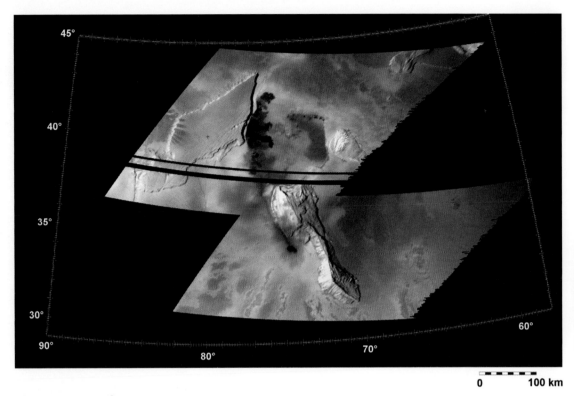
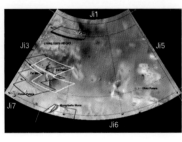

Plate Ji2.1.2 **Zal Montes**

Encounter: *Galileo* I27   Resolution: 350 meters/pixel
Orthographic Projection   Map Scale: 1 cm = 63 km

Two regional mapping mosaics were acquired of the Zal Montes region, permitting stereo mapping of these features. Zal Montes consists of two discrete parts, a 2-kilometer-high tabular mesa to the north and a scarp-bounded craggy 7-kilometer-high peak to the south. A low plateau at 34°N, 87°W in the I25 mosaic has been highly fractured, possibly during uplift. In the I25 view, a low scarp extends from the northern mesa to a prominent mountain peak at 25°N, 83°W. Several lobate or blocky deposits can also be seen at the base of these mountains.

Zal Patera, the large circular feature between Zal Montes, is the site of volcanic activity. Dark flows emanate from the western edge of the patera near where it abuts the edge of the tabular mesa and correspond to a hotspot. The small dark patera at 34°N, 74°W is the center of an old lava flow field extending to the southwest. Diffuse dark streaks radiate from the center of the small dark spot at 41°N, 84°W. The fact that these dark deposits fall on the tabular mesa to the east of the dark spot shows that they were not lava flows but rather were dark "airborne" ash deposits that fell on both high and low terrains during eruption.

(A fragmentary image to the northeast of Zal Montes, part of the I27 terminator mapping sequence, has been incorporated directly into the global mosaic at 60°N, 60°W. The view is oblique and detail is somewhat blurred. It shows the southern half of an unnamed mountain and featureless volcanic plains. The southern fragments of these mosaics are shown in Plates Ji6.1 and Ji7.4.)

Lambert Conformal Conic Projection
Map Scale: 1 cm = 143 km

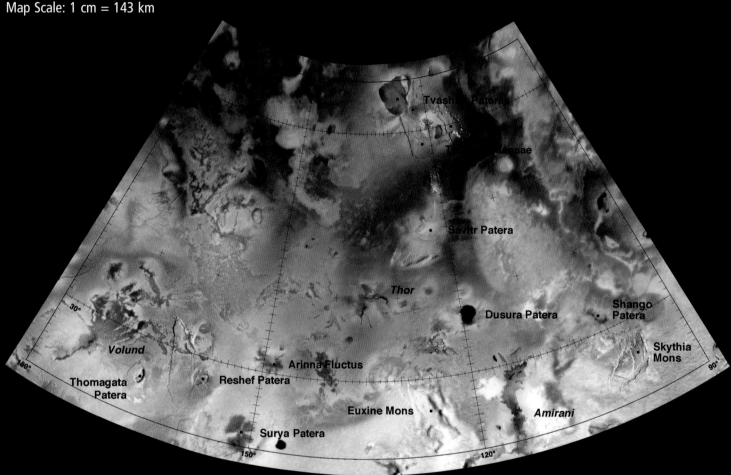

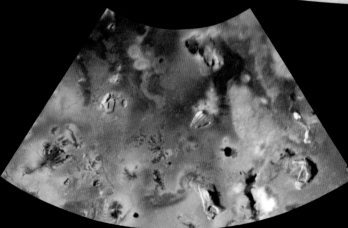

Plate Ji3/Ji3-m Amirani Quadrangle

Volcanic centers such as Thor, Amirani, and Tvashtar dominate this region. *Galileo* first observed an eruption at Tvashtar in 2000, when a giant reddish ring was seen. Although not shown in the earlier higher-resolution images used here, the ring is visible in the low-resolution view in Plate Ji3.2.5. This ring was very similar to the one observed around Pele (Plate Ji9). *New Horizons* observed a spectacular eruption of Tvashtar in 2007, so evidently it is frequently though not continuously active. The Amirani site was observed several times by *Galileo*. The mountains in this region also display a provocative range of morphologies, from flat-topped mesas to smaller sharp craggy peaks to large rounded degraded massifs. Such changes probably reflect the age of the mountain and the degree to which the uplifted landmass creeps down-slope over time due to gravity. At least one volcanic center, Savitr Patera, appears to have cut into the side of a mountain. The transition from Io's reddish poles to its more yellowish equatorial region is also quite obvious.

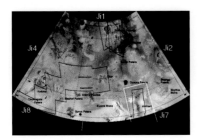

Plate Ji3.1.1a Amirani

Encounter: *Galileo* I27   Resolution: 210 meters/pixel
Orthographic Projection   Map Scale: 1 cm = 31 km

The Amirani volcanic flow field stretches over 300 kilometers and consists of numerous overlapping bright and dark lava flows. The small dark lobes are the most recent flows. On Io, brightness may reflect age, as older lava flows are progressively covered with brighter sulfurous plume deposits. The center of the flow field appears to be near 21°N, 115°W. Flow fields such as these may be Io's equivalent of flood basalts on Earth. The half-moon-shaped dark spot is a shallow volcanic caldera and site of the Maui plume observed by *Voyager*.

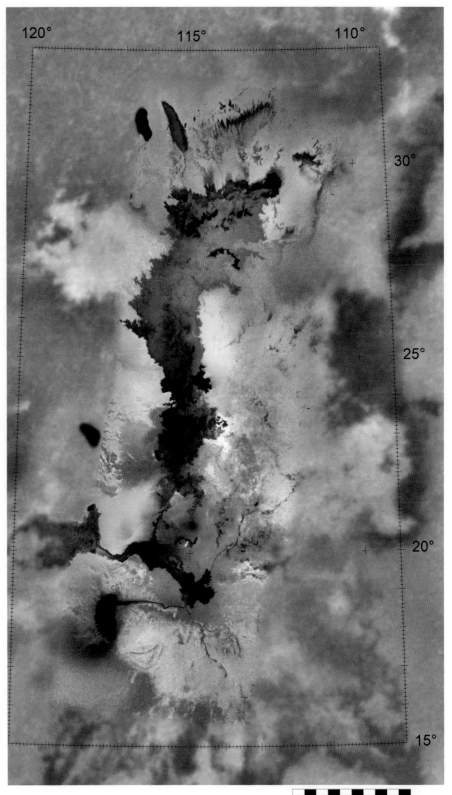

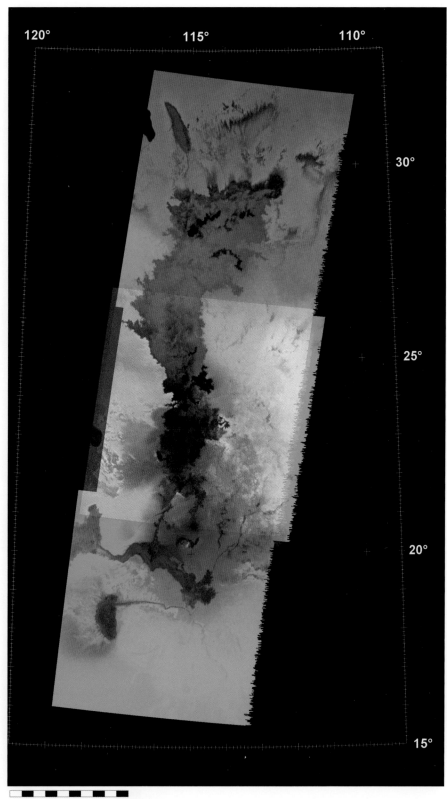

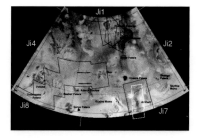

Plate Ji3.1.1b **Amirani: Color**
Encounter: *Galileo* I27
Resolution: 210 meters/pixel
Orthographic Projection
Map Scale: 1 cm = 31 km

This is the same image sequence as shown in Plate Ji8.1.1, except that here only the partial high-resolution I27 color mosaic is used, rather than the lower-resolution global color map.

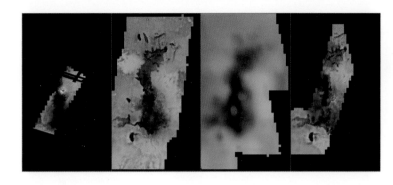

Plate Ji3.1.2-N **Amirani: NIMS Thermal Maps**

Encounter: *Galileo* I24   Resolution: 6.3 kilometers/pixel
Encounter: *Galileo* I27   Resolution: 6.3 kilometers/pixel
Encounter: *Galileo* I31   Resolution: 8.5 kilometers/pixel
Encounter: *Galileo* I31   Resolution: 3.3 kilometers/pixel

Galileo NIMS observed Amirani on three separate occasions, October 1999, February 2000, and (twice) in August 2001 (from left to right), as shown in this 4-image panel. Here low-resolution NIMS 3-color thermal maps are combined with high-resolution SSI mosaics obtained at similar times. Hotspots within the flow field (visible as red spots in this rendering) correlate with some of the darkest and most recent lava flows, indicating they were active at the time. Indicated temperatures may exceed 1000 K. Several changes are seen in the locations of the primary hotspots, indicating shifting or growing lava formations. (The I31 thermal maps may be slightly distorted.)

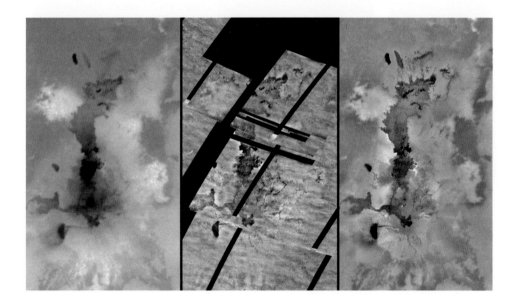

Plate Ji3.1.3 **Changes at Amirani**

Encounter: Galileo I21   Resolution: 1290 meters/pixel
Encounter: Galileo I24   Resolution: 510 meters/pixel
Encounter: Galileo I27   Resolution: 210 meters/pixel

Galileo observed Amirani three times at moderate- to high-resolution over a span of approximately 8 months. Changes are apparent in the shapes of several small dark flows near the northern end of the flow complex, as well as in the diffuse shading centered near the southern end of the flow complex. No further observations were acquired later in the mission.

Atlas of the Galilean Satellites

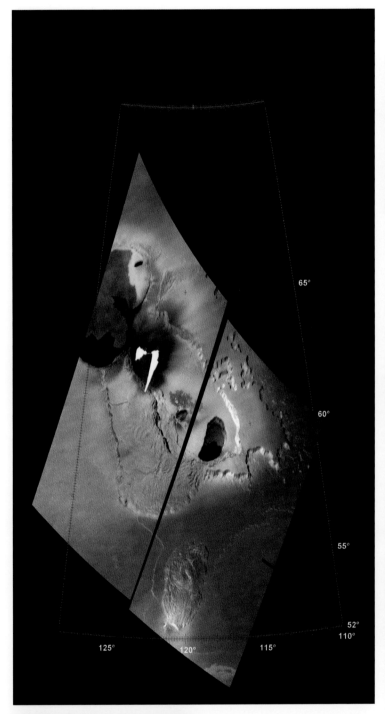

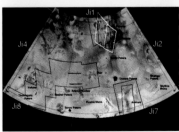

Plate Ji3.2.1a **Tvashtar**
Encounter: *Galileo* I25   Resolution: 190 meters/pixel

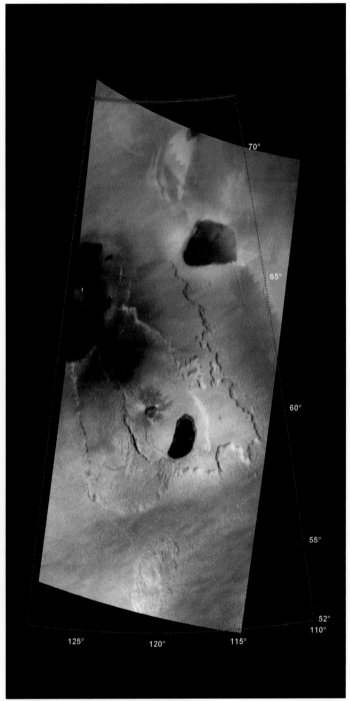

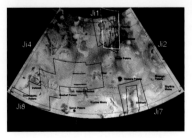

Plate Ji3.2.1b **Tvashtar**
Encounter: *Galileo* I27   Resolution: 320 meters/pixel

Plate Ji3.2.1c Tvashtar

Encounter: *Galileo* I32   Resolution: 200 meters/pixel
Orthographic Projection   Map Scale: 1 cm = 49.8 km

Io is not very predictable. For the I25 encounter with Io, images were targetted at what appeared to be several large calderas. That is debatable, but the images fortuitously captured an intense ongoing eruption near the center of these features. This hot feature saturated the camera CCD, forming the oddly shaped bright streak shown here. This has been interpreted as a fire fountain of hot lava erupting along a linear fissure.

Two more observations were then planned in later orbits to monitor this new eruption, a site not previously known as a hotspot. During I27, the original fire fountain was quiet, though still hot, and the eruption center had moved to the large patera to the northwest. Distant observations during I31 showed that Tvashtar had suddenly grown a large reddish plume very similar to Pele.

The large "ring" of mountains surrounding the eruption site is actually two: a flat-topped 1- to 2-kilometer-high mesa to the east, and massif to the west, roughly 6 kilometers high. The mesa is deeply incised along its margins. These reentrants are characteristic of scarp erosion by sublimation and collapse. The flat plain between these edifices includes a 2-kilometer-deep caldera (inactive?), a small dark patera surrounded by radial deposits, and the fire fountain site. Colors are from the global mosaic.

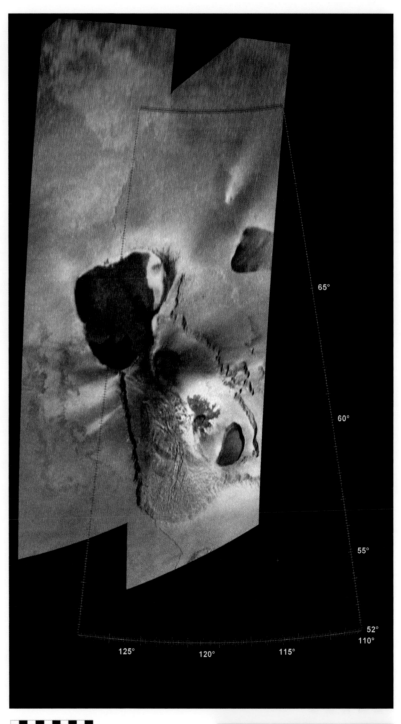

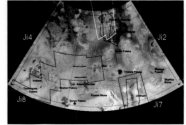

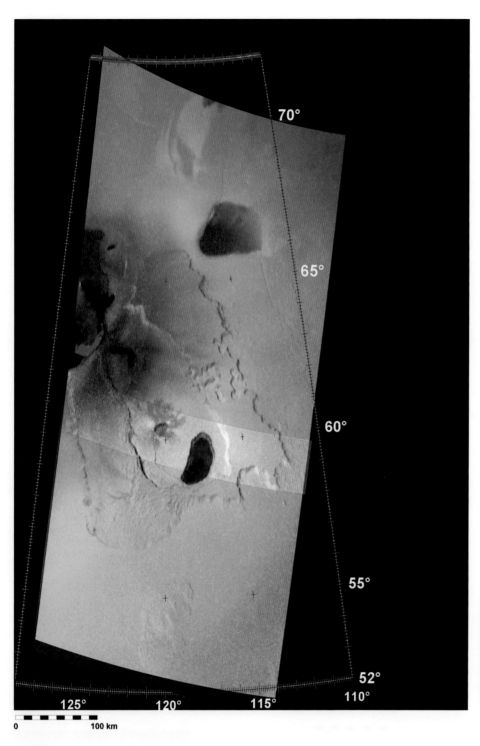

Plate Ji3.2.2 **Tvashtar: Color**
Encounter: *Galileo* I27
Resolution: 320 meters/pixel
Orthographic Projection
Map Scale: 1 cm = 49.8 km

This I27 color sequence is the highest-resolution mapping of Tvashtar in the infrared. In this sequence, the eruption has shifted from the earlier fire-fountain site (the white hook-shaped feature seen in Plate Ji3.2.1a) to the fish-tail-shaped dark lava flows to the northwest. Much of this feature appears to be active, including the distal ends of the "tail." This sequence includes images at 0.97 (partially returned), 0.76, and 0.4 microns. Compare this view to the NIMS views in Plate Ji3.2.3-N. The fire-fountain site was warm (Plate Ji3.2.3-N), but not warm enough to be detected in the 0.97-micron SSI filters, which is sensitive only to the higher temperatures.

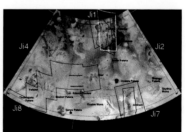

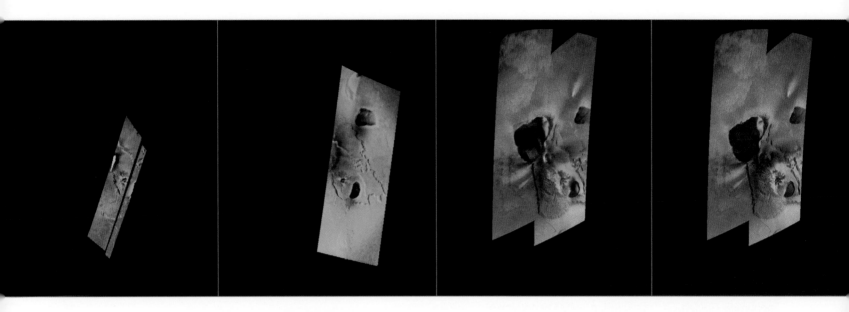

Plate Ji3.2.3-N **Tvashtar: NIMS Thermal Maps**

Encounter: *Galileo* I25　　Resolution: 4.6 kilometers/pixel
Encounter: *Galileo* I27　　Resolution: 4.8 kilometers/pixel
Encounter: *Galileo* I31　　Resolution: 1.7 kilometers/pixel
Encounter: *Galileo* I31　　Resolution: 1.0 kilometers/pixel

Galileo NIMS also observed Tvashtar on three separate occasions, November 1999, February 2000, and (twice) in August 2001, as shown left-to-right in this 4-image panel. Here low-resolution NIMS 3-color thermal maps are combined with high-resolution SSI mosaics obtained at similar times. In the first two observations, NIMS resolution is too low to resolve details, but the I31 mosaics show that the dark fish-tail-shaped flow within the large patera and the original fire-fountain site to the south (glowing red) were all thermally active at the time. Minimum NIMS temperatures are ~1060 K, SSI temperatures 1300–1500 K, and ground-based estimates up to 1300–1900 K!

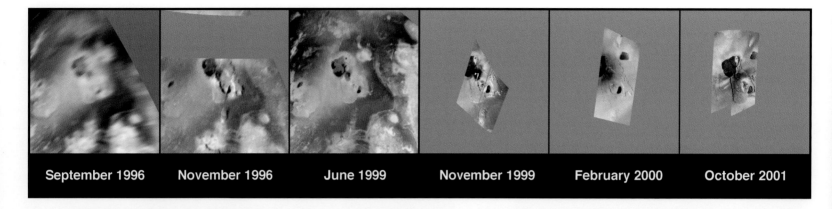

Plate Ji3.2.4 **Changes at Tvashtar**

From left to right:

Encounter: *Galileo* G2   Resolution: 4900 meters/pixel
Encounter: *Galileo* C3   Resolution: 3460 meters/pixel
Encounter: *Galileo* C21   Resolution: 1560 meters/pixel
Encounter: *Galileo* I25   Resolution: 190 meters/pixel
Encounter: *Galileo* I27   Resolution: 320 meters/pixel
Encounter: *Galileo* I32   Resolution: 200 meters/pixel

This time-sequence of 6 images shows changes that occurred at Tvashtar during the course of the *Galileo* mission from September 1996 to October 2001. They vary in resolution, viewing perspective, and even in the color filter used, but are still a useful if incomplete record. The most dramatic change occurred between C21 and I25, when a new eruption began. The I25 image records the fire fountain shown in Plate Ji3.2.1a. By I27, the intensity at the fire fountain had decreased and the main eruption had shifted to the north, and by I32 a dark red plume deposit had formed, first seen at low resolution in G29 (see Plate Ji3.2.5). During I32, the eruption scene appeared relatively quiet, except for the new dark streaks formed earlier. *New Horizons* observed an active plume again in 2007 (Figure 5.4.5).

Plate Ji3.2.5 **Red Ring at Tvashtar**

Encounter: *Galileo* G2   Resolution: 4900 meters/pixel
Encounter: *Galileo* G29   Resolution: 10300 meters/pixel

Although the Tvashtar eruption was first detected on November 26, 1999, the eruption surprised observers when a large 1200-kilometer-wide dark red plume ring deposit was observed in December 2000. These low-resolution enhanced-color views show the anti-Jovian hemisphere before and after the Tvashtar eruption. The top view was acquired in September 1996, the bottom view in December 2000, at least 13 months after the main eruption. The newly formed dark reddish plume ring from Tvashtar is evident at upper right in this sinusoidal map projection. The reddish color indicates a sulfur-rich composition similar to that of the long-lived and much loved Pele plume ring (visible at lower left and in quadrangle Ji9). *New Horizons* witnessed a large active plume at Tvashtar in 2007 (Fig. 5.4.5).

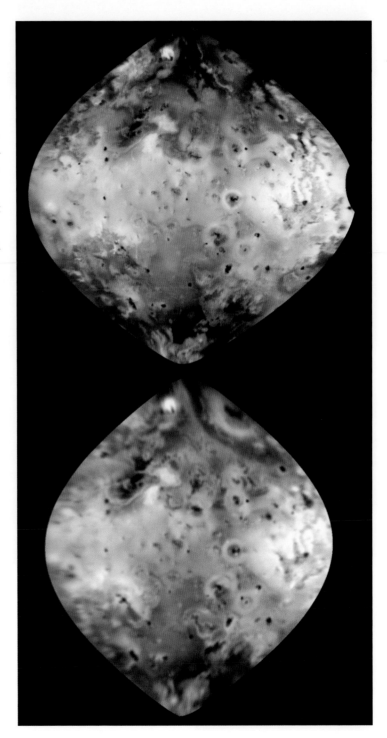

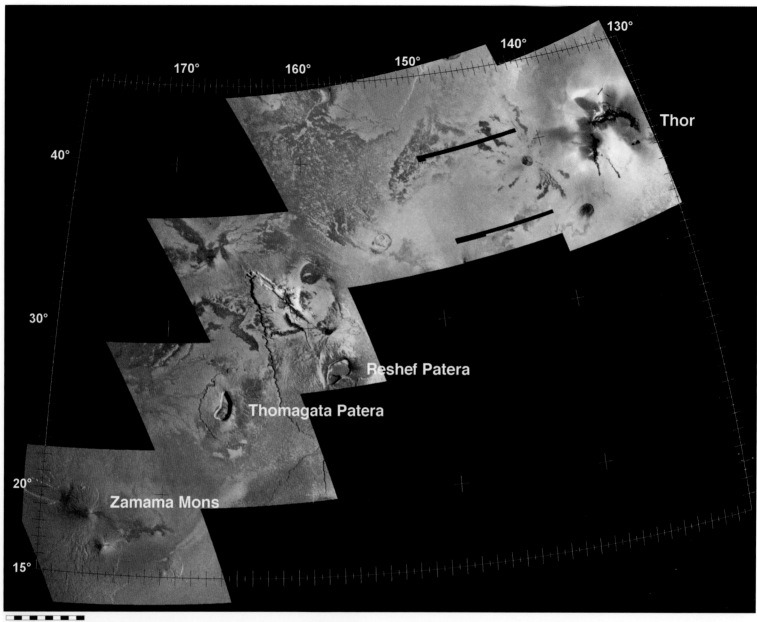

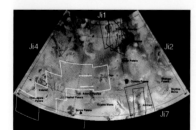

Plate Ji3.3 From Zamama to Thor: Regional Mapping Mosaic

Encounter: *Galileo* I32   Resolution: 340 meters/pixel
Orthographic Projection   Map Scale: 1 cm = 47.2 km

This is the northern half of a regional mapping mosaic along the sunrise terminator during I32 (see also Plate Ji8.7). The two Zamama volcanoes (Plate Ji8.4) sit on the western edge of the mosaic, which extends eastward to the Thor volcanic complex at 40°N, 135°W. Thor became active sometime between May (I30) and October 2001 (I32). Although the region has always been colorful, the very dark flows, and dark diffuse deposits and whitish SO_2-rich ring surrounding them, are entirely new. *Galileo* also saw a 500-kilometer-high plume over Thor during I31, the tallest plume ever observed. Between Thor and Zamama lie several low plateaus and depressed calderas. This monochrome mosaic has been combined with the lower-resolution global color map.

Plate Ji3.3-N From Zamama to Thor: NIMS Thermal Map

Encounter: *Galileo* I32
Resolution: 8.2 kilometers/pixel

The newly active volcanic center Thor glows brightly in this thermal map, which unfortunately does not include Zamama. No other activity is detected in these data. Thor was one of the hottest regions on Io during this encounter. Minimum temperatures are estimated at 800 K but are likely higher.

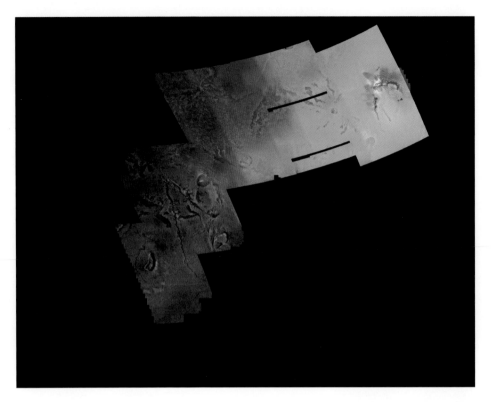

Plate Ji3.4 Changes at Thor

Encounter: *Galileo* I24
Resolution: 1470 meters/pixel
Encounter: *Galileo* I32
Resolution: 340 meters/pixel
Orthographic Projection
Map Scale: 1 cm = 40.2 km

When Thor erupted violently in 2001, it completely remade the region. None of the dark lava flows or dark and bright diffuse deposits are apparent in views obtained in 1999. NIMS detected no hotspots here as of C30 (May 2001) but did in its global map in August. Little is known about the details of this eruption, which occurred very late in the mission, except for the high temperatures (>800 K), extended volcanic deposits and large plume.

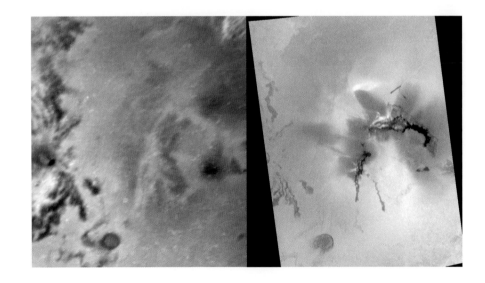

Lambert Conformal Conic Projection
Map Scale: 1 cm = 143 km

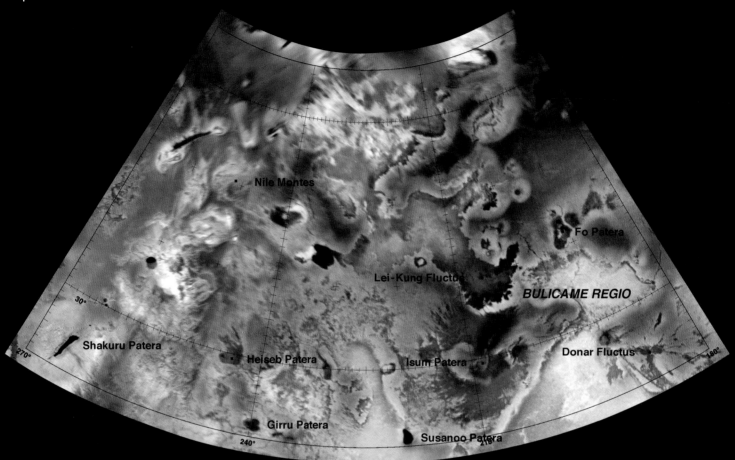

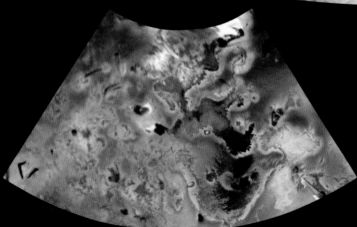

Plate Ji4/Ji4-m Lei-Kung Fluctus Quadrangle

This region consists of volcanic plains noteworthy for their exaggerated reddish, yellowish and grayish colors. Vast dark lava flows (Lei-Kung Fluctus) cover large areas in the central portions of this quadrangle. The whitish deposits are most likely sulfur dioxide frosts. The dark black spike at Shakuru is probably a molten lava lake, which is flanked to the west by a 7-kilometer-high plateau. Additional mountains can be identified in the shadow image, including those at 58°N, 192°W, 52°N, 254°W and 58°N, 254°W, and a machete-shaped plateau at 23°N, 190°W.

Plate Ji4.1 Terrain Sapping: Very High Resolution
Encounter: *Galileo* I27   Resolution: 6 meters/pixel
Orthographic Projection   Map Scale: 1 cm = 2 km

The original objective of this observation was to examine areas thought to be modified by sapping, or ground collapse. These are our highest-resolution images of Io, but context images planned for later orbits were never acquired, and the geologic context is unknown and precise location on the surface is uncertain. The smooth dark gray unit at 30.7°N, 192.8°W is the top of a relatively smooth and perhaps lava-covered mesa casting shadows to the west. The convoluted bright and dark patterns do not resemble other areas of Io and their origins are unknown. The smoother material could be either small lava flows of unconsolidated debris formed as the mesa scarp (possibly) retreated due to erosion. Without topographic or context information, the geology here remains open to interpretation.

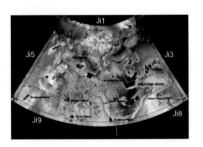

Plate Ji4.1.X Terrain Sapping: Original Perspective

This view shows part of Plate Ji4.1 in its original oblique perspective, and simulates the view an astronaut might have on approach to the surface. Part of a smooth dark-topped mesa is visible in the foreground. The shadowy edge of the mesa is to the left. Beyond are eroded and resurfaced lava plains.

Lambert Conformal Conic Projection
Map Scale: 1 cm = 143 km

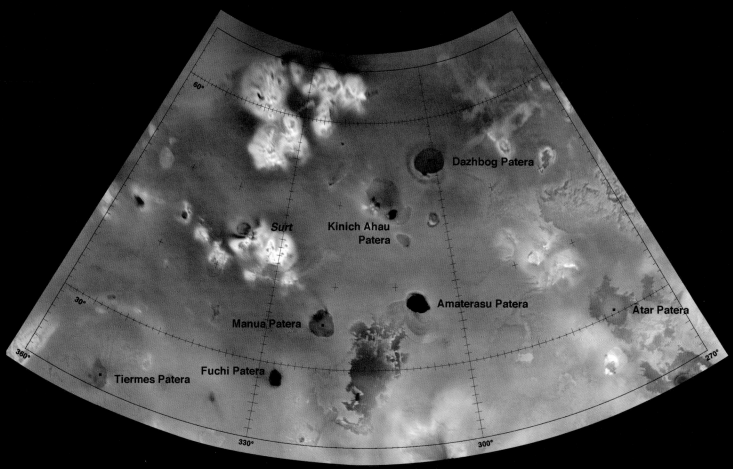

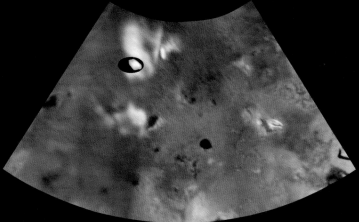

Plate Ji5/Ji5-m **Amaterasu Patera Quadrangle**

Amaterasu Patera quadrangle is dominated by relatively featureless volcanic plains. The dark round spots are volcanic calderas (called patera) and are or were molten lava lakes similar to those seen at Kilauea in Hawai'i. Amaterasu Patera is one of the darkest features on Io and may be relatively hot. Dazhbog was the site of a major Pele-style plume eruption sometime between November 1999 and August 2001. Neither were observed closely by *Galileo*. Irregular bright features, called aureoles, are prominent in this quadrangle. Their origin is not well understood, although they are most likely composed of sulfur dioxide frosts. Mountains can be found near the center of several of these features. In general, however, there are relatively fewer mountains here compared with other regions. One of these bright spots is also associated with Surt, an active volcanic center during *Voyager* days.

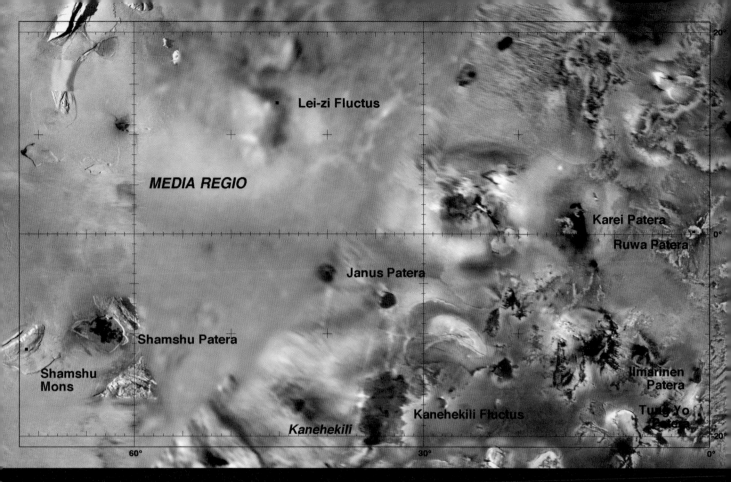

Plate Ji6/Ji6-m Sp: Kanehekili Quadrangle

Abundant mountains in the west (observed by *Galileo*) give way to numerous volcanic calderas and flow fields towards the east (seen by *Voyager*). The concentration of mountains surrounding Shamshu Patera is unusual and suggests that in some cases mountain building and volcanic caldera formation may be linked in some way. Sp: Kanehekili (~35°S, 17°W) eruption site has been monitored since at least 1990, though never observed at high resolution.

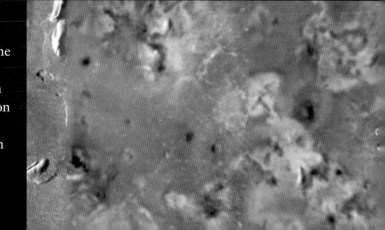

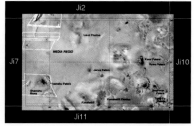

Plate Ji6.1 From Shamshu Mons to Mongibelli Mons: Regional Mapping Mosaic

Encounter: *Galileo* I27
Resolution: 340 meters/pixel
Orthographic Projection
Map Scale: 1 cm = 68.9 km

These images are part of a mosaic (only half of which was returned) of the sunset terminator from Shamshu Mons to Zal Montes (see Plate Ji2.1.2). This area has one of the highest spatial density of mountains anywhere on Io. Several prominent mountains surround Shamshu Patera (10°S, 62°W), a large volcanic caldera partially covered with dark lava flows. Shamshu Mons itself is ~3 kilometers high, while the partially imaged mountain at 15°S, 60°W rises at least 5 kilometers. Many of these mountains have fractured or broken into several sections, either during or after uplift.

The unusual structure to the north is the double-ridged Mongibello Mons, rising 6 kilometers high. The origin of the double ridge structure is uncertain. It may reflect layering in Io's crust, exposed when the mountain block was uplifted and rotated, or it could be the expression of parallel faults formed when the mountain was uplifted. Low-lying layered plains occur on several sides, noted for their crenulated scarps. Just south of Mongibello Mons is an elongate low-rising mountain noteworthy for its whitish color. Several smaller massifs lie due east of Mongibello, including a broken massif 1.65 kilometers high and a 4-kilometer-high steep-sided peak at 17°N, 55°W, rising like Mont St. Michel from the (volcanic) plains (see also Plate Ji7.4). Shown at 60% scale to fit page.

Map Scale at Equator: 1 cm = 121 km

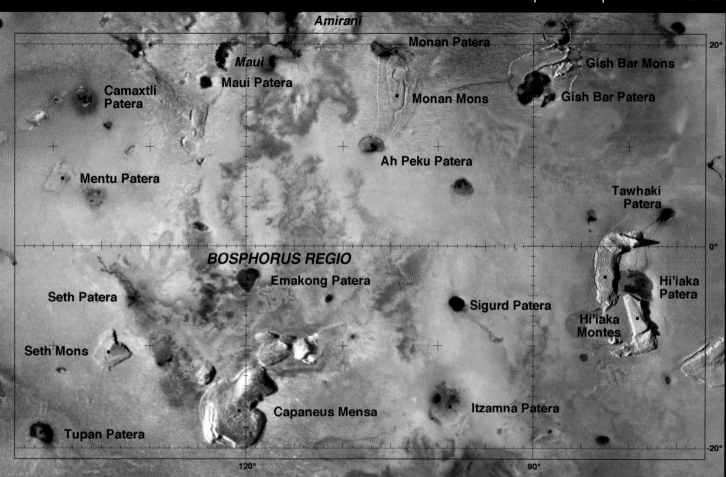

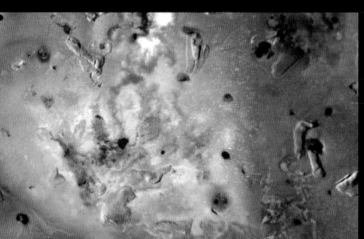

Plate Ji7/Ji7-m Emakong Patera Quadrangle

Emakong quadrangle may be one of the busiest places on Io. The region features a high concentration of mountains and dark calderas, but is also dominated by the extensive Emakong and Seth Paterae volcanic flow complexes that make up Bosphorus Regio. The whitish and yellowish deposits and low temperatures have led to suggestions that the lavas here are composed mostly of sulfur, a rare and exotic lava on Earth. The Maui and Amirani eruption sites lie to the north. The Hi'iaka Montes complex of mountains is also one of the most varied and complex on Io. Note that unscrambled I24 images were used along the northern margin (from Gish Bar Mons to Amirani), giving the plains a misleading speckled texture. Bright speckles south of Capaneus Mensa are due to radiation noise in some of the regional mapping images used in this hemisphere.

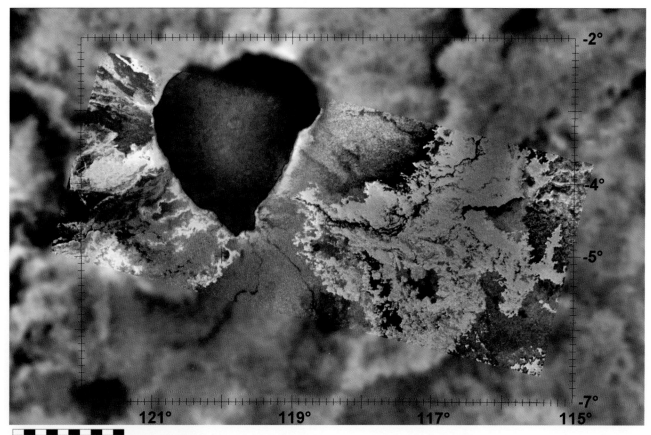

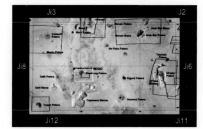

Plate Ji7.1.1 Emakong Patera

Encounter: *Galileo* I25   Resolution: 144 meters/pixel
Orthographic Projection   Map Scale: 1 cm = 16.5 km

This view shows the heart of the vast Emakong volcanic flow field. Most flows appear to emanate from the central heart-shaped caldera, which exhibits little or no topographic relief. The colors of Emakong and low temperatures may be an indication of sulfur volcanism but this has not been observed directly. This view also shows a relatively bright lava flow to the east of Emakong Patera (shown in Plate Ji7.1.3).

Plate Ji7.1.2-N Emakong Patera: NIMS Thermal Map

Encounter: *Galileo* I32   Resolution: 4.0 kilometers/pixel

Emakong Patera consistently shows relatively low thermal emission, with temperatures of only 300K or so. A faint glow was seen within and along the edge of the patera, consistent with exposure of hot sulfur lava along the edge of a lava lake, or with prolonged cooling of older silicate flows.

Plate Ji7.1.3 **Emakong Lava Channels**

Encounter: *Galileo* I25   Resolution: 32 meters/pixel
Orthographic Projection   Map Scale: 1 cm = 5.8 km

The lava channel seen in Plate Ji7.1.1 is shown here to be a complex structure. This flow is relatively narrow at its origin at the edge of the dark caldera and then broadens considerably down-slope as it flows east. A dark lava channel runs through the center of this flow. This channel splits apart and comes back together several times, forming "islands." This braiding is common in lava channels on Earth as well, and typically occurs in lavas that are very runny, including basalt but also possibly molten sulfur. Several episodes of channel formation can be identified. Mosaic shown at 65% scale to fit page.

Plate Ji7.1.4-N **Emakong Lava Channels: NIMS Thermal Map**

Encounter: *Galileo* I32   Resolution: 0.8 kilometers/pixel

The NIMS view shows no significant hotspots within the eastern Emakong lava flow field as observed by *Galileo*.

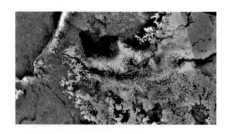

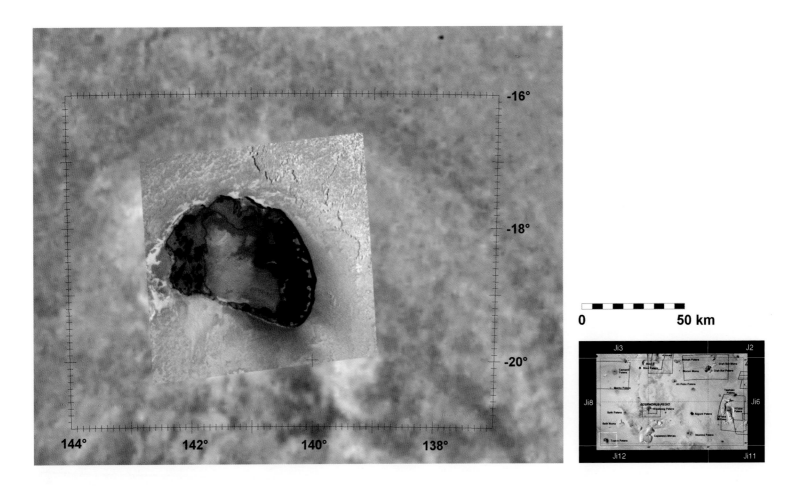

Plate Ji7.2.1 **Tupan Patera**

Encounter: *Galileo* I32   Resolution: 135 meters/pixel
Orthographic Projection   Map Scale: 1 cm = 18.6 km

The Tupan caldera is a persistent hotspot and the reddish and greenish color seen in the quadrangle indicate that it has been active recently. The shadows on the eastern margin tell us that the depression is 1 kilometer deep, but no shadows can be seen around the brighter central "island," indicating it has relief of no more than 60 meters. In fact, the dark "bath-tub" ring that snakes across the orange deposits on the floor of the caldera is not a shadow but may be the edge of a crusted-over lava lake. The green colors along the western floor may be due to iron contamination or presence of olivine or pyroxene. Etched plains just to the northeast of Tupan Patera are part of the volcanic plains observed across most of Io.

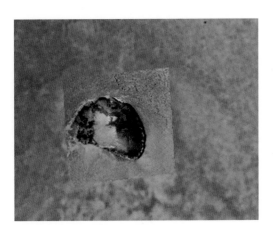

Plate Ji7.2.2-N **Tupan Patera: NIMS Thermal Map**

Encounter: *Galileo* I32   Resolution: 4.0 kilometers/pixel

The hottest regions at Tupan correspond to the dark eastern portion of the caldera floor, a possible lava lake. A weaker thermal emission emanates from the dark spotted area in the western part of the caldera, but the bright "island" at center is cold. Minimum temperatures are at least 750 K and may range to 1100 K at times.

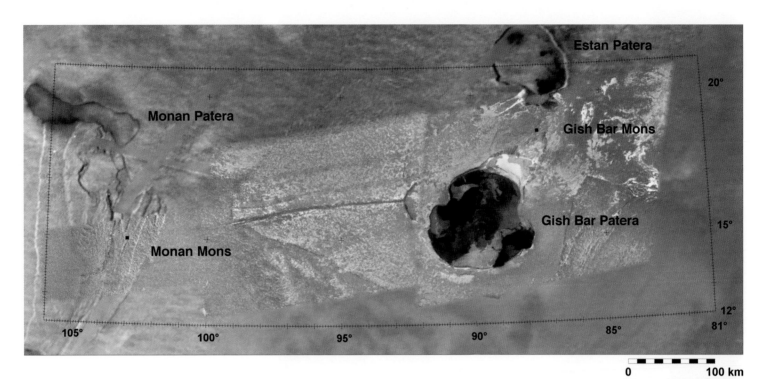

Plate Ji7.3.1 **Gish Bar Patera**

Encounter: *Galileo* I32   Resolution: 250 meters/pixel
Orthographic Projection   Map Scale: 1 cm = 21.9 km

This mosaic is the only surviving half of a planned stereo sequence. The Gish Bar caldera (15°N, 9°W) and overlapping lava flows on the caldera floor, as well as the reddish-floored patera to the north at 22°N, 87°W, are very evident, but the local noontime Sun conditions make the prominent mountains between the two large dark-floored paterae all but disappear. Most mountains on Io are difficult to recognize under such viewing conditions, perhaps due to the extensive blanketing plume deposits that cover all terrains. These mountains can be seen in the quadrangle maps (Plate Ji7M). The smaller northern peak rises up to 4 kilometers and is cut into by both Gish Bar Patera and Estan Patera to the north (20°N, 87°W). The taller peak to the east rises ~11 kilometers.

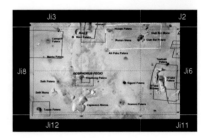

Plate Ji7.3.2-N **Gish Bar Patera: NIMS Thermal Map**

Encounter: *Galileo* I31   Resolution: 4.4 kilometers/pixel

Gish Bar Patera is very active thermally, and changes have been observed in the dark flows on the floor of the caldera. At least three hotspots can be identified on the dark floor itself, including the dark spot in the southeast section. Minimum temperatures of ~500K are associated with these flows. A significant increase in thermal output after October 1999 may be associated with new dark lavas seen in Plate Ji7.3.1. Dark flows in Estan Patera, north of Gish Bar Mons, may be even warmer.

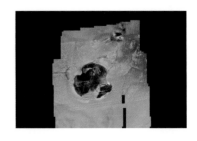

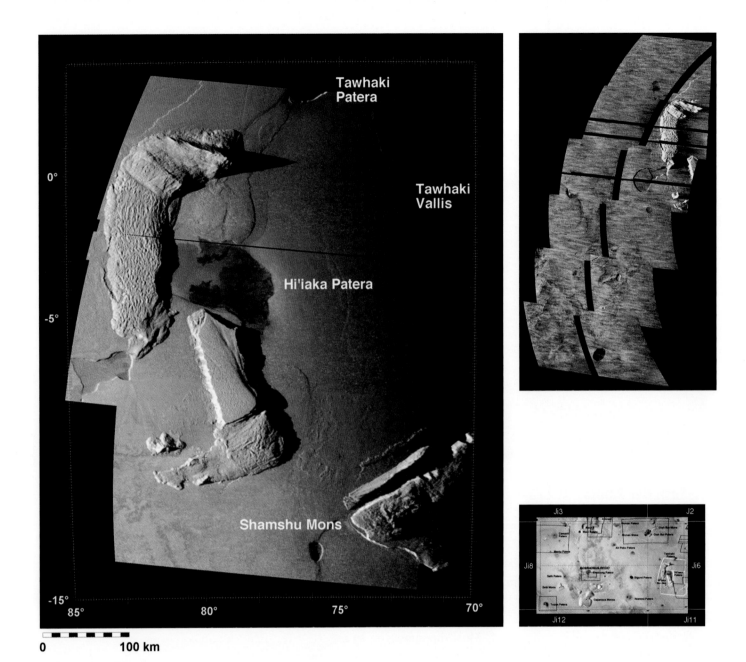

Plate Ji7.4 Hi'iaka Montes and Tawhaki Vallis

Encounter: *Galileo* I25   Resolution: 270 meters/pixel
Orthographic Projection   Map Scale: 1 cm = 42.3 km

Hi'iaka Montes is split into two parts. The southeastern section is a flat-topped mesa, 2.5 kilometers high. Small mounds of debris can be seen along the edge, indicating the edge is slowly crumbling. The northwestern section is a complex structure featuring a ridged plateau 3 to 5 kilometers high tilting 4° to the west and several rugged promontories on the north end, the tallest of which reaches 9 kilometers high. The ridges may have formed due to crumpling of the surface, either before or after uplift of the mountain. Also visible is part of Shamshu Montes to the southeast (12°S, 73°W; see also Plate Ji6.1).

In the volcanic plains to the east of Hi'iaka Montes is the narrow channel, Tawhaki Vallis. Similar to channels seen in Hawai'i, Tawhaki Vallis is 40 to 60 meters deep and was carved not by running water but by hot flowing lava from nearby Tawhaki Patera just to the north, an active hotspot. The only other observed lava channel is at Emakong (Plate Ji7.1.1). These images were obtained as part of a sequence that includes Zal Montes (Plate Ji2.1.1). Hi'iaka Patera itself glows at 450K. Inset version is based on I2Y descrambled images.

Mercator Projection
Map Scale at Equator: 1 cm = 121 km

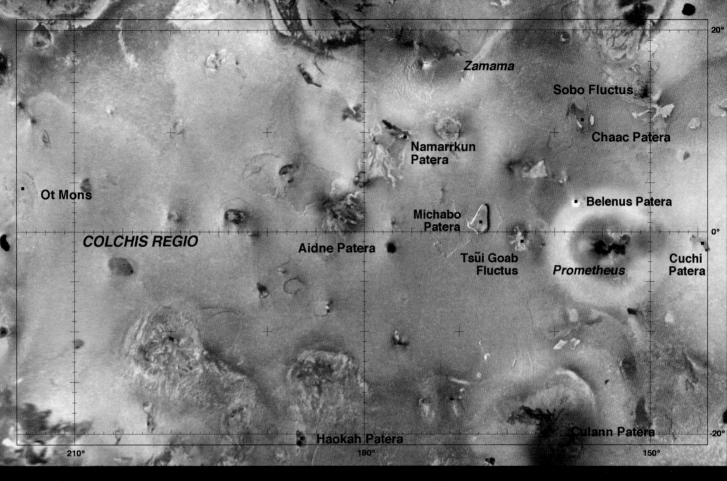

Plate Ji8/Ji8-m Prometheus Quadrangle

The extensive smooth volcanic plains of Prometheus quadrangle contrast starkly with the complex morphologies of Emakong quadrangle (Plate Ji7). Layered plains and flat-topped mesas occur, but mountains are much less common here than other locations (quadrangle Ji7, for example). Rather, we see numerous volcanic pits or calderas (paterae), many of which are 1 to 3 kilometers deep. These represent sites where volcanic centers have collapsed. The extensive relatively white plains of Colchis Regio are sulfur dioxide deposits.

Three volcanic centers stand out: Zamama, Culann, and Prometheus, all of which were *Galileo* high-resolution targets. Although the reddish and greenish colors are an indication of recent volcanic activity, all three volcanic centers are distinctly

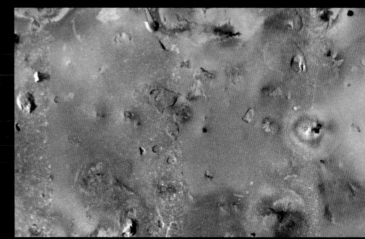

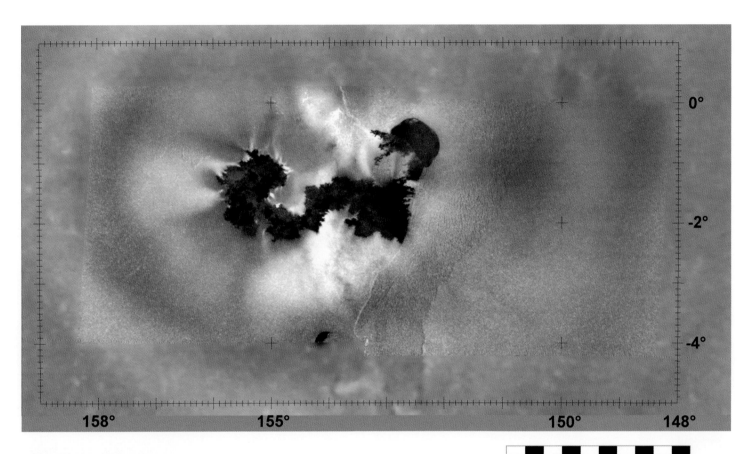

Plate Ji8.1.1a Prometheus
Encounter: *Galileo* I27   Resolution: 170 meters/pixel
Orthographic Projection   Map Scale: 1 cm = 20.1 km

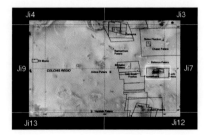

Prometheus was a favorite target of the *Galileo* mission. First observed by *Voyager* in 1979 as a ~140-kilometer-high plume of gas and dust (Figure 5.4.2), Prometheus is one of the most consistently active volcanoes on Io. It is the type example of a promethean eruption style, characterized by relatively slow but prolonged eruption rates, compound overlapping flow fields, and SO_2-driven plumes (<200 kilometers high). *Galileo* obtained its first close-up views in 1999, but had already discovered that the plume had moved 80 km to the west (Plate Ji8.1.7). Here we see more clearly the source of the plume, several bright jet-like features along the edges of the western lobe of a new dark lava flow field. The volcanic plume forms around the edge of the flow where hot lava volatilizes sulfur dioxide in the surrounding bright plains, creating jets of gas and dust. The lava flows apparently emanate from the dark kidney-shaped patera at 0.7°S, 152.5°W. A ridged plateau roughly 100 meters high surrounds the patera to the north, south and east and blocks lava flow in those directions.

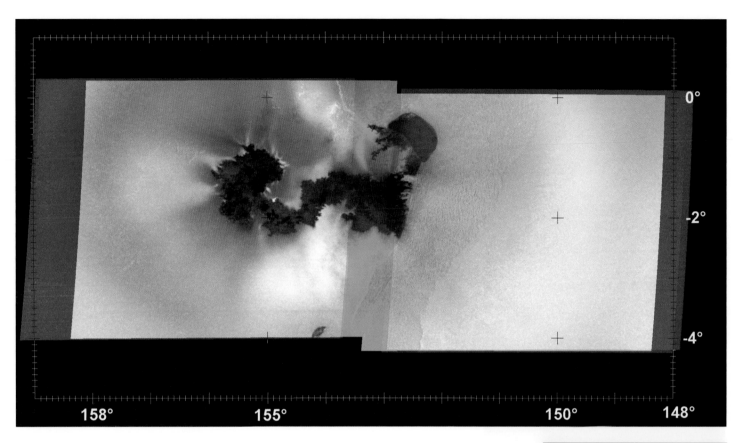

Plate Ji8.1.1b **Prometheus in Color**
Encounter: *Galileo* I27   Resolution: 170 meters/pixel
Orthographic Projection   Map Scale: 1 cm = 20.1 km

These images are from the same sequence as shown in Plate Ji8.1.1a, except that here only the I27 partial color mosaic is used.

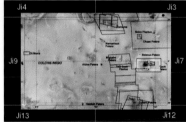

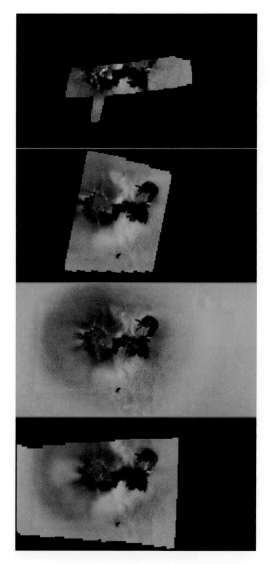

Plate Ji8.1.2-N **Prometheus: NIMS Thermal Maps**

Encounter: *Galileo* I24   Resolution: 1.4 kilometers/pixel
Encounter: *Galileo* I24   Resolution: 3.8 kilometers/pixel
Encounter: *Galileo* I25   Resolution: 9.5 kilometers/pixel
Encounter: *Galileo* I27   Resolution: 3.3 kilometers/pixel

Several targeted infrared observations were made of Prometheus. The best of these, obtained during orbit I24 and shown in the first panel, reveals discrete hotspots across the western dark lobe, and likley correlating with dark flow features within the lobe. A small hotspot also occurs at the eastern end of the main flow body (at 1°S, 152.4°W) near where it connects with the dark patera. This spot was persistent during the three encounters and may be the source vent for the main flow field. Resolution was poorer in subsequent encounters but each shows a similar thermal distribution. Minimum temperatures of ~900 K are indicated, although other estimates range from 1250 K up to 1480 K for discrete areas within the flow field.

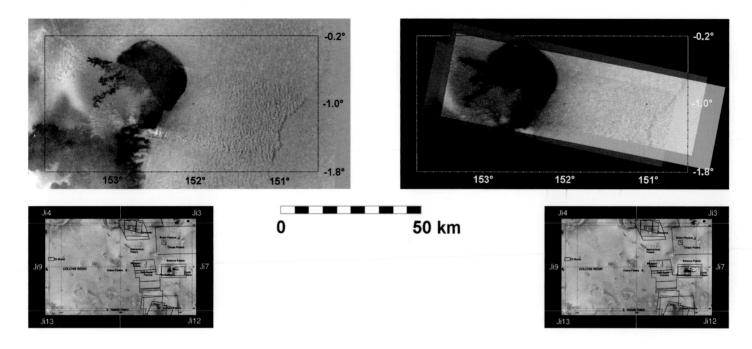

Plate Ji8.1.3a **Prometheus Caldera**

Plate Ji8.1.3b **Prometheus Caldera: Color**
Encounter: *Galileo* I24   Resolution: 120 meters/pixel
Orthographic Projection   Map Scale: 1 cm = 14.2 km

The vast Prometheus lava flow field originates from this fairly ordinary looking dark kidney-shaped volcanic caldera. No relief is apparent along the edges of the dark patera. The ridged plains to the east are actually part of a roughly 200-meter-high plateau surrounding the caldera to the east and north that block lava flow and are visible in the regional view. The ridges are common on Io's plains but their origins are unknown. Gray-tone and I24-only color views are shown. These images were not corrupted by the I24 anomaly.

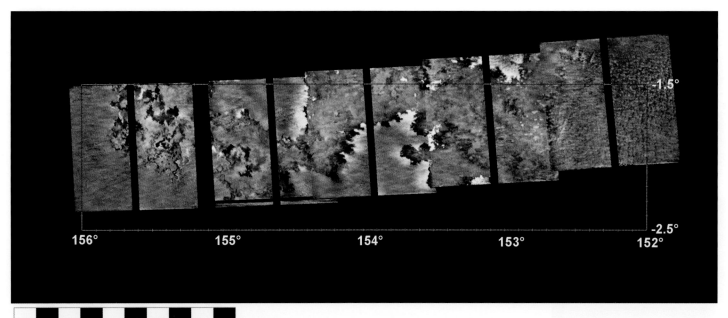

0   50 km

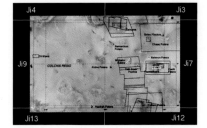

Plate Ji8.1.4 Prometheus Lava Flows: High Resolution (descrambled)

Encounter: *Galileo* I24   Resolution: 70 meters/pixel
Orthographic Projection   Map Scale: 1 cm = 8.3 km

Two high-resolution views of the western flow field were acquired by *Galileo*. Although scrambled, the I24 mosaic reveals numerous small overlapping and intermingled dark and gray lava flows within the main body. Although likely to be a few meters or less, flow thicknesses are unknown. Color from global mosaic.

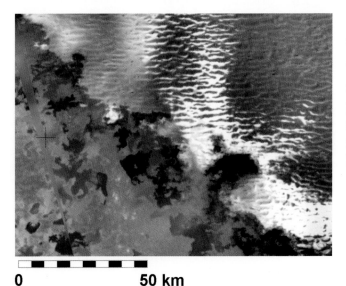

Plate Ji8.1.5.X Prometheus Lava Flows: Close-up

Orthographic Projection   Map Scale: 1 cm = 1.4 km

This enlargement from Plate Ji8.1.5 more clearly shows the interaction between the advancing lavas and the ridged bright plains. We cannot measure flow thicknesses directly in these images. Lava clearly flows between ridges, indicating flows are not much thicker than ridges, once we determine ridge heights.

0   50 km

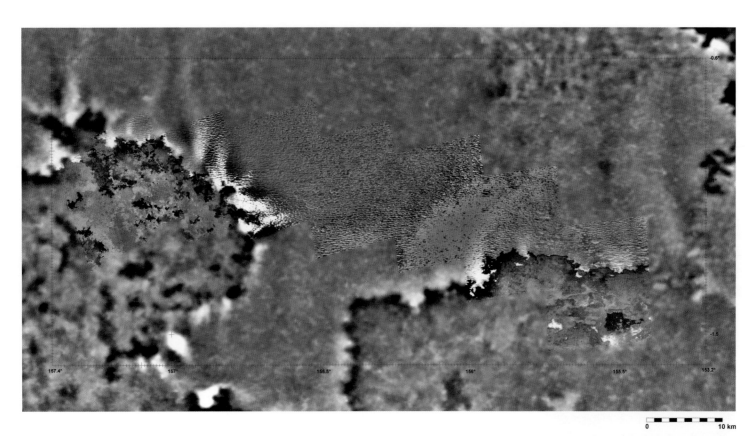

Plate Ji8.1.5 **Prometheus Lava Flows: High Resolution**
Encounter: *Galileo* I27   Resolution: 12 meters/pixel
Orthographic Projection   Map Scale: 1 cm = 4.71 km

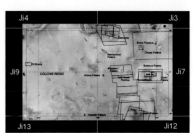

When the I24 mosaic was returned scrambled, a second very high resolution mosaic was planned. New dark flow lobes (since I24) can be identified in this mosaic, indicating widespread small changes routinely occur within the flow field. The mosaic also clearly shows numerous dark tongues of lava extruding into the surrounding plains. Many of these dark lobes are surrounded by jets of bright material, probably sulfur dioxide frosts and the source(s) of the main Prometheus plume. The plains themselves feature tightly packed short parallel ridges. The origin of these ridges, seen in many other locations, is unknown. The bright frosts are deposited only on ridge sides facing the lava flows, another indication that the plume venting is directionally focused from the lava–plains contact. Color from global mosaic. Mosaic shown at 30% scale to fit page. See also Plate Ji8.1.5.X for enlargement.

Plate Ji8.1.6-N Prometheus Lava Flows: NIMS Thermal Map
Encounter: *Galileo* I27   Resolution: 0.3 kilometers/pixel

This observation, coinciding with high-resolution imaging (Plate Ji8.1.5) confirms that the western flow lobe is generally hot, whereas the eastern lobe glows only faintly and in only a few spots.

Plate Ji8.1.7 Changes at Prometheus
From left to right:

Encounter: *Voyager* 1	Resolution: 4660 meters/pixel
Encounter: *Galileo* G2	Resolution: 4900 meters/pixel
Encounter: *Galileo* C3	Resolution: 3460 meters/pixel
Encounter: *Galileo* E4	Resolution: 5900 meters/pixel
Encounter: *Galileo* C10	Resolution: 3850 meters/pixel
Encounter: *Galileo* E14	Resolution: 2600 meters/pixel
Encounter: *Galileo* C21	Resolution: 1290 meters/pixel
Encounter: *Galileo* I24	Resolution: 1470 meters/pixel
Encounter: *Galileo* G29	Resolution: 10300 meters/pixel

This sequence shows how Prometheus changed from *Voyager* days (1979) through the *Galileo* mission (late 2000). The first two frames show the plume source had expanded and moved ~80 kilometers to the west sometime during the 17-year gap. Although a dark feature is visible within the plume ring in the first (*Voyager*) image, it does not correspond directly to the dark flow field seen in later *Galileo* images. Minor changes to the shape of the plume can be seen during the *Galileo* mission, probably reflecting changes in the advancing flow front, but otherwise no further lateral movement was observed.

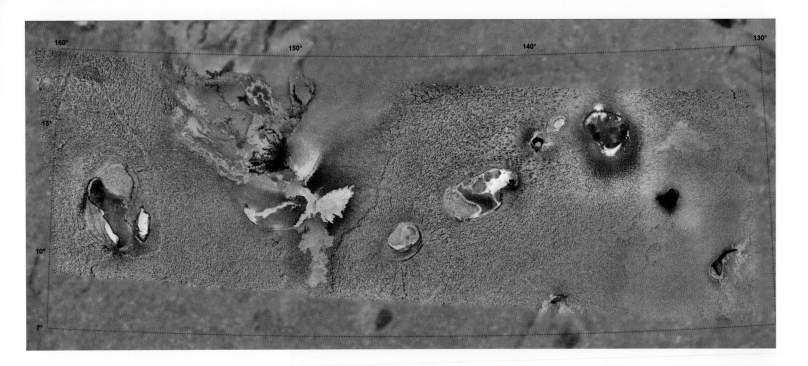

Plate Ji8.2.1 Sobo Fluctus Region

Encounter: *Galileo* I27   Resolution: 185 meters/pixel
Orthographic Projection   Map Scale: 1 cm = 47.2 km

This mapping mosaic, due north of Prometheus, was targeted to examine a suite of volcanic morphologies. Here we see ovoid patera of different sizes, shapes and colors, dark and bright lava flow fields at Sobo Fluctus, and ridged volcanic plains throughout. A green halo surrounds Camaxtli Patera. Several concentric scarps form the southern margin of Grannos Patera. The ridged plains have a corrugated or ridged texture that changes orientation in different parts of the mosaic. Chaac Patera is featured in high resolution in Plate Ji8.3. Feature names in this crowded scene can be found in the Gazetteer. Mosaic shown at 45% scale to fit page.

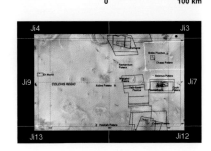

Plate Ji8.2.2-N Sobo Fluctus Region: NIMS Thermal Maps

Encounter: *Galileo* I27   Resolution: 4.5 kilometers/pixel
Encounter: *Galileo* I32   Resolution: 5.3 kilometers/pixel

The Camax region was observed by NIMS during orbits I27 (top) and I32 (bottom). Several darker features are thermally active. Sobo Fluctus and Camaxtli Patera are both prominent hotspots, as well as Ruaumoko Patera to the west and and Tien Mu Patera to the southeast of Camaxtli Patera. Both of these small spots have very dark floors. No major changes in hotspot distribution or intensity was observed. Chaac Patera, although somewhat distinctive in color during I27, was not active during either encounter.

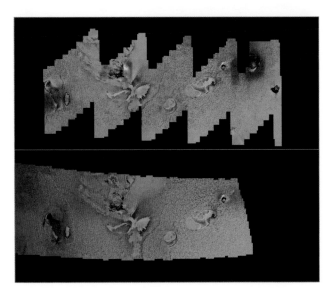

Plate Ji8.3.1 Chaac Patera: Very High Resolution
Encounter: *Galileo* I27   Resolution: 7–8 meters/pixel
Orthographic Projection   Map Scale: 1 cm = 0.9 km

Only fragments of this 5-frame mosaic were returned to Earth, but they are still very fascinating. The caldera itself is 2.75 kilometers deep along the northern wall cliff. The caldera floor is patterned in dark, bright and gray amoeboid shapes that appear to consist of rounded or pitted mesas and shallow interconnected depressions. This may be characteristic of lava flow deflation, whereby the frozen surfaces of lava flows or sheets partly collapse after lava withdraws, but other interpretations are possible.

These are the highest-resolution images of Io for which we have context images. Their locations on the floor of Chaac Patera are shown in Plate Ji8.3.2; they have been moved closer together here to fit the page. An alternate view of the caldera walls is shown in Plate Ji8.3.3-X. Mosaic shown at 75% to fit page.

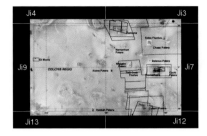

Plate Ji8.3.2 Chaac Patera: Context

Encounter: *Galileo* I27
Orthographic Projection
Map Scale: 1 cm = 5 km

Here the five high-resolution images are placed in context and in their approximate locations on the Chaac Patera floor. Only the locations of the last, southernmost, image and perhaps the two on the north rim image are known with confidence. See Plate Ji8.3.1 for enlargement.

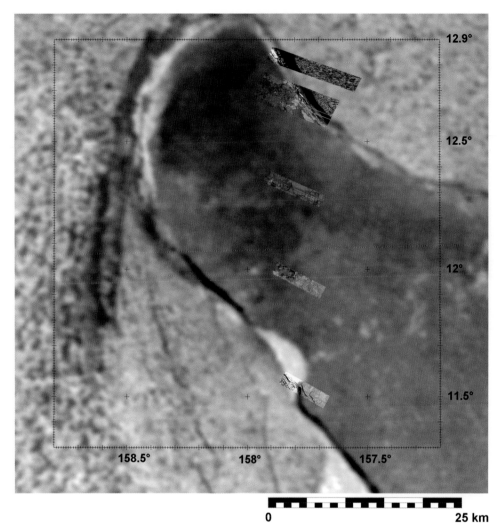

Plate Ji8.3.4-N Chaac Patera: NIMS Thermal Map

Encounter: *Galileo* I27   Resolution: 1.5 kilometers/pixel

The reddish tint on the floor of Chaac Patera in this rendition is not due to thermal emission but an unusual and unresolved composition. Balder Patera, the smaller oval depression just east of Chaac Patera, is noteworthy for its apparent high concentration of SO_2 (showing as blue in this rendition).

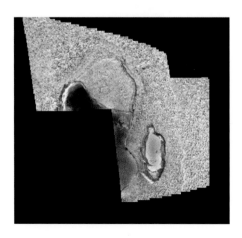

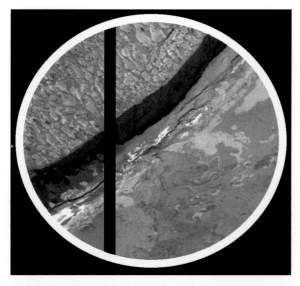

Plate Ji8.3.3-X Chaac Patera: Backlit Cliffs of Io
Encounter: *Galileo* I27   Resolution: 7 meters/pixel

The two northernmost images of this sequence have been rotated and brightened to show the sheer cliffs of Chaac Patera from the perspective of an approaching astronaut. The shadowed scarp is backlit by reflected light from the caldera floor. It is a puzzle that these cliffs do not highlight multiple thin layers of lava, as do some volcanic cliffs on Earth and Mars. This could be due to compression or solidification of lavas on Io into thicker coherent layers, or to the absence on Io of the type of erosion that makes resistant layers stand out on those two planets. Indirect evidence suggests most lavas on Io are no more than a few meters thick; perhaps the resolution was still not sufficient to resolve individual lava layers in this cliff.

Parts of the caldera wall are sheer, indicating the rim materials are relatively strong; other parts are cluttered with slumped debris blocks. The fact that such sheer cliffs remain standing on Io has long been noted to indicate that weak materials like sulfur are not the dominant lava type on Io: such cliffs would collapse. Stronger silicates dominate the lava compositions, although sulfur is an important volatile component on Io.

The stubby ridged pattern of Io's volcanic plains (see also Plate Ji8.1.5-X), north of the caldera rim scarp, are very evident. The ridges have surprising complex shapes, although their origins remain unclear

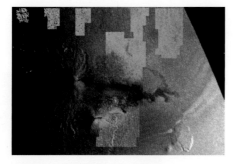

Plate Ji8.4.2-N Zamama: NIMS Thermal Map
Encounter: *Galileo* I24   Resolution: 4.9 kilometers/pixel

Although incomplete, this map indicates that the upper parts of the long dark flow of Zamama are warm. Minimum temperatures are estimated to be 1100 K.

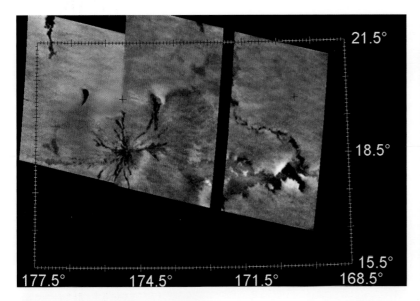
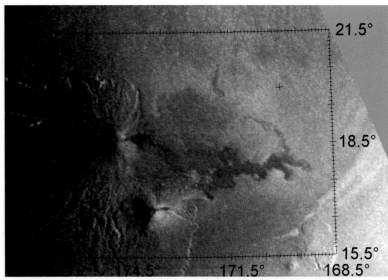
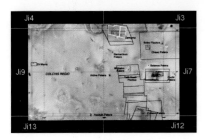
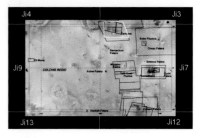

Plate Ji8.4.1a Zamama: High-Sun View
Encounter: *Galileo* I24   Resolution: 380 meters/pixel
Orthographic Projection   Map Scale: 1 cm = 32.1 km

This descrambled I24 mosaic showing Zamama in regional context is of low quality, but reveals that Zamama is comprised of a radial field of dark narrow lava flows similar to those associated with conical shield volcanoes on Earth. The large dark flow extending to the east emanates from the center of this flow field and is the source of observed thermal and plume activity. It is interesting to note that there is no trace of this long dark flow in the 1979 *Voyager* images. Faint dark squiggly patterns in the eastern section may be eroded or partly buried older lava flows. The mosaic is part of a series that extends west to Isum Patera.

Plate Ji8.4.1b Zamama: Low-Sun View
Encounter: *Galileo* I32   Resolution: 340 meters/pixel
Orthographic Projection   Map Scale: 1 cm = 32.1 km

This low-Sun view of Zamama confirms that the radial lava flows are centered on a 60-kilometer-wide shield volcano (see also Figure 5.4.3). A second similar volcano lies just south of Zamama, although it does not appear to have been active recently and its lava flows have faded. Both volcanoes are approximately 2 kilometers high. Few of Io's hundreds of volcanoes exceed 2 kilometers in elevation; these two examples are the most Earth-like of Io's volcanoes, at least in terms of what humans think of when they hear the word "volcano." The great Mauna Loa volcano (9 kilometers), and the Mars volcanoes of Tharsis (9 to 27 kilometers) tower much higher and are much larger by volume. Low volcano volumes on Io could be explained if deep magma sources find alternative outlets after some limited period of time, or if magma has difficulty in general in reaching the surface. A much smaller similar volcano can be seen at 15.3°N, 169.7°W.

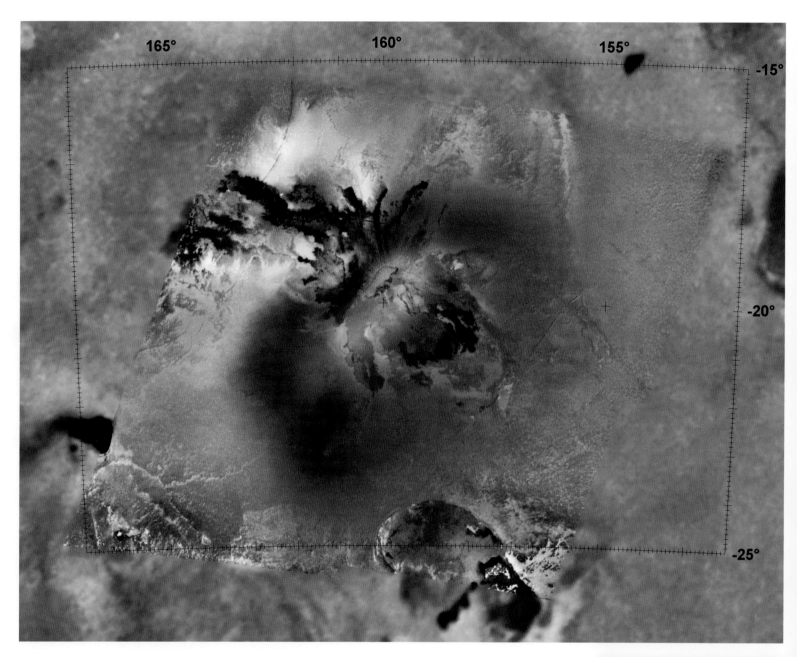

Plate Ji8.5.1a Culann Patera

Encounter: *Galileo* I25 Resolution: 205 meters/pixel
Orthographic Projection Map Scale: 1 cm = 26.1 km

Culann Patera features reddish and greenish flow units and plume deposits. These colors are associated elsewhere on Io with recent and ongoing volcanic activity and NIMS observed a persistent hotspot and occasional plume at Culann (see Plate Ji8.7.2-N and Figure 5.4.4). The green color may be due to the presence of olivine or sulfur. The volcanic flows seen here are centered on an oblong central patera. No relief is evident in the low-Sun images shown Plate Ji8.7, indicating that Culann has very low surface slopes and no conical peak similar to Zamama (Plate Ji8.4.1b). Eruption temperatures are estimated at 1000 K but could have exceeded 1200 K.

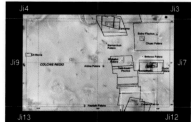

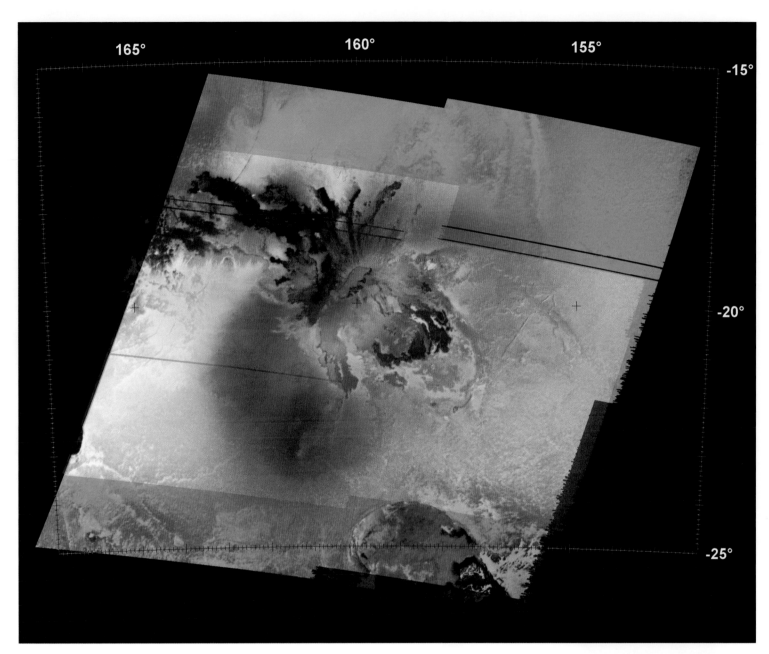

Plate Ji8.5.1b **Culann Patera: Color**
Encounter: *Galileo* I25   Resolution: 205 meters/pixel
Orthographic Projection   Map Scale: 1 cm = 26.1 km

These images are from the same sequence as shown in Plate Ji8.5.1a, except that here only the I25 partial color mosaic is used rather than the lower-resolution global color mosaic.

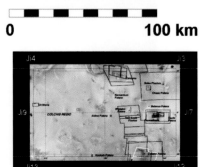

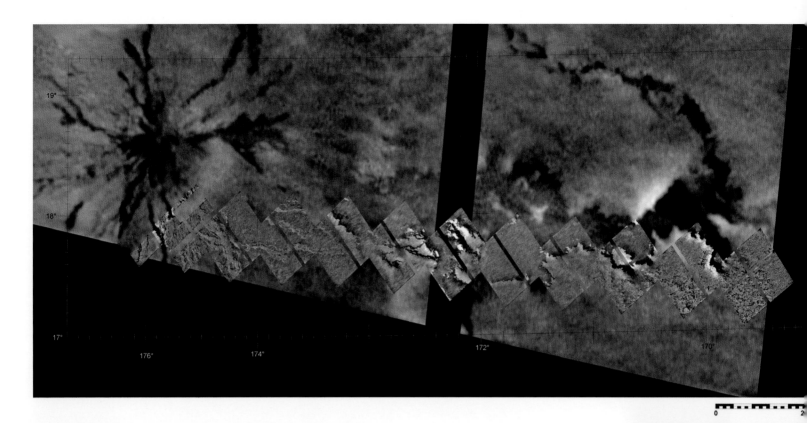

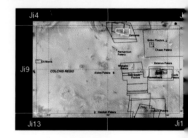

Plate Ji8.4.3 Zamama: High-Sun, High-resolution View
Encounter: *Galileo* I24   Resolution: 40 meters/pixel
Orthographic Projection   Map Scale: 1 cm = 10 km

Like Prometheus, the Zamama plume does not originate from the main vent, but instead from the edges of lava flows where they interact with SO_2-frost-rich surrounding terrains. This descrambled high-Sun I24 view shows the dark lava flows clearly, but also sets of shorter dark and bright radial flows close to the main vent at the western end of the mosaic. This view shows the high-resolution mosaic together with a low-resolution context view. Mosaic shown at 45% to fit page.

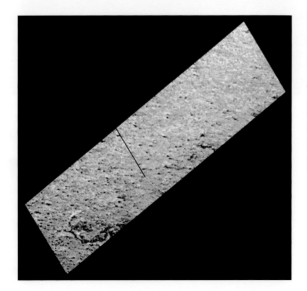

Plate Ji8.4.4 Zamama: High-Sun, Highest-resolution View
Encounter: *Galileo* I24   Resolution: 20 meters/pixel
Orthographic Projection   Map Scale: 1 cm = 10 km

This fragmentary image was the last in the series shown in Plate Ji8.4.3. It was returned in an unscrambled mode. The frame is located beyond the eastern end of the mosaic, in a region of ridged plains. The dark sinuous feature is of unknown origin, although it may be an old partially buried lava flow.

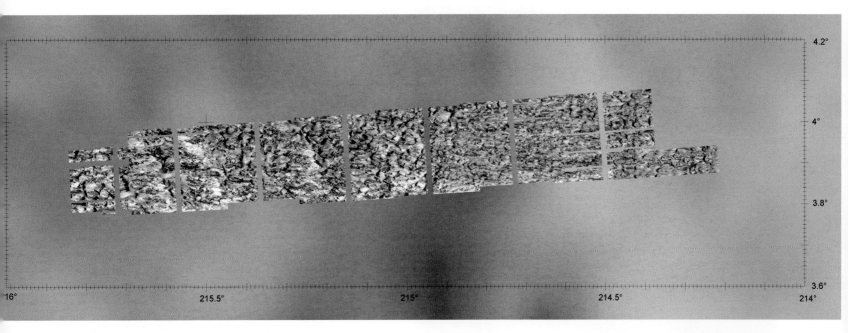

Plate Ji8.6.1 **Ot Mons**
Encounter: *Galileo* I24   Resolution: 18 meters/pixel
Orthographic Projection   Map Scale: 1 cm = 3.9 km

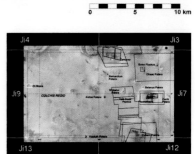

This is the highest-resolution view of Io's mountains we have from *Galileo*. Context imaging does not exist and the exact location of these unscrambled I24 images on Ot Mons is rather uncertain (although the mosaic spans roughly half the width of the mountain). The lack of context and the high-Sun illumination combine to make interpretation difficult. The odd bright and dark patterns are consistent with down-slope creep of material, but other interpretations are possible. We see no direct evidence for layered rock units in this mosaic, expect perhaps in subtle shading differences from east to west. Compare this view with the low-Sun views at lower resolution on Tohil Mons (Plate Ji12.3). Ot Mons rises only 3.5 kilometers above the surface and appears to have been heavily modified by mass wasting.

Plate Ji8.6.2 **Ot Mons: Highest Resolution**
Encounter: *Galileo* I24   Resolution: 9 meters/pixel
Orthographic Projection   Map Scale: 1 cm = 2.4 km

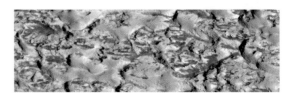

The westernmost partial frame of the Ot Mons mosaic was returned unscrambled. More clearly seen in this small view are the numerous oblong depressions, some enclosed, some forming open-ended amphitheatres similar to those formed by landslides. Dark material appears to coat some of the ridge crests between depressions. Some of the larger depressions appear to be partly filled with a smooth material. These features may be consistent with erosion or collapse of the mountain's flanks, although the lack of context and stereo imaging makes any interpretation suspect.

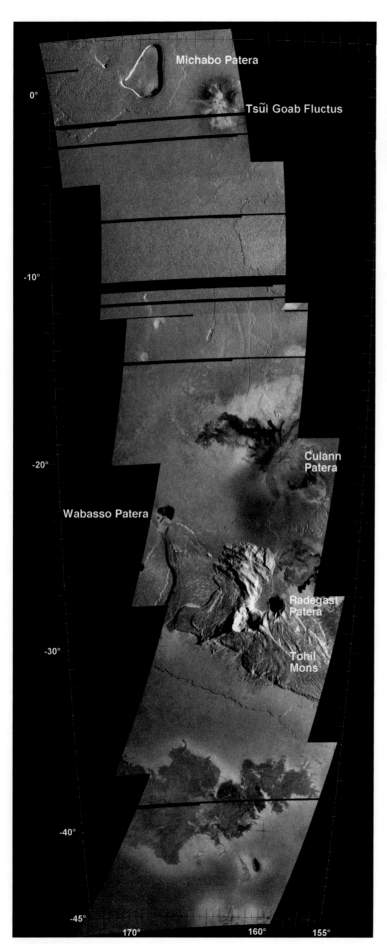

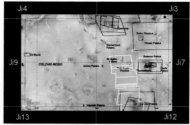

Plate Ji8.7.1 From Michabo Patera to Tohil Mons: Regional Mapping Mosaic

Encounter: *Galileo* I32   Resolution: 325 meters/pixel
Orthographic Projection   Map Scale: 1 cm = 64 km

This mapping mosaic along the sunrise terminator includes Tohil Mons (Plate Ji12.3), Culann Patera (Plate Ji8.5), as well as the large caldera, Michabo Patera, a bright and dark radial flow complex and small shield volcano near 0°N, 163°W, and a variety of pits, lineations and scarps within the intervening volcanic plains. Michabo Patera is 1.4 kilometers deep and surrounded by a low mesa ~100 meters thick. Whether the layering here and in the volcanic plains to the south (especially at 8°S, 160°W) is due to erosion or is evidence of massive sheet lavas is unclear at these resolutions.

Plate Ji8.7.2-N From Michabo Patera to Tohil Mons: NIMS Thermal Map

Encounter: *Galileo* I32   Resolution: 8.0 kilometers/pixel

NIMS obtained infrared data along with the imaging swath. These data do not cover the entire mosaic, and only show thermal activity at Culann Patera itself (Plate Ji8.5).

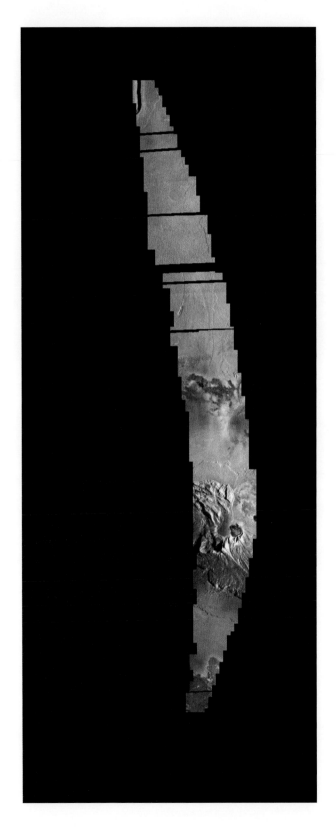

Mercator Projection
Map Scale at Equator: 1 cm = 121 km

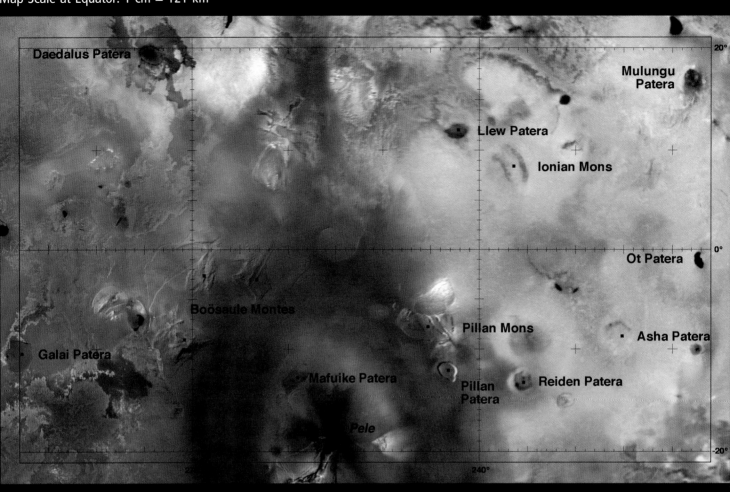

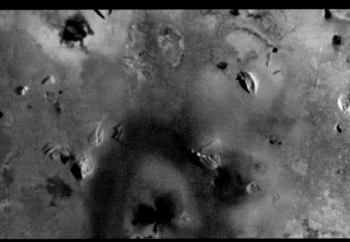

Plate Ji9/Ji9-m Pele Quadrangle

This is a region of superlatives. It features the largest volcanic feature and the tallest mountain on Io. The large reddish ring is fallout of the Pele volcanic plume, observed with minor variations by HST and *Galileo* since 1979. Despite the vast size of the eruption deposit, the volcanic center is not a towering volcano but rather a narrow caldera containing an active lava lake. This region was also the site of a spectacular eruption in 1997 at nearby Pillan (Plate Ji9.2; this map shows the surface as of 1979), a site observed by *Galileo*. Extensive lava flows of different colors and ages cover the southwestern portion of the scene. The small patera just west of Boösaule Montes has a distinctly yellowish color. Daedalus Patera is a prominent hotspot and changes have been detected here, but *Galileo* did not get a close look.

Although it does not look imposing here, due to the geometry of *Voyager*'s arrival, the southern edifice of Boösaule Montes (at 10°S, 271°W) towers 17 kilometers above the plains and is among the tallest known structures in the Solar System. The northern plateau rises 8 kilometers, while the northeastern part rises 7 kilometers. Pillan Mons may be a source for the Pillan eruption.

Io | 317

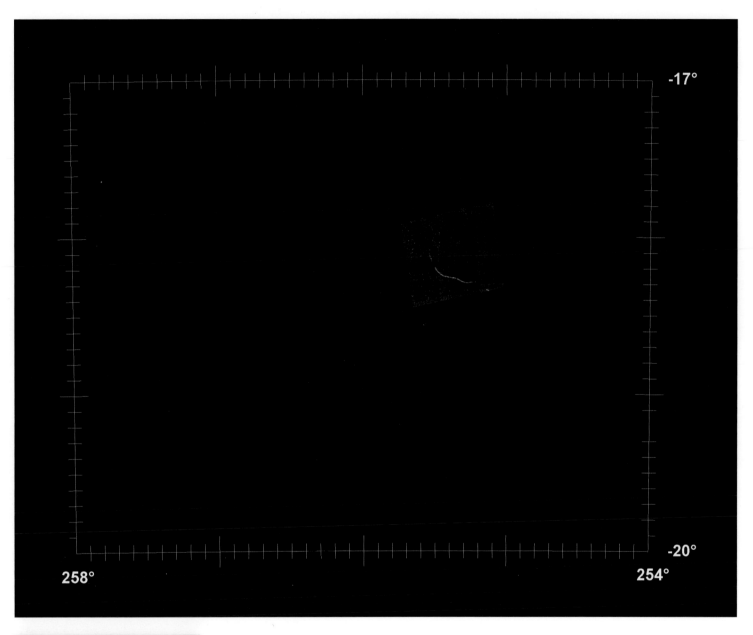

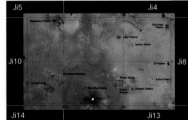

Plate Ji9.1.1a
Encounter: *Galileo* I24   Resolution: 28 meters/pixel

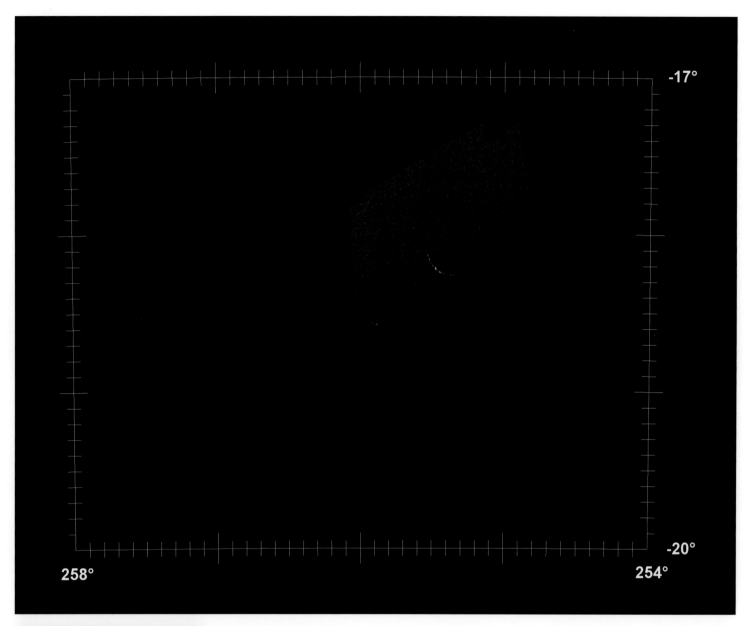
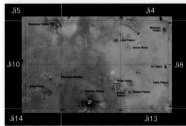

Plate Ji9.1.1b
Encounter: *Galileo* I27   Resolution: 18 meters/pixel

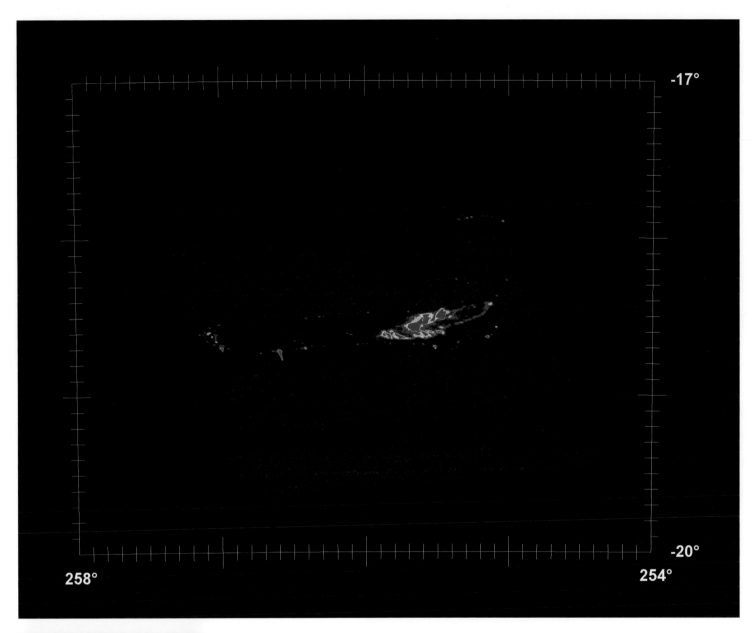
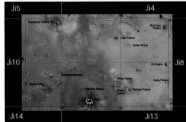

Plate Ji9.1.1c
Encounter: *Galileo* I32   Resolution: 65 meters/pixel

Plate Ji9.1.1 **Pele at Night**
Orthographic Projection   Map Scale: 1 cm = 14.2 km

Pele was in darkness during each of *Galileo*'s successful close encounters until late in the mission, when spacecraft anomalies and budgetary decisions scuttled imaging plans. Nonetheless, both SSI and NIMS observed hot glowing lava at Pele in the cold Ionian night (and once in daylight at very low resolution) on three orbits. The best view, showing the entire caldera in one view, was obtained in I32 (the image here is an average of three images made to reduce compression noise). A thin beaded ribbon of glowing hot lava traces the oblong outer edges of a lava lake. Hotspots along this trace are intense enough to saturate the CCD. Inside this narrow outer ring lies a smaller but more intense oblong loop of glowing lava.

The smaller I24 and I27 views both feature a single small section of a narrow ribbon of bright exposed lava. This feature is a good match to a faint ribbon of glowing lava just north of the hot central region in the I32 image, within a northern extension of the patera. (An alternative fit would be to the narrow ribbons of lava at the far western end of the Pele feature.) Narrow threads of glowing lava are common on Earth in molten lava lakes confined to calderas, where the edge of the chilled lava crust grinds against the caldera wall, causing small chunks to founder or break away. The intensely glowing center section is likely an area of lava fountains and/or foundering of the crust in the center of the lake. The giant reddish Pele plume originates from either the center fire fountains or the edge of the lava lake.

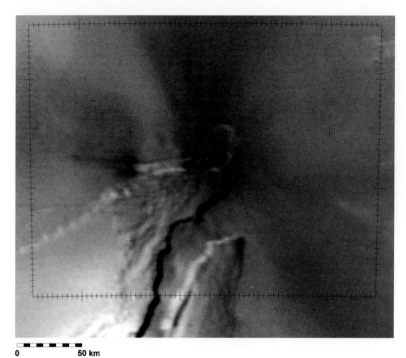

Plate Ji9.1.3 **Pele in Daylight**
Encounter: *Voyager* 1   Resolution: 630 meters/pixel
Encounter: *Galileo* I32   Resolution: 65 meters/pixel
Orthographic Projection   Map Scale: 1 cm = 51 km

The best views of Pele in daylight were obtained by *Voyager* in 1979 at ~0.6 km resolution. Although *Galileo* did observe glowing lava at Pele in daylight, this was at very low resolution. In the absence of resolved daylight imaging, I combine here the best *Galileo* night view (shown in bright red in this rendition) with the daylight *Voyager* view. The pattern of hot glowing lava matches surprisingly well the lower-resolution *Voyager* view of the Pele caldera, which forms a golf-club-shaped walled depression on the northern edge of Danube Planum, which rise 2 to 3 kilometers high. There is also a strong correspondence between the inner area of hot fountaining lava and the dark feature in the eastern part of the caldera. These similarities indicate that while the Pele eruption has been nearly continuous since 1979, no major changes occurred over the 20-year period, consistent with the notion of an active lava lake confined within a walled patera.

Plate Ji9.1.2-N Pele: NIMS Thermal Maps

Encounter: *Voyager* 1   Resolution: 0.65 kilometers/pixel
Encounter: *Galileo* I27   Resolution: 0.6 kilometers/pixel
Encounter: *Galileo* I31   Resolution: 1.3 kilometers/pixel
Encounter: *Galileo* I32   Resolution: 0.9 kilometers/pixel

Estimated eruption temperatures at Pele range from 1400K to 1750K and probably correspond to the hot fountaining lava, likely basaltic or ultramafic in composition. No lateral lava flows are associated with Pele. The distribution of thermal sources is relatively consistent from encounter to encounter, although there appear to be some variations in the distribution of hotspots along the outer edge of Pele.

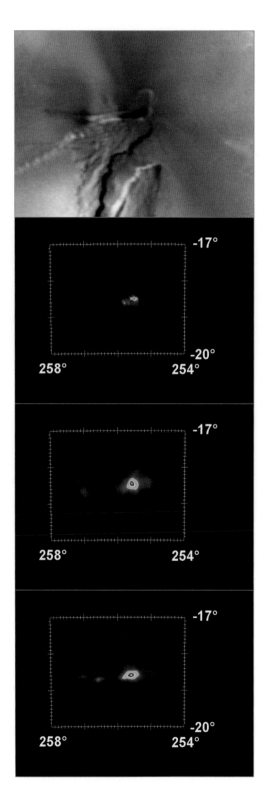

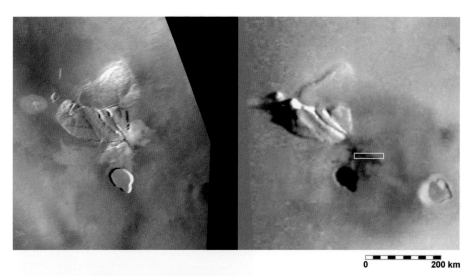

Plate Ji9.2.1 Pillan: Regional View
Encounter: *Voyager* 1   Resolution: 1000 meters/pixel
Encounter: *Galileo* E14   Resolution: 2600 meters/pixel
Orthographic Projection   Map Scale: 1 cm = 94.5 km

The first great eruption witnessed by *Galileo* occurred in Spring 1997, when a large new diffuse dark spot 450 kilometers across was observed at low resolution at Pillan. Pillan is the type example of pillanian eruption, noted for relatively short but intense eruptive phases, with high heat and lava output. Plumes may be associated, and one was observed briefly at Pillan. At its heart was a new large dark flow complex. These two views show the regional relief and geologic context of the Pillan flow field. The Sun is to the left in the *Voyager* view (left) and to the right in the *Galileo* view (right). The white rectangle marks the preferred location of the high-resolution mosaic (Plate Ji9.2.2). The area is dominated by the 85- by 175-kilometer-wide mountain at center. This deeply fractured mountain rises ~5 kilometers above the plains. A low tongue-shaped deposit a few hundred meters thick extends from the northern face of the mountain. The dark flows appear to have formed at or near a fissure at the southern end of the mountain and flowed toward and into the 1- to 2-kilometer-deep caldera, 65 kilometers to the south. Sadly, the cliff over which the lavas poured into the caldera were not imaged at high resolution. Active or not, these lava falls would have been some of the most amazing images returned from space and may have resembled the spectacular lava falls at Kilauea in Hawai'i in 1969. These new flows also covered the floor of the caldera, turning it dark.

0 25 km

Plate Ji9.2.2 Pillan: Lava Fields, High Resolution

Encounter: *Galileo* I24 Resolution: 18 meters/pixel
Orthographic Projection Map Scale: 1 cm = 4.5 km

The new flows at Pillan were an obvious imaging target during the first close pass of Io in October 1999. These are among the highest-resolution views we have of Ionian lava flows. Despite the I24 camera anomaly, considerable detail is apparent in the restored images. We appear to be viewing part of a large contiguous flow sheet. This is not surprising since most of the Pillan flows appear to have been emplaced over a period of a few months, if not much shorter. These flows contrast with the Prometheus flows, shown at similar resolution in Plate Ji8.1.5, which have been emplaced much more slowly. Many of the features seen here, including large pits, rounded depressions, flow margins, and wide channels with raised rims (probably flow levees) are commonly observed on basaltic lava flow fields here on Earth and on Mars. Shadows on flow margins indicate that the thickest flows are 8 to 10 meters thick.

The best context imaging for this site is high-Sun imaging at only 1.3-kilometer resolution (also obtained during I24). An educated guess suggests that the mosaic lies north of the center of the flow field as shown here. An alternate guess puts the channel feature at longitude 242.1°W, about one frame-width south and west of the location shown here. The transition from smoother and darker material to more rugged material at longitude 241°W appears to correlate with a change in brightness in the available context imaging and may be the boundary of the new dark flow units. See also Plate Ji9.2.4-N.

Plate Ji9.2.3 Pillan: Lava Fields, Highest Resolution
Encounter: *Galileo* I24   Resolution: 9 meters/pixel
Orthographic Projection   Map Scale: 1 cm = 1.1 km

The final image in the Pillan sequence (Plate Ji9.2.2) is a quarter frame returned intact in a nonscrambled imaging mode. It is at full resolution and shows a relatively smooth surface, with some areas of pitting, scoring, and several shallow irregular depressions. There are no obvious leveed channels or volcanic pits in this image, however.

Plate Ji9.2.4-N Pillan: NIMS Thermal Map
Encounter: *Galileo* I24   Resolution: 0.2 kilometers/pixel

A simultaneous NIMS track was obtained during the high-resolution Pillan imaging mosaic (Plate Ji9.2.2). Although incomplete, it clearly shows that the dark, smoother deposit in the middle of the track is warmer than the more rugged materials to the east. The hottest part of the flow (observed) was near the vicinity of the leveed channels and pits. The precise location of the NIMS swath with respect to the imaging mosaic is uncertain to within half an image frame. Minimum eruption temperatures were estimated at 1870 K (!) but have been recently revised downward to 1600 K. These values are consistent with basaltic or ultramafic lavas, depending on eruption conditions. Pillan's color and albedo suggest that the lavas may include magnesium-rich orthopyroxenes, a very common mafic lava mineral (another being olivine). The full composition eludes us, however.

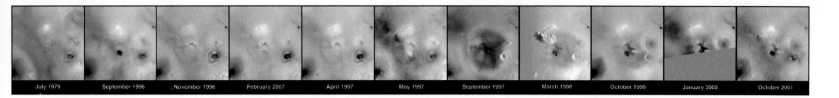

Plate Ji9.2.5 **Changes at Pillan**

From left to right:

Encounter: *Voyager* 1	Resolution: 3000 meters/pixel
Encounter: *Galileo* G2	Resolution: 4940 meters/pixel
Encounter: *Galileo* C3	Resolution: 3000 meters/pixel
Encounter: *Galileo* E6	Resolution: 4100 meters/pixel
Encounter: *Galileo* G7	Resolution: 6100 meters/pixel
Encounter: *Galileo* G8	Resolution: 10500 meters/pixel
Encounter: *Galileo* C10	Resolution: 5100 meters/pixel
Encounter: *Galileo* E14	Resolution: 2600 meters/pixel
Encounter: *Galileo* I24	Resolution: 1490 meters/pixel
Encounter: *Galileo* E26	Resolution: 3450 meters/pixel
Encounter: *Galileo* I32	Resolution: 5070 meters/pixel

Like Tvashtar (Plate Ji3.2), *Galileo* observed dramatic changes at Pillan. The big change occurred between orbit G8 and C10, when a new dark lava flow and dark plume ring appeared. The eruption had apparently begun by May 1997 and was in full vigor one month later during C10. Although the dark lava flow was fully formed, only two small lobes had entered the caldera floor to the south at this time. The caldera floor was not completely covered by dark flows until E14. Post-eruption changes include a general fading of the dark plume ring.

Plate Ji9.3-N **Pele and Pillan: SSI Eclipse Thermal Map**
Encounter: *Galileo* E11   Resolution: 13.8 kilometers/pixel

The *Galileo* imaging camera was also sensitive to very high temperatures (although it could not detect the fainter hotspots better seen in NIMS). These hotspots were best seen when Io was eclipsed by Jupiter, darkening the surface. Here, Pele and Pillan (to the right) glow intensely in the Ionian darkness. Marduk glows faintly to the east of Pele and Pillan. The glow along the western edge of Io is a plume over Acala Fluctus. Red represents hot in this view. Io glows faintly in this image, perhaps due to residual surface heat. A small faint ring near the north pole could be an Ionian aurora (which has been documented in other eclipse images).

Mercator Projection
Map Scale at Equator: 1 cm = 121 km

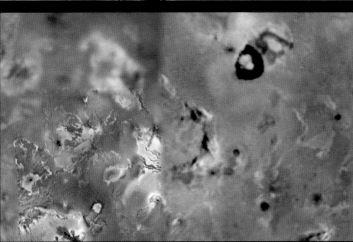

Plate Ji10/Ji10-m Loki Quadrangle

The giant dark horseshoe is Loki Patera, site of what may be a gigantic lava lake. The other great volcano is Ra Patera. Despite extending 250 kilometers, Ra Patera itself rises less than 2 kilometers high. The narrow radial reddish lava flows, like those at Emakong Patera, are considered to be candidate sulfur flows. Despite observed activity here, eruption temperatures have never been very high. The views shown here are those of *Voyager* in 1979, due to their much higher resolution. Ra Patera has seen extensive changes since *Voyager* days (Plate Ji10.2.2). Surprisingly few changes have been observed at Loki despite the high level of activity. Acala Fluctus also became a plume site during the *Galileo* mission. The bright sulfur dioxide deposits that characterize equatorial regions elsewhere are limited here to the area of Ra Patera.

Voyager's noontime views used here reveal little in the way of relief. Topographic features in this region are revealed in *Voyager* stereo images and a few low-Sun *Galileo* views. The ridged material immediately east of Nyambe Patera (at 2°N, 341°W) is a mountain ~7 kilometers high, easily visible in *Voyager* stereo images. Flanking Ra Patera and extending in an arc to the northeast toward Carancho Patera is an arcuate mountain range of varying height (Plate Ji10.1.2). The circular mesas at Inachus and Apis Tholus are related to volcanism in some way: both are ~2 kilometers high and have 1- to 3-kilometer-deep volcanic calderas at their center.

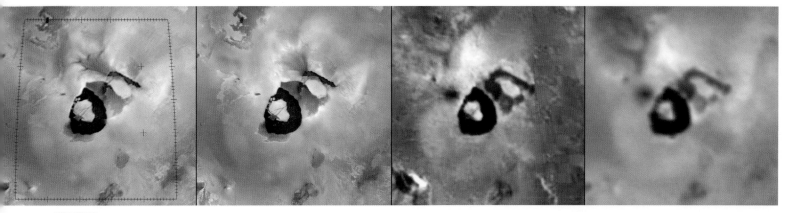

Plate Ji10.1.1 **Changes at Loki**

Encounter: *Voyager* 1	Resolution: 2900 meters/pixel
Encounter: *Voyager* 1	Resolution: 1150 meters/pixel
Encounter: *Galileo* G7	Resolution: 6100 meters/pixel
Encounter: *Galileo* C22	Resolution: 9550 meters/pixel
Orthographic Projection	Map Scale: 1 cm = 181 km

Our best views remains that from *Voyager* in 1979, limited to ~1 kilometer resolution. Loki consists of three major parts: a dark horseshoe patera, a less dark polygonal patch extending from the northeast margin, and a dark gash-like feature to the northeast. The horseshoe encloses a bright polygonal feature that resembles an island and is divided by several narrow fractures. The "island" has remained unchanged during the past 28 years and is likely a fixed feature. Numerous kilometer-wide bright spots of unknown origin are scattered across the dark floor. The dark horseshoe has been interpreted as an overturning lava lake or a caldera floor subject to repeated flooding by lavas.

Voyager observed a large volcanic plume rising above the western end of the dark fissure and a smaller plume from the eastern end. These dark plumes can be seen changing in the two *Voyager* images (at left), but this is due almost entirely to the changing viewing geometry as *Voyager* passed Io in March 1979. A peculiar feature of the first *Voyager* view is the bright northwest edge of the dark deposit. This suggests the presence of a low scarp, given the viewing angle of the spacecraft from the southeast (see also Plate Ji10.1.2). Earth-based and *Galileo* observations indicate that most of Loki's heat output was from the dark horseshoe, however.

The horseshoe itself appears little changed between 1979 and 2001, although repeated eruptions have been monitored from Earth within its borders. Dark material, on the other hand, appears to have been extruded from both ends of the dark gash, as shown in the two *Galileo* images (at right). Some of these dark materials have flowed into the polygon north of the horseshoe. Some of these dark flows may have entered Loki Patera proper. Most of these changes took place before *Galileo*'s arrival in 1996, with little obvious change occurring since. Diffuse new dark material is also centered on the small dark patera due west of the horseshoe, further evidence this is an active volcanic center. There is no evidence for dark plume jets in *Galileo* images, suggesting that the dark gash has been quiet.

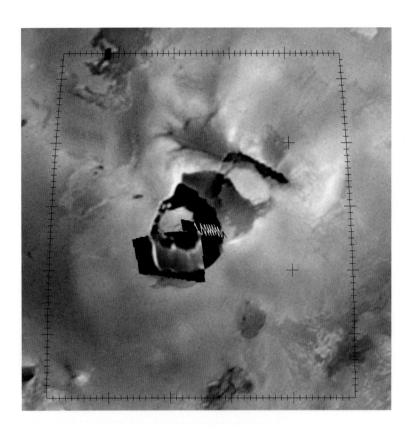

Plate Ji10.1.3-N **Loki: NIMS Thermal Maps**
Encounter: *Galileo* I24   Resolution: 0.8 kilometers/pixel
Encounter: *Galileo* I32   Resolution: 2.3 kilometers/pixel

NIMS acquired two thermal views of Loki. Both were at night and neither form a complete map. Warm areas in the thermal maps correspond closely to the dark patera-floor material in the visible images. *Voyager* saw no evidence for topography in the central bright "island," but NIMS saw hot material faintly glowing in one of these narrow fractures during its I24 sweep across central Loki.

The darkest part of the horseshoe itself, as viewed by *Voyager*, is a narrow section along the southwest wall of the patera. The infrared view of the southern half of Loki during I32 similarly shows an enhanced heat flow from this southwestern wall, glowing white in this rendition. The remainder of the patera floor is also warm. Unfortunately, the northern portion of the Loki region was not observed. Temperatures during these observations were 260–450 K over most of the dark floor, rising to 450–550 K along the dark southwestern margin. Loki's thermal output has been variable over timescales of weeks or days and temperatures in excess of 950 K have been detected at other times. The *Galileo* photopolarimeter also recorded that the hottest parts of the patera floor shifted from the southwest to the southeast quadrant.

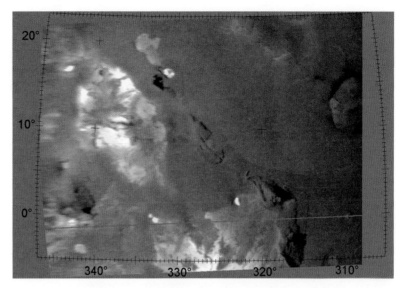

Plate Ji10.1.2(G) **Loki: Best *Galileo* View**
Encounter: *Galileo* I32   Resolution: 1150 meters/pixel

Plate Ji10.1.2(V) **Loki: *Voyager* Context View**
Encounter: *Voyager* 1   Resolution: 1150 meters/pixel
Orthographic Projection   Map Scale: 1 cm = 137 km

Loki is one of the largest volcanic features and probably the most intense volcanic site on Io, releasing at least 10% to ~25% of the total heat radiated from the surface at any given time. Despite high interest, *Galileo*'s only "high"-resolution view of Loki (G) shows the western portion of Loki at sunset. This image was planned to search for evidence of topography along the edge of Loki Patera in the form of shadows. The image does indeed show a long shadow but only along the northern half of Loki Patera (15°N, 310°W). The shadow length indicates a scarp height of roughly 2.5 kilometers, decreasing to near zero to the south. This shadow also correlates with the bright scarp in one of the *Voyager* images (Plate Ji10.1.1) and with a low shield volcano to the northwest (centered on the dark spot at 16°N, 313°W). Evidently, Loki Patera cut into the outer flank of this volcano. The southwestern edge of Loki has no apparent relief, even though this is where thermal activity appears to have been most intense (see Plate J10.1.3-N).

The floor of Loki Patera itself is bright in the *Galileo* view (G), in contrast to the very dark color when viewed at midday (V). This may be consistent with a smooth glassy lava surface. Dark areas elsewhere in the *Galileo* view are mountain shadows, including Carancho Montes (longitude 316°W), which rises 7 to 8 kilometers. Other mountains are located at 15°N, 332°W and 1°N, 342°W, and a dissected mesa at 9°N, 327°W.

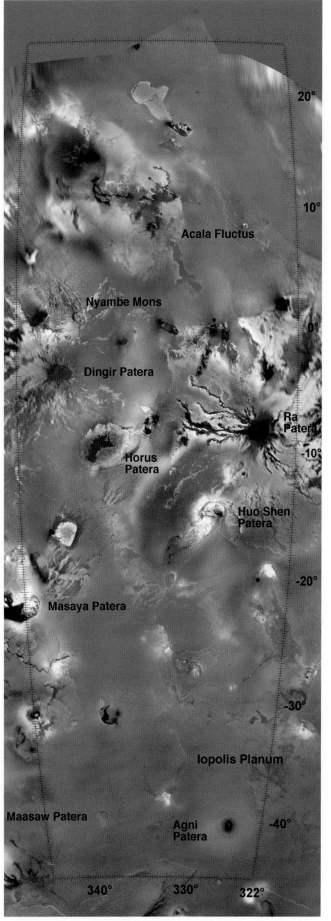

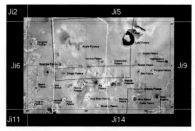

Plate Ji10.2.1 **From Maasaw Patera to Ra Patera: Regional Mapping Mosaic**

Encounter: *Voyager* 1   Resolution: 570 meters/pixel
Orthographic Projection   Map Scale: 1 cm = 94.5 km

Voyager 1 acquired this 5-frame mosaic extending north–south along the 0° prime meridian (the sub-Jovian longitude). The Ra Patera volcanic center dominates the scene. The long narrow flows extending from the center of Ra may be composed of sulfur, although this has not been confirmed. Ra Patera was the site of a major eruption sometime between 1994 and 1995, but is shown here as in 1979 (Plate Ji10.2.2). Acala Fluctus has been a major recurring eruption site as well. Iopolis Planum and Nyambe Mons are the prominent mountains standing 3 to 7 kilometers high. Maasaw Patera was seen at high resolution by *Voyager* (Plate Ji14.1). The dark deposit extending west from Masaya Patera lies on top of a 1- to 2-kilometer-high plateau, indicating that it was an ash deposit.

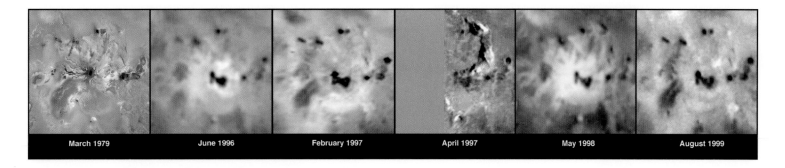

March 1979 | June 1996 | February 1997 | April 1997 | May 1998 | August 1999

Plate Ji10.2.2 **Changes at Ra Patera**

Encounter: *Voyager* 1   Resolution: 1000 meters/pixel
Encounter: *Galileo* G1   Resolution: 15400 meters/pixel
Encounter: *Galileo* E6   Resolution: 5600 meters/pixel
Encounter: *Galileo* G7   Resolution: 6100 meters/pixel
Encounter: *Galileo* E15   Resolution: 12500 meters/pixel
Encounter: *Galileo* C22   Resolution: 9550 meters/pixel

In 1995, John Spencer and colleagues working with Hubble Space Telescope images discovered that the Ra Patera region had brightened considerably between 3/1994 and 3/1995, indicating an eruption had occurred. Thermal outbursts were observed from Earth in March and September of 1995. When *Galileo* first resolved Io in June 1996, roughly 2 years later, it discovered a large new dark lava flow south of the Carancho plateau and surrounded by an extended zone of diffuse bright material. *Galileo* also observed a volcanic plume, indicating that the eruption was still ongoing, although a plume was not subsequently observed and no hotspot was ever detected.

The new lavas flowed south and east, around the large arcuate mountain complex just east of Ra (and visible in the April 1997 image). Much of the *Voyager*-era lava flow field was covered or obliterated by the new eruption. During *Galileo*'s mission, the old *Voyager*-era features began to reemerge through the plume fallout. The bright plume deposits formed in 1994 were evidently rather thin and ephemeral, and fade with time. Planned higher-resolution late-mission *Galileo* images were never acquired due to safing and budgetary cancellations.

Lambert Conformal Conic Projection
Map Scale: 1 cm = 124 km

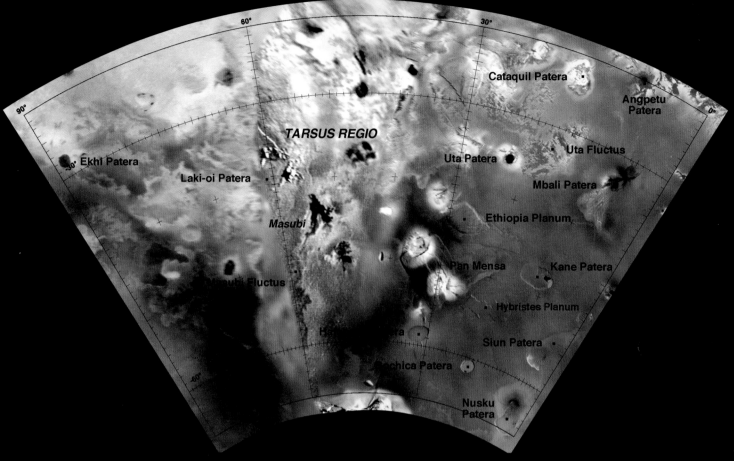

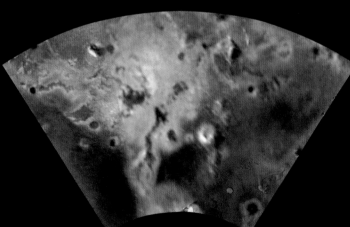

Plate Ji11/Ji11-m Masubi Quadrangle

The extensive large whitish and grayish deposits of Tarsus Regio dominate this striking area. The diffuse whitish deposits are probably sulfur dioxide or sulfur allotropes that escaped or were driven off during volcanic eruptions. At the heart of Tarsus lies the Masubi volcanic complex. This area has been the site of intermittent volcanic activity, and has been a frequent plume site since *Voyager* days. All attempts to image the area at high resolution with *Galileo* were unsuccessful, although some indications of new lava flows were seen in long-range images.

The Pan Mensa plateau is unusual, even for Io. The mesa is flanked by two volcanic calderae and may represent a failure of the crust or lithosphere in the zone between them. Fracturing along the crest of the mesa probably formed during uplift in the center of this elongate dome, splitting the rock. Mountains at 28°S, 30°W and 26°S, 68°W were identified by limb images or stereo imaging. In general, mountains are uncommon in this area.

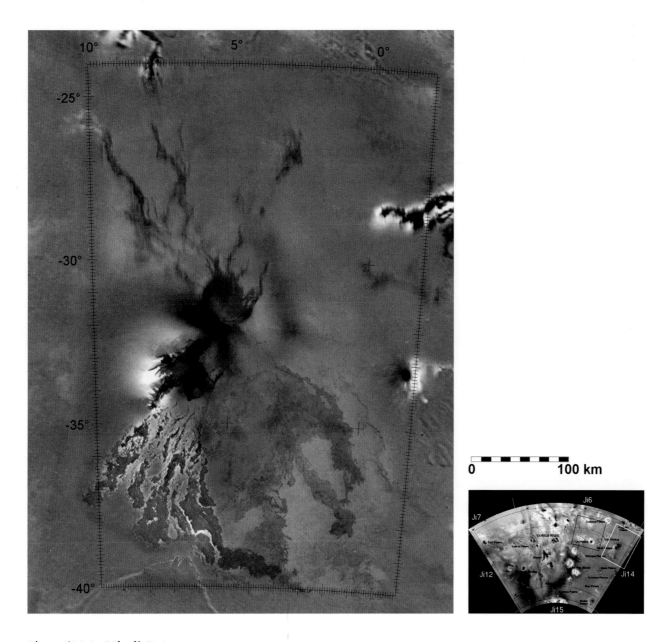

Plate Ji11.1 **Mbali Patera**

Encounter: *Voyager* 1 Resolution: 310 meters/pixel
Orthographic Projection Map Scale: 1 cm = 36.6 km

Mbali is a lava flow field emanating from near the dark circular spot. Little is known about this complex flow field. There is little known topography associated with the flow field, although a small shield volcano could be present. This view is supplemented by *Voyager* color, which lacks infrared sensitivity. No hotspots have been detected here and Mbali is probably dormant or extinct.

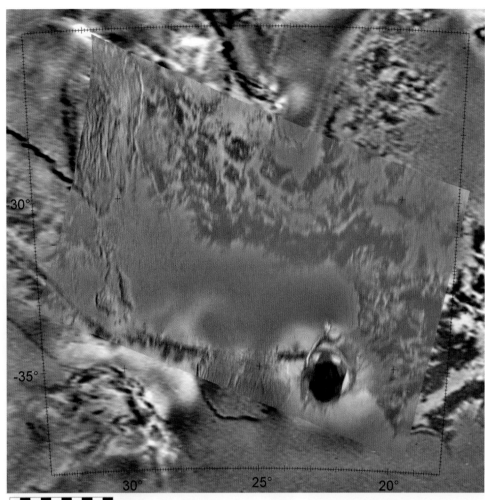
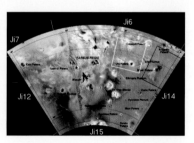

Plate Ji11.2 **Uta Patera**

Encounter: *Voyager* 1   Resolution: 300 meters/pixel
Orthographic Projection   Map Scale: 1 cm = 35.4 km

Voyager acquired this image of Io in 1979, but due to a radiation-induced timing anomaly some of its images were smeared, including this one. A dark arcuate fissure extends to the west of Uta Patera, the dark circular caldera at lower right, past an oval mesa-like plateau.

Lambert Conformal Conic Projection
Map Scale: 1 cm = 143 km

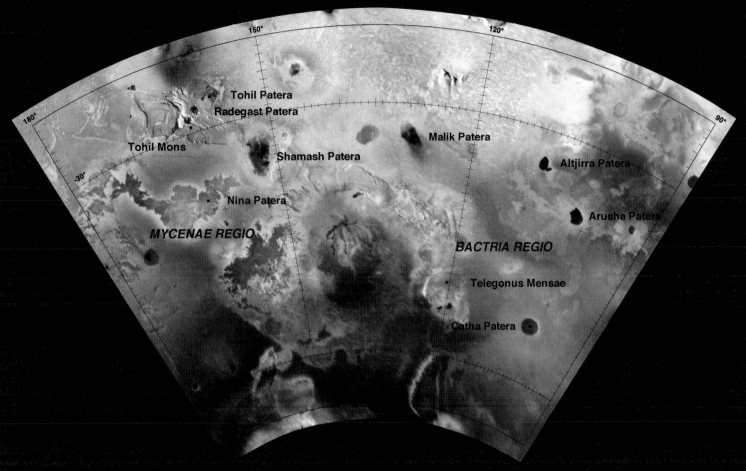

Plate Ji12/Ji12-m Tohil Mons Quadrangle

Tohil Mons is the highest feature of this region topographically, and was observed several times by *Galileo* (Plate Ji12.3). This complex structure towers 8 to 9 kilometers above the plains, and appears to be most rugged part of 1200-kilometer-long linear topographic feature that extends out to the elongate bright frosted plateau at 45°S, 125°W. A variety of lava flow types occur throughout the region, including the extensive dark lava flow in Mycenae Regio, and the apparently bright lava flows surrounding Arusha Patera. Low mesas are common in the south. Bright sulfur dioxide aureoles surround the mesa at Telegonus Mensae (see Plate Ji12.2) and mountains to the north, as well as several lava flows to the west. Reddish deposits to the north and west are the southern parts of the Culann Patera eruption site (see Plate Ji8).

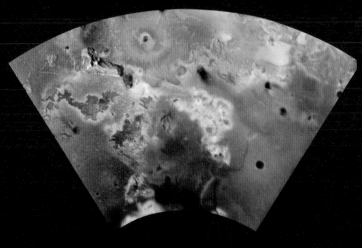

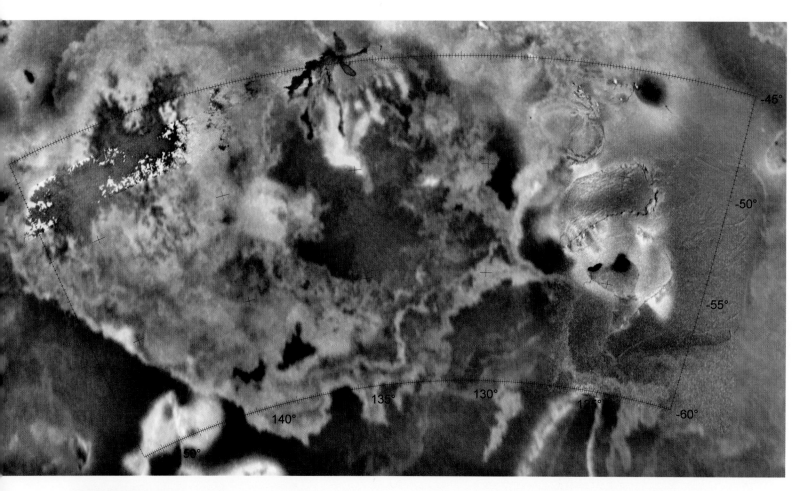

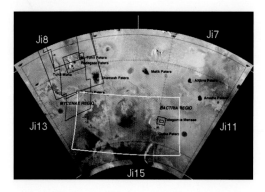

Plate Ji12.1 Southern Latitudes: Regional Mapping Mosaic

Encounter: *Galileo* I27   Resolution: 350 meters/pixel
Orthographic Projection   Map Scale: 1 cm = 55 km

Little of this 4-frame mosaic was returned to Earth. It was intended to allow mapping of the extensive flow fields in this region. The southeast corner of Telegunis Mensae, the large mesa-like plateau at 51°S, 120°W, was targeted for high-resolution observation (Plate J12.2). From a distance, Telegonus Mensae does not appear impressive. It lies on the northern margin of a large multi-colored patera. It was chosen due mostly to the geometry of the I32 flyby in order to investigate erosional processes on Io. Whitish deposits along the margin are probably SO_2 sapping from the mesa scarp.

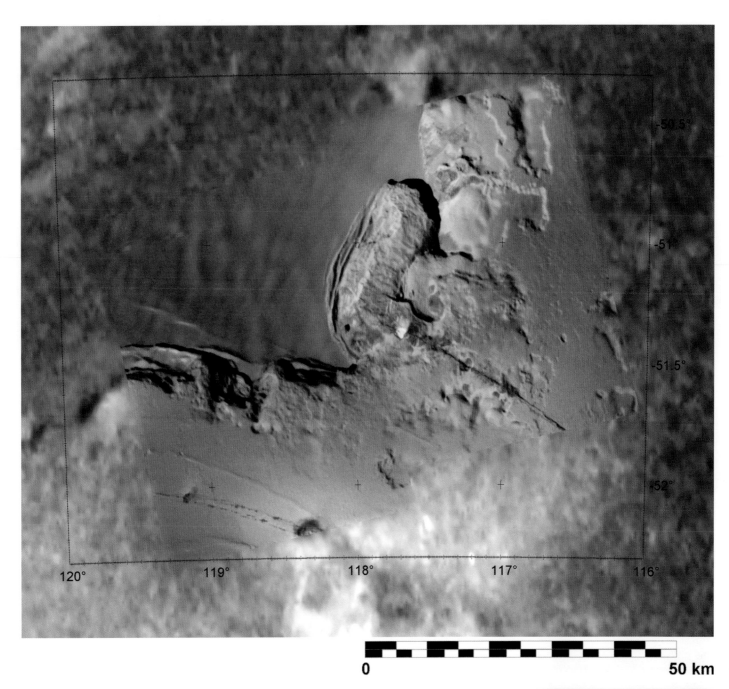

Plate Ji12.2.1 Telegonus Mensae: High Resolution

Encounter: *Galileo* I32   Resolution: 42 meters/pixel
Orthographic Projection   Map Scale: 1 cm = 5.8 km

This high-resolution view of the southeast corner of Telegunis Mensae shows several alcoves that have been cut into the edge of the mesa by mass wasting. Terraces along the edge of the mesa are an indication of incomplete slumping of material as the edge collapses. The cliff face stands ~1.5 kilometers high. The narrow dark linear feature at 51.5°S, 117°W is the focus of the very high resolution view in Plate Ji12.2.2.

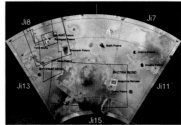

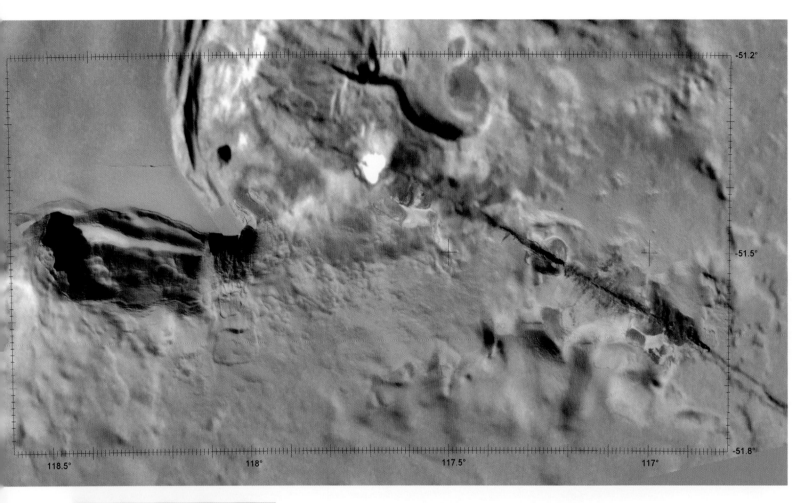

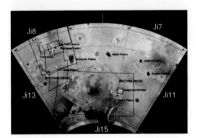

Plate Ji12.2.2 **Telegonus Mensae: Very High Resolution**
Encounter: *Galileo* I32 Resolution: 10 meters/pixel
Orthographic Projection Map Scale: 1 cm = 1.8 km

These are among our highest-resolution images of Io. The linear feature visible in lower-resolution images appears to be a fracture, from which small amounts of dark material have escaped. The images reveal several processes are involved in scarp retreat on Io. Only minor amounts of debris from the retreating scarps remain on the plains. Several large intact blocks, at 51.5°S, 118.3°W, have broken free and moved downhill. A thin deposit covers much of the scarp, including the large blocks. The wrinkled surface of this surface is characteristic of slow down-slope creep of a soil-like layer. Several small landslides have broken from the scarp face and flowed southward onto the plains, forming overlapping lobate deposits at 51.65°S, 118°W.

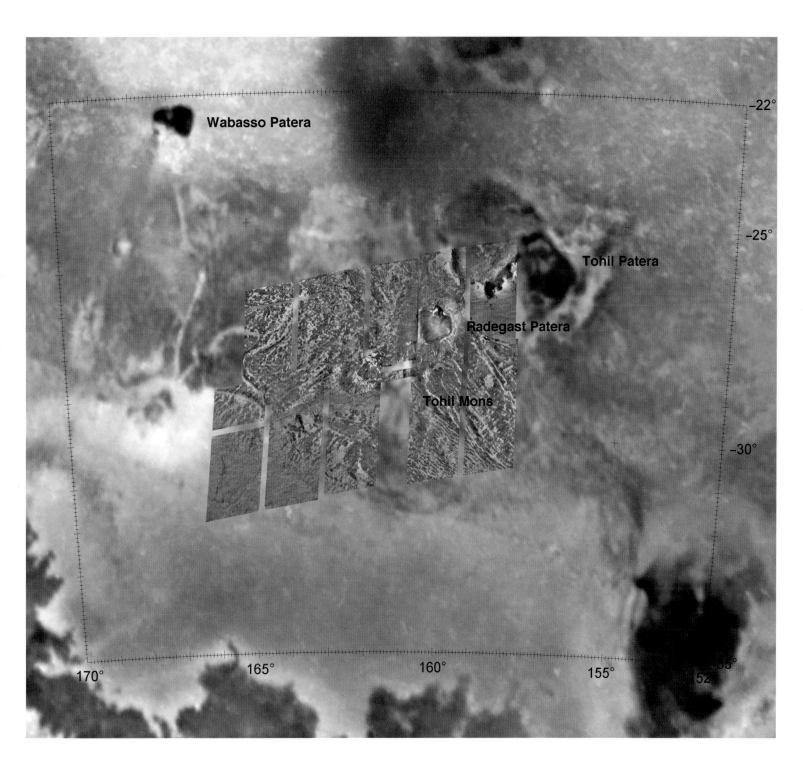

Plate Ji12.3.1a **Tohil Mons: High-Sun View**
Encounter: *Galileo* I24   Resolution: 184 meters/pixel

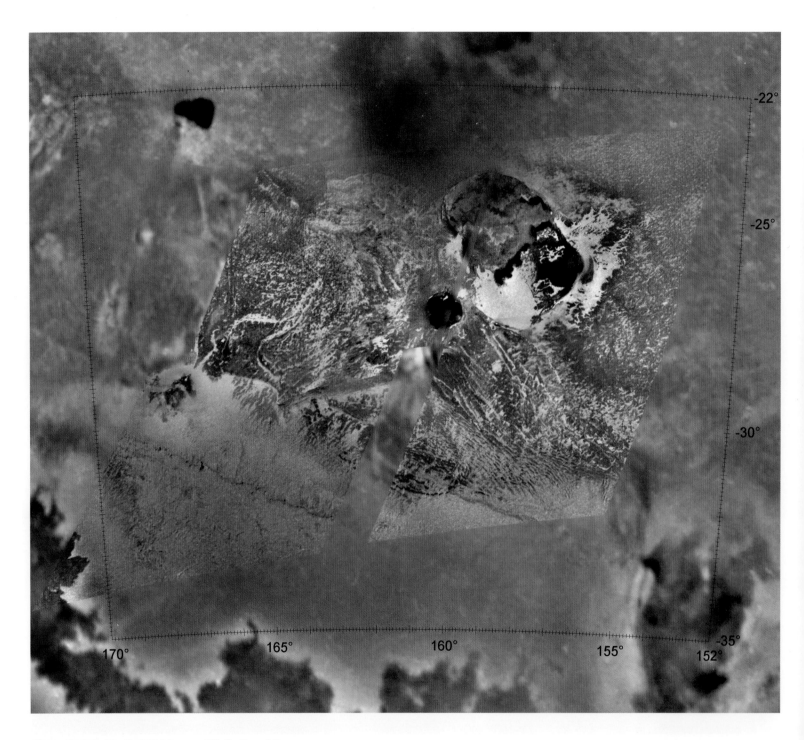

Plate Ji12.3.1b **Tohil Mons: High-Sun View**
Encounter: *Galileo* I27   Resolution: 165 meters/pixel

Io | 341

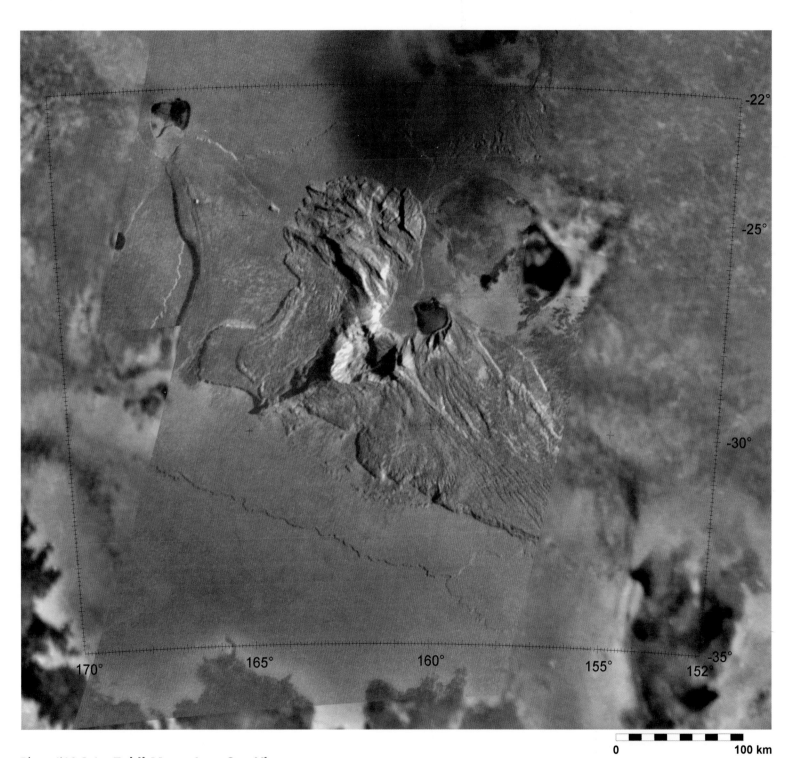

Plate Ji12.3.1c **Tohil Mons: Low-Sun View**
Encounter: *Galileo* I32   Resolution: 340 meters/pixel

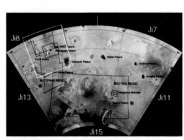

Plate Ji12.3.1 Tohil Mons: Three Views
Orthographic Projection   Map Scale: 1 cm = 28.7 km

Tohil Mons is the best observed and one of the most distinctive mountains on Io. Three image mosaics were obtained. The I24 and I27 images form a high-Sun stereo pair, despite the scrambling of the I24 images. The lighting makes topographic relief almost impossible to discern. The low-Sun I32 regional mapping images, as well as the high-resolution mosaic (Plate Ji12.3.3), show the structure of the mountain in great detail.

Towering 9.5 kilometers above the plains, Tohil is comprised of several distinct parts. A rectangular eastern plateau rises gently toward the west. The lobate outlines of this plateau suggest it has undergone partial collapse. An intensely fractured northern plateau, at 25°S, 161°W, rises ~3.5 kilometers. The two components converge along a peculiar linear alignment of paterae, amphitheatres and ridges. Tohil Patera lies immediately north of Tohil Mons, at 26°S, 158°W, and includes dark and bright deposits of various colors. The smaller, dark-floored Radagast Patera, a hotspot, lies between the northern mountain and the southeastern plateau. Along the crest of Tohil Mons is a large quasi-circular amphitheatre whose floor sits roughly 2 kilometers above the plains. A small caldera might lie on the floor of this feature. Several steep-sided promontories form the southern crest of this amphitheatre. A narrow ridge extends along the crest to the southeast. The alignment of these features suggests that a major fault may form the backbone of Tohil Mons.

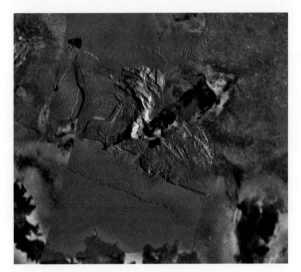

Plate Ji12.3.2-N Tohil Mons: NIMS Thermal Map
Encounter: *Galileo* I32   Resolution: 1.2 kilometers/pixel

The limited NIMS view shows a small hotspot within Radagast Patera, along the western edge and corresponding to the darker northern part of the caldera. Minimum temperatures of this spot are in the range of 325 to 400 K. Small discrete hotspots of this type may be more common on Io than the limited *Galileo* mapping shows.

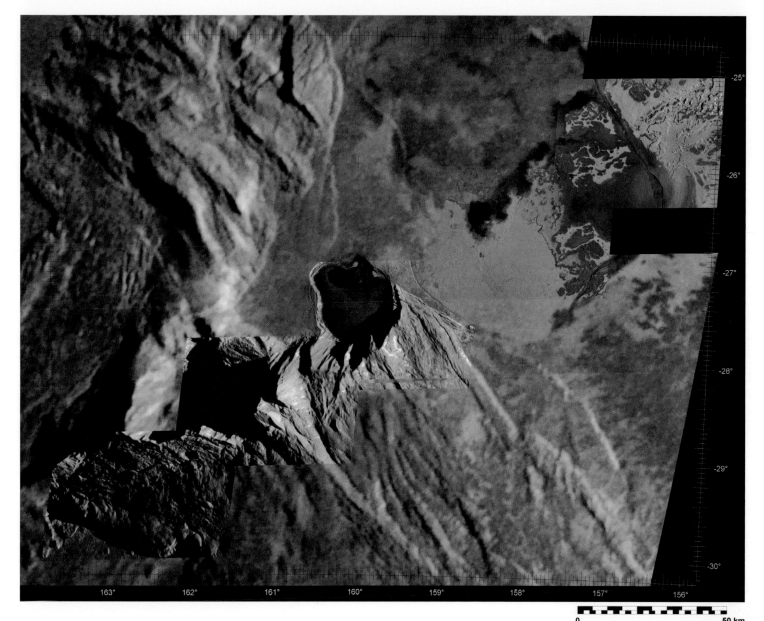

Plate Ji12.3.3 **Tohil Mons: High-resolution Mosaic**
Encounter: *Galileo* I32   Resolution: 52 meters/pixel
Orthographic Projection   Map Scale: 1 cm = 12.3 km

At high resolution we see additional mass wasting and deformation features on Tohil Mons. A few small lobate landslides can be seen on the plains below to the southwest, and the face of the mountain here is crumpled or wrinkled, suggesting that much of the flank is creeping downhill or is intensely folded. A hint of layering can be seen on the slopes of the steep promontory at 28.6°S, 161.1°W, but otherwise evidence for layering is scarce.

The dark-floored Radagast Patera is surrounded by a scarp 300–500 meters high. The darker northwestern feature within Radagast is the likely site of a hotspot (Plate Ji12.3.2-N). No relief is evident across multi-colored Tohil Patera. Low-lying mesas less than 100 meters high cross the plains to the east of Tohil Patera (25.5°S, 156.2°W). These may be due to the erosion of volatile-rich layers within the caldera, or the erosion of the caldera floor by hot lava. Shown at 50% scale fit page.

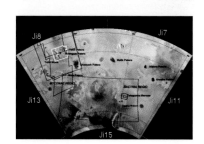

Lambert Conformal Conic Projection
Map Scale: 1 cm = 143 km

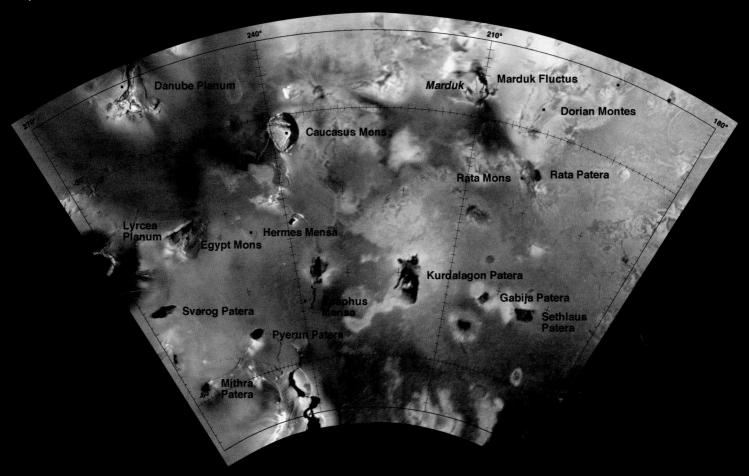

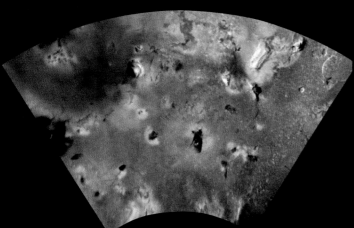

Plate Ji13/Ji13-m Caucasus Mons Quadrangle

The large reddish ring to the northwest is part of the Pele eruptive plume (seen in Plate Ji9). The 2- to 3-kilometer-high Danube Planum plateau forms the southern boundary of the Pele caldera (also in Ji9). Several prominent mountains are apparent: Egypt Mons, Dorian Montes, Caucasus Mons, and the etched plains at 62°S, 250°W. The conical peak west of Caucasus Mons (30°S, 245°W) is a 2-kilometer-high shield volcano very similar to Zamama (see Plate Ji8.4), and is one of the rare expressions of volcanic relief on Io. Dark reddish volcanic deposits at 30°S, 210°W mark the Marduk eruption site. Layered plains found to the south are better expressed in Plate Ji15.

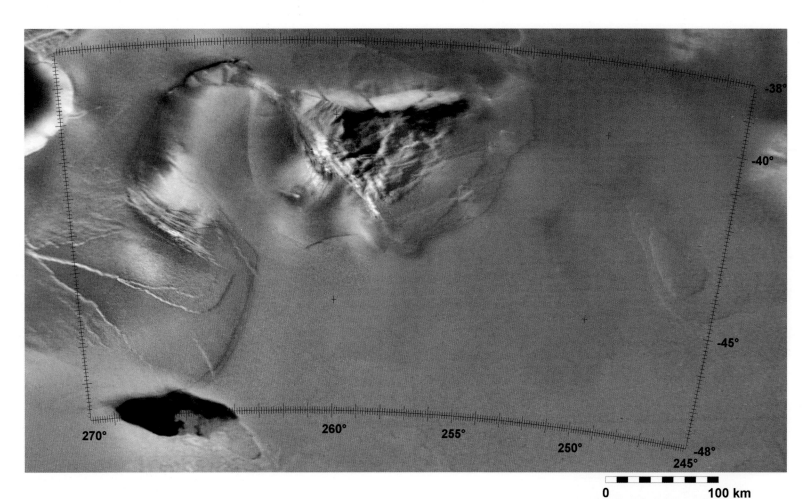

Plate Ji13.1 **Egypt Mons**

Encounter: *Voyager* 1 Resolution: 220 meters/pixel
Orthographic Projection Map Scale: 1 cm = 32.5 km

One of five high-resolution views from *Voyager* 1, Egypt Mons is a distinctive asymmetric triangular (or flatiron) massif rising 10 kilometers above the plains. The combination of steep northern scarp and gentle sloping southern flank indicates that it is probably an intact crustal block that tilted southward as it was uplifted (Euboea Montes, Plate Ji14, is another example). The steep scarp on the northern flank is probably a thrust fault scarp. Faulting and uplift probably occurred as the crust of Io undergoes compression due to global volcanic burial. Several small bright spots on the southwestern flank may be fumarole-like deposits of sulfur dioxide. Eroded layered plains tens of meters thick surround Egypt Mons and extend to the west.

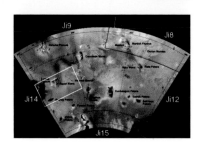

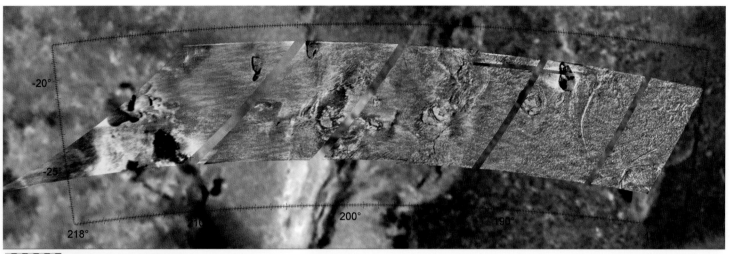

0   100 km

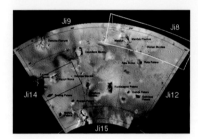

Plate Ji13.2 Dorian Montes

Encounter: *Galileo* I24   Resolution: 450 meters/pixel

Orthographic Projection   Map Scale: 1 cm = 66.4 km

This I24 regional view of the Dorian Montes (27°S, 202°W) was scrambled and partly restored. Although it does reveal several low scarps within the plains of Io, it is not easy to interpret and is shown here for completeness. Rata Mons (at 23°S, 194°W and visible in the quadrangle maps) is all but invisible in this high-Sun view.

Lambert Conformal Conic Projection
Map Scale: 1 cm = 143 km

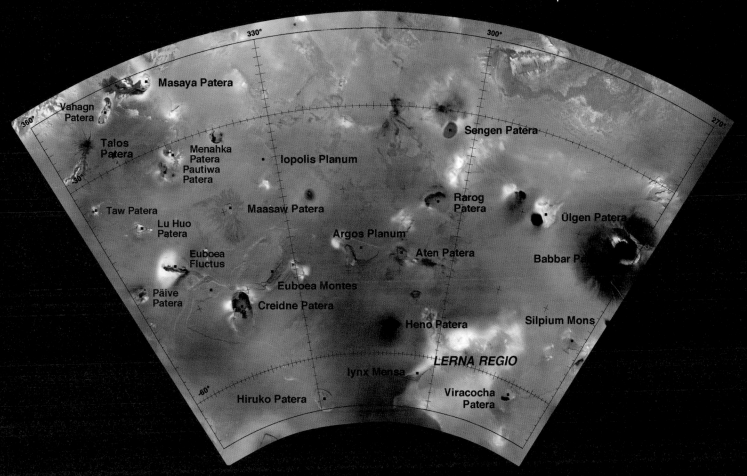

Plate Ji14/Ji14-m Creidne Patera Quadrangle

Many of the best *Voyager* 1 observations are centered in this region due to the geometry of the 1979 flyby. This region is dominated by small eruptive volcanic centers. Highlights include the lava flows surrounding Maasaw Patera and Talos Patera. The most prominent mountain in the region is the asymmetric 13-kilometer-high Euboea Montes. This tilted crustal block features an enormous landslide on its northern flank and figured prominently in helping to understand the origin of Io's tall mountains. Another curiosity is the dark deposit extending westward from Masaya Patera. This deposit appears to have been dropped on top of a mesa-like plateau immediately to the south (revealed in stereo images of the region), indicating the deposit was lofted explosively into space. The persistent dark ring surrounding Babbar Patera is also rare for Io, another prominent example being the 1997 eruption at Pillan (Plate Ji9.2).

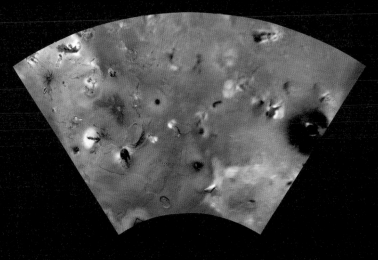

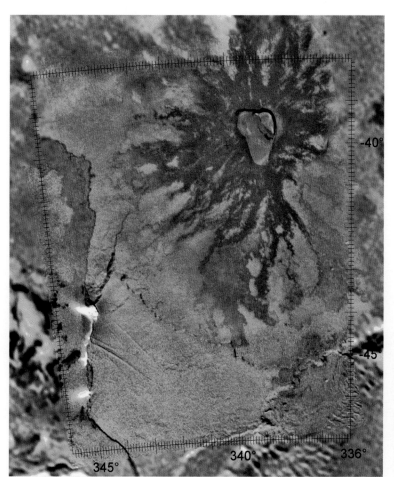
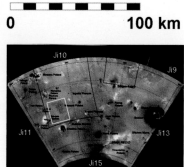

Plate Ji14.1 **Maasaw Patera**
Encounter: *Voyager* 1   Resolution: 250 meters/pixel
Orthographic Projection   Map Scale: 1 cm = 28.3 km

This impressive set of radiating lava flows is centered on the 32- by 44-kilometer-wide, 2-kilometer-deep Maasaw Patera volcanic caldera. The steep walls indicate that Io's crust must be strong enough to support such relief and is not composed of thick layers of sulfur. There is almost no relief on the lava-covered flanks of Maasaw, however, indicating that the lavas had a very low viscosity.

Bright deposits emanating from at least three discrete spots (e.g., 43°S, 345°W) on the low scarps along the southwestern edge of the mosaic are related to scarp retreat. The bright material is probably sulfur dioxide released from within the scarp face, which stands less than 100 meters high. The thicker deposit in the southeast corner of the plate is the northern edge of the Euboea Montes landslide deposit, one of the largest landslides by volume in the Solar System. The edge of this landslide is roughly 2 kilometers thick, and formed when a layered deposit slid off the top of the 13-kilometer-high Euboea Montes (off the bottom of the plate) as it was or after it had been uplifted. This view uses *Voyager* color data, which do not extend into the infrared.

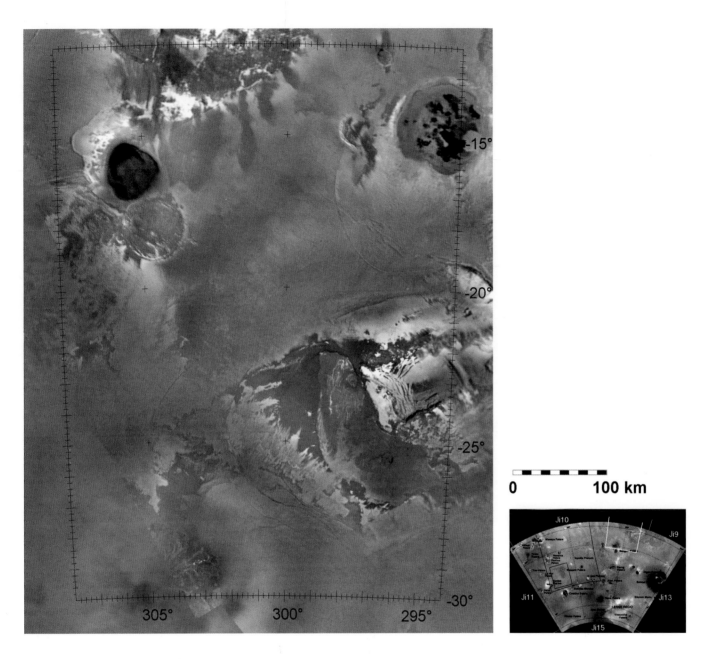

Plate Ji14.2 **Mesa Complex**
Encounter: *Voyager* 1   Resolution: 330 meters/pixel
Orthographic Projection   Map Scale: 1 cm = 39 km

Another *Voyager* 1 high-resolution view, this area includes a large rectangle at center right flanked by dark flows and volcanic calderas. The odd whitish and orange, 40- by 65-kilometer-wide rectangle, centered at 23°S, 295°W, is actually a 5- to 6-kilometer-high flat-topped mesa. A large reddish boomerang-shaped patera abuts the western margin of this mesa, surrounded by darker lava flows. Similar dark and bright flow patterns abut the eastern margin. The circular paterae to the north are of interest for the variety of albedos and colors they display. These differences may or may not relate to the ages of most recent volcanic activity, or to the composition of the lavas.

Polar Stereographic Projection
Map Scale: 1 cm = 130 km

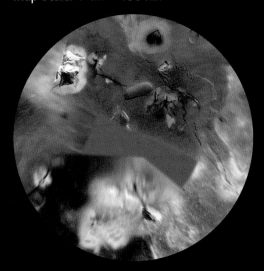
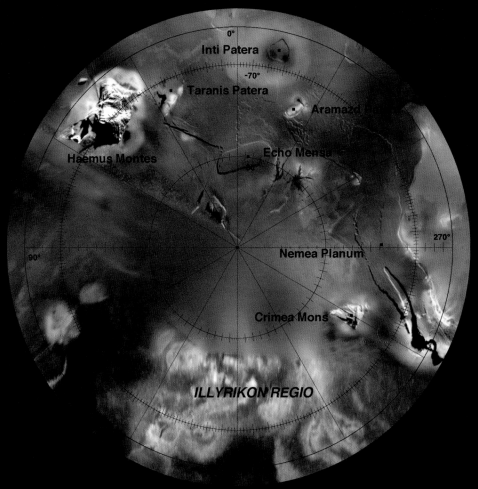

Plate Ji15/Ji15-m Haemus Montes Quadrangle

Voyager 1 passed almost directly over Io's southern pole, providing much of the sharply resolved details seen here. The region is populated with low mesa-like plateaus a few hundred meters high. Examples include Echo Mensa and Nemea Plaunum (the taller peaks at 65°S, 242°W stand roughly 4 to 6 kilometers high). Their origins are unclear but appear to be formed by erosional scarp retreat, perhaps like the mesas of the desert southwest of the United States. Here the mechanism is sublimation of sulfur or sulfur dioxide within the layers and the weakened scarps collapse.

The most prominent mountain in the scene, the double-peak Haemus Montes towers 9–10 kilometers above the surrounding plains. Crimea Mons stands a somewhat less-lofty 4 kilometers high. Black diffuse deposits emanate from two spots on the rim of Inti Patera and are at the center of a large whitish deposit. Bright whitish aureole deposits also surround Haemus Montes, in Illyrikon Regio, and at discrete spots along the edges of layered plains. At least one mountain lies within Illyrikon Regio, consistent with the idea that bright aureoles are often associated with topographic relief and volcanic eruptions.

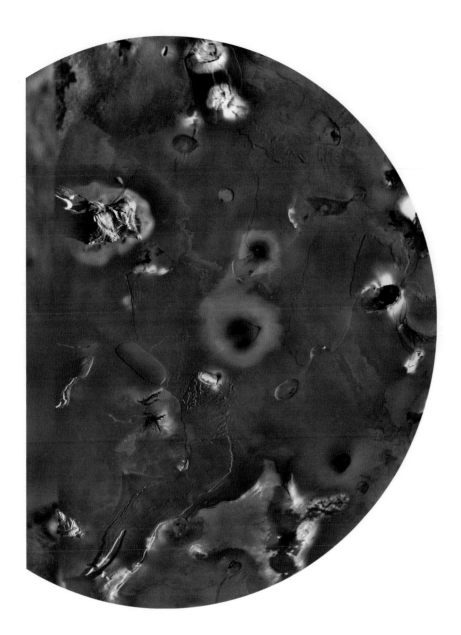

Plate Ji15.1 South Polar Region
Encounter: *Voyager* 1   Resolution: 440 meters/pixel
Orthographic Projection   Map Scale: 1 cm = 140 km

Voyager obtained a 9-frame mosaic of the south polar region at 440-meter resolution, a level of detail that is not shown in the global mosaic. Aside from a few prominent mountains (see Plate Ji15.2, for example), the area is dominated by extensive tracks of layered plains. These plains are bounded by scarps typically a hundred meters high or less. In some cases, more than one layer is evident. In some areas the scarps are highly dissected, evidence of scarp retreat through the loss of a volatile material in the layers, leading to scarp weakening. Here on Io, the rock is dominantly volcanic, and the volatile is not water but most likely sulfur dioxide or other sulfur compounds.

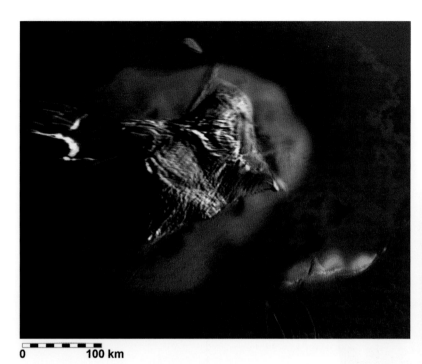
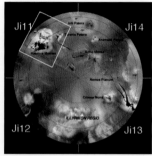

0 100 km

Plate Ji15.2 **Haemus Montes: Close-up**
Encounter: *Voyager* 1   Resolution: 440 meters/pixel
Orthographic Projection   Map Scale: 1 cm = 46.8 km

Rather than using low-resolution *Galileo* color mosaics, this view of Haemus Montes includes *Voyager* 3-color data. Although insensitive to the near-infrared, the higher-resolution simultaneous *Voyager* color data clearly show that parts of the eastern flank of Haemus Montes are not covered by the whitish, SO_2 deposits. This "bare" patch is on the flank and is bounded by a scarp. White frost material covers the upper layer, but not the lower. Apparently, some layers within the mountain contain less sulfur dioxide than others.

Appendix 1: Glossary

Albedo   the relative amount of sunlight reflected by a planetary surface. Low-albedo objects are dark and absorb most incident sunlight.

Anti-Jovian hemisphere   the side of a satellite always facing away from Jupiter (see *Synchronous rotation*).

Apex (and **antapex**)   the point (or region) on a planet or satellite that is most forward as it orbits (usually centered at 0°N, 90°W).

Basalt   a common silicate volcanic rock formed from cooled lavas. Lower in silica than granite, basalts tend to be enriched in magnesium, iron and calcium, and include minerals such as pyroxene, olivine and plagioclase.

Breccia   a rock formed during impact or volcanic eruption consisting of broken fragments of older rocks and refrozen melt, reaggregated to form a new unit of rock.

Caldera   a circular, oval, or polygonal volcanic depression formed by collapse due to eruption. Often forms a walled depression, which can sometimes be filled by lava or lava flows.

Catena   see *Crater chain*.

Central dome   a mound of fractured deep-crustal material in the center of a crater uplifted during the impact event. This distinct variation on icy satellites takes the form of a broad rounded mound in larger craters.

Central peak   a mound of fractured deep-crustal material in the center of a crater uplifted during the impact event. Usually takes form of conical peak(s).

Chaos   a region of disrupted crust on Europa, often formed from fragmented remnants of the original crust as well as new material.

Comet   a small (<100km across) irregularly shaped object orbiting the Sun, usually at great distances, and composed of ices, organic (carbon-rich) materials, and silicates. Comets can occasionally be disturbed into highly elliptical orbits approaching the Sun.

Convection   movement and flow within a liquid mass or a rock layer behaving in a liquid manner, usually involving vertical overturn.

Core   the innermost (and usually most dense) part of a planet, often composed of iron or iron compounds.

Crater chain a linear arrangement of roughly equal-sized closely spaced craters, usually formed simultaneously.

Creep slow (usually imperceptible) movement of a crustal or lithospheric mass under the influence of gravity.

Crust an outer layer of a planet with a distinctly different composition from that which lies beneath (e.g., mantle).

Cycloidal a repeating pattern of connected arcs. The arcs form sharp bends at the connection points.

Diapir/diapirism vertical movement or exchange of rock layers by creep in the form of multiple large rising and descending "blobs."

Differentiation the process whereby a planet of roughly uniform composition separates into a dense (usually iron-rich) core and less-dense outer layers, such as an icy crust. This process often releases a large quantity of planetary heat.

Dilational opening or pulling apart of the crust or lithosphere across a fracture, creating a new space usually filled by other materials.

Diurnal daily (in the context of the referenced body).

Ductile able to be deformed or altered in shape without breaking or fracturing.

Ejecta (blanket) the roughly continuous concentric blanket or layer of material thrown out of an impact crater during formation. Commonly consists of fragmented and melted crustal material.

Embay partial flooding of a depression by lavas or other liquids up to a uniform topographic level.

Facula a large bright quasi-circular spot.

Fault a planar break within a planetary crust or lithosphere, usually involving motion or sliding along its surface.

Fire fountain an energetic eruption of molten lava displaying fountain-like sheets of lava jetting upward into space (or the air).

Flexus ridge, typically curved.

Flood basalt a sheet of multiple lava flows covering a large contiguous region.

Fluctus a large associated field of lava flows, sometimes layered.

Fossae a set of long narrow depressions (see also *Graben*).

Fumarole a small vent emitting gases or vapors.

Furrow see *Trough* and *Graben*.

Graben a linear or curved fault-bounded valley, formed by one or two fault-scarps and a down-dropped linear block of crust.

Hydrated mineral a mineral that has been altered by bonding with water.

Hydrocarbon   a complex organic compound consisting of carbon and hydrogen.

Impact crater   a large circular (or elliptical) walled depression formed by the impact of a meteorite, asteroid or comet.

Infrared   that part of the electromagnetic spectrum longward of red. Infrared is invisible to humans, but usually sensed as heat. Many geologic materials have diagnostic features in this range of the spectrum.

K(elvin)   a measure of temperature equivalent to 1.8° Fahrenheit or 1° Celsius.

Kilometer   unit of length equal to ~0.62 miles (100 km = 62 miles).

Lava   molten rock, be it silicates or ice, erupted on the surface.

Lava lake   a molten pool of lava confined within a depression.

Leading hemisphere   the side of a synchronously rotating planet or satellite that is most forward as it orbits (centered at 0°N, 90°W).

Lenticula   an oval- or lens-shaped feature on Europa smaller than 50 kilometers across, characterized by a coarse texture and replaces older terrains.

Levee   a raised ridge lining the edge of a channel or lake formed by liquids such as water or lava.

Linea   narrow elongate linear or curved marking.

Lithosphere   the outermost cooler (solid) layer of a planet. The layer is defined by its mechanical and thermal properties and is not necessarily equivalent to the crust.

Macula   a large dark quasi-circular spot.

Mafic   a rock or mineral group characterized by high iron and magnesium and low (<52%) silica content. Basalt is a common example.

Magma   molten rock, be it silicates or ice, that remains underground.

Mantle   a thick global layer beneath the crust, usually composed of material with a composition and density intermediate between that of the outer crust and inner core.

Mass wasting   the down-slope movement of soil or regolith due to gravity.

Massif   a large irregular mountain or promontory.

Matrix   a material unit on Europa characterized by coarse disorganized surface texture. Usually replaces older units and has a rubble-like appearance.

Mensae/Mesa   flat-topped plateau.

Meter   a unit of length equal to ~3.3 feet (100 m = 328 ft).

Mons/Montes   mountain.

NIMS   Near-Infrared Mapping Spectrometer, the infrared imaging instrument onboard *Galileo*.

Nonsynchronous rotation a rotation state faster (or slower) than synchronous. In the context of the Galilean satellites, this discrepancy is very slight and may take several thousand years to complete.

Oblique to view or act upon an object at a large angle. In perspective: to view objects from the side rather than from above.

Occultation passing of one planetary body in front of another, blocking it from view as seen by an observer at a distant location.

Olivine a magnesium iron silicate mineral common in mafic or ultramafic rocks and noted for its green color.

Palimpsest a very degraded impact crater formed on icy satellites, noted for its lack of relief.

Patera see *Caldera*.

Penedome crater an impact crater formed on icy satellites, noted for its prominent central dome and lack of rim relief.

Penepalimpsest a degraded impact crater formed on icy satellites, noted for its relatively low relief.

Phase angle the angle between the Sun and an observer, as measured on the surface of a planetary body. Low phase angle implies the observer and the Sun are in the same general region of the sky as measured from the surface (the surface of the Moon when it is at full phase as viewed from Earth, for example).

Phase function an equation describing the change in brightness of a planetary surface as the phase angle changes.

Polar wander the process by which a planet or its outer layers shift orientation while the planetary body maintains the original orientation of its spin axis. By this mechanism a polar terrain can be slowly moved to more equatorial climates.

Polymer a large molecule composed of identical smaller molecules bonded together, usually in chains. Sulfur on Io may combine in this way frequently.

Planum plateau.

Plume high-velocity cloud of gas and/or dust seen arching into naked space over a volcanic vent or source.

Pyroxene a magnesium iron aluminum silicate mineral common in mafic or ultramafic rocks. Other ions are also commonly bonded with the mineral.

Ray a linear alignment of bright (or dark) ejecta material, radiating from the center of an impact crater. A ray may consist of secondary craters and surface coatings.

Regio large continuous region with a related geologic history.

Regolith the loose disaggregated "soil" of a planetary surface, formed by any combination of meteorite bombardment, wind action, precipitation, chemical alteration, or other geologic processes.

Relaxation see *Creep*.

Resonance (orbital) a regular periodicity between two or more planets or satellites such that their orbital periods are simple integer multiples, leading to repeated regular encounters. This can lead to subsequent distortions of their orbital shapes, enhanced through regular gravitational pumping.

Rim the steep inward-facing scarp marking the morphologic edge of an impact crater.

Rock solid aggregate of minerals or related constituents. Composition can be silicates, ices, or any other naturally occurring minerals (although "rock" is often considered equivalent to silicate rock).

Salt a mineral formed by reaction between acids and bases. The most commonly known are those involving combinations of metal ions with bases such as chlorine, carbonate, hydronium, sulfate, among others.

Scarp a tilted topographic surface, typically elongated and generally continuous and uniform in slope.

Secondary craters smaller craters formed by the impact of solid or coherent blocks of material ejected from a larger crater during impact. Usually formed beyond the ejecta blanket.

Shear deformation of rock due to divergent or unequal lateral displacement. Shear is often associated with sliding, crumpling or grinding of rock along or near discrete fault zones.

Shield volcano a volcanic pile forming a large conical mountain with relatively low slopes ($<5°$).

Silica a mineral component consisting of one silicon and two oxygen atoms.

Silicates a class of minerals dominated by silicon and oxygen bonding, usually in combination with various metallic atoms or ions.

Spectrometer an optical instrument designed to sense the reflected spectrum of an object.

SSI Solid-state imager, the CCD "camera" aboard *Galileo*.

Strike-slip fault a planar break within a planetary crust or lithosphere involving lateral (sideways) motion or sliding along its surface.

Sublimation transformation of a solid (water ice, for example) to a gas or vapor directly.

Sulci/Sulcus a region characterized by sets of parallel ridges, furrows, or faults.

Sulfate salt of sulfuric acid. The base molecule consists of one sulfur and four oxygen atoms.

Synchronous rotation when a planet's or satellite's rotation period matches its revolution about the primary planet (or star) so as to keep one side always facing that primary.

Talus an accumulated pile of loose rubble at the base of a cliff or scarp.

Tectonics the structure and deformation of planetary lithospheres and surfaces.

Terrace a narrow bench or platform formed by a coherent block of material that has slid part way down a slope.

Tholis circular domed or flat-topped mound.

Tide the daily rise and fall of the surface of a planet or satellite in response to cyclical gravitational forces or fluctuations.

Trough an elongate linear or curved walled depression.

True polar wander see *Polar wander*.

Ultramafic a rock or mineral group characterized by very low (<45%) silica content, lower than basalt.

Trailing hemisphere the side of a synchronously rotating planet or satellite that faces backward as it orbits (centered at 0°N, 270°W).

Vallis a channel or valley, typically formed by fluid flow, such as water or lava.

Viscosity the property of a rock or magma that defines its ability to deform or flow under pressure or gravity. Low viscosity implies ease of flow.

Viscous relaxation see *Creep*.

Volcanism the processes of melting and mobilizing melted crustal or mantle rocks.

Volcano an accumulated pile of lava flows, forming a mountain or layer.

Appendix 2: Supplemental readings

Supplemental and recommended readings related to features or ideas described in the *Atlas* are listed first by topic, and then grouped as follows. Books and collections of journal articles are listed first and in chronological order. Many additional references can be found mentioned in these compilations. These listings are followed by single journal articles of specific importance, also in chronological order. This list of readings includes many of my own articles, and is by no means complete. My apologies for any I may have left off by oversight or need.

Comet Shoemaker-Levy 9 disruption and Jupiter impact

Books
The Great Comet Crash, ed. J. Spencer and J. Milton, Cambridge University Press, New York, 1995.
The Collision of Comet Shoemaker-Levy 9 and Jupiter, ed. K. Noll, H. Weaver and P. Feldman, Cambridge University Press, New York, 1996.

Journal articles
Melosh, H.J. and Schenk, P. (1993). Split comets and the origin of crater chains on Ganymede and Callisto, *Nature*, **365**, 731–733.
Schenk, P.M., Asphaug, E., McKinnon, W.B., Melosh, H.J., and Weissman, P. (1996). Cometary nuclei and tidal disruption: The geologic record of crater chains on Callisto and Ganymede, *Icarus*, **121**, 249–274.

Voyager and *Galileo* missions and results

Voyager and before

Journal compilations and books
Science, **204**, 945–1008, 1979, and *Science*, **206**, 925–996, 1979.
   First results from the two *Voyager* encounters.
Voyage to Jupiter, D. Morrison and J. Swan, NASA Special Publication 439, 1980.
   A personal recounting of the *Voyager* encounters.
Satellites of Jupiter – Special Section, *Icarus*, **44**(2) (Nov.), 1980.
   First results from the *Voyager* encounters with Jupiter.
Satellites of Jupiter, University of Arizona Press, Tucson, AZ, 1982.
   First detailed scientific analyses of the *Voyager* results.

Satellites, University of Arizona Press, Tucson, AZ, 1986.
   The Jupiter chapters provide a reassessment of the Galilean satellites.
Satellites of the Outer Planets, D. Rothery, Clarendon Press, Oxford, 1992.
The Giant Planet Jupiter, J. Rogers, Cambridge University Press, New York, 1995.
   Focus is on the history of telescopic observations of Jupiter and its satellites.
Icy Galilean Satellites – Special Section, *J. Geophys. Res.*, **100**(E9) (Sept. 25), 1995.

Galileo and beyond

Journal compilations and books

Galileo: Exploration of Jupiter's System, ed. C. Yeates, NASA Special Publication 479, 1985.
   A description of the original *Galileo* mission plan.
Galileo Remote Sensing – Special Section, *J. Geophys. Res.*, **135**(1) (Sept.), 1998.
Jupiter Odyssey, D. Harland, Springer-Praxis, London/New York, 2000.
Mission Jupiter, D. Fischer, Springer-Verlag, New York, 2001.
   Two books on the *Galileo* mission from the perspective of mission history.
Jupiter, ed. F. Bagenal, T. Dowling and W. McKinnon, Cambridge University Press, Cambridge/New York, 2004.
   Probably the best scientific compilation of the *Galileo* results.
Schenk, P. (2004). Ice moons of Sol? In *Icy Worlds of the Solar System*, ed. P. Dasch, Cambridge University Press, New York, pp. 110–134.
New Horizons at Jupiter – Special Section, *Science*, **318**(5848) (Oct. 12), 2007.
   Articles on *New Horizons*' pass of Jupiter and its moons, with emphasis on Io.
Collins, G.C., McKinnon, W.B., Moore, J.M., Pappalardo, R.T., Prokter, L.M., and Schenk, P.M. Tectonics of the outer planet satellites. In *Planetary Tectonics*, ed. T. Watters and R. Schultz. Cambridge University Press, Cambridge/New York, 2010.

Io

Journal compilations and books

Time-variable Phenomenon in the Jovian System, NASA Special Publication 494, 1987.
   Several post-*Voyager* Io-related articles.
Io Volcanism in the *Galileo* Era – Special Section, *Geophys. Res. Lett.*, **24**(20) (Oct. 15), 1997.
Galileo: Io Up Close – Special Section, *Science*, **288**(5469) (May 19), 2001.
Geology and Geophysics of Io – Special Section, *J. Geophys. Res.*, **106**(E12) (Dec. 25), 2001.

Io After *Galileo* – Special Section, *Icarus*, **169**(1) (May), 2004.
Io after Galileo, ed. R. Lopes and J. Spencer, Praxis, Chichester, UK, 2007.
Volcanism on Io, A. Davies, Cambridge University Press, New York, 2007.
   Two excellent and comprehensive summations of the *Galileo* results.

Journal articles
Spencer, J. and Schneider, N. (1996). Io on the eve of the *Galileo* mission, *Ann. Rev. Earth Planet. Sci.*, **24**, 125–190.
   Our best pre-*Galileo* perspectives on Io.
Spencer, J., McEwen, A., McGrath, M., Sartoretti, P., Nash, D., Noll, K., and Gilmore, D. (1997). Volcanic resurfacing of Io: Post-repair HST imaging, *Icarus*, **127**, 221–237.
   Pre-*Galileo* images of Io, 1997.
Schenk, P., McEwen, A., Davies, A., Davenport, T., and Jones, K. (1997). Geology and topography of Ra Patera, Io, in the *Voyager* era: Prelude to eruption, *Geophys. Res. Lett.*, **24**, 2467–2470.
Schenk, P.M. and Bulmer, M.H. (1998). Origin of mountains on Io by thrust faulting and large-scale mass movements, *Science*, **279**, 1514–1518.
Schenk, P. and Williams, D. (2004). A potential thermal erosion lava channel on Io, *Geophys. Res. Lett.*, **31**, L23702.

Europa

Journal compilations and books
Europa – Special Section, *Nature*, **391**(3) (Jan. 22), 1998.
Galileo Mission Results – Special Section, *J. Geophys. Res.*, volume 106, number E9 (Sept. 25), 2000.
Europa: Special Section, *Astrobiology*, **3**(4) (Dec.), 2003.
   Focus on astrobiology and composition of Europa.
Europa's Icy Shell – Special Section, *Icarus*, **177**(2) (Oct.), 2005.
Europa, ed. R. Pappalardo, W. McKinnon and K. Khuruna, University of Arizona Press, Tucson, AZ, 2010.
   An excellent comprehensive and *unbiased* summary of *Galileo*'s finding.

Journal articles
Schenk, P.M. and Seyfert, C.F. (1980). Fault offsets and proposed plate motions for Europa, *EOS*, **61**, 286.
McEwen, A., Schenk, P.M., and McKinnon, W.B. (1989). Fault offsets and lateral crustal movement on Europa: Evidence for a mobile ice shell, *Icarus*, **79**, 75–100.
Pappalardo, R., Head, J., and Greeley, R. (1999). Hidden ocean of Europa, *Sci. Am.*, **281**(4), 34–43.

Prockter, L. and Pappalardo, R. (2000). Folds on Europa: Implications for crustal cycling and accommodation of extension, *Science*, **289**, 941–944.

Hussmann, H. and Spohn, T. (2004). Thermal-orbital evolution of Io and Europa. *Icarus*, **171**, 391–410.

Figueredo, P. and Greeley, R. (2004). Resurfacing history of Europa from pole-to-pole geological mapping, *Icarus*, **167**, 287–312.

Schenk, P.M. and Pappalardo, R. (2004). Topographic variations in chaos on Europa: Implications for diapiric formation, *Geophys. Res. Lett.*, **31**, L16703, doi:10.1029/2004GL019978.

Prockter, L. and Schenk, P. (2005). Origin and evolution of Castalia Macula, an anomalous young depression on Europa, *Icarus*, **177**, 305.

Nimmo, F. and Schenk, P. (2006). Normal faulting on Europa, *J. Struct. Geol.*, **28**, 2194–2203.

Dalton, J.B. (2007). Linear mixture modeling of Europa's non-ice material based on cryogenic laboratory spectroscopy, *Geophys. Res. Lett.*, **34**, L21205, doi:10.1029/2007GL031497.

Schenk, P., Matsuyama, I., and Nimmo, F. (2008). Evidence for true polar wander on Europa from global scale small circle depressions, *Nature*, **453**, 368–371.

Ganymede and Callisto

Journal compilations and books
See *Galileo and beyond* listing.

Journal articles
Schenk, P.M. and McKinnon, W.B. (1985). Dark halo craters and the thickness of grooved terrain on Ganymede, *J. Geophys. Res.*, **90**, C775–C783.

Schenk, P. (1995). The geology of Callisto, *J. Geophys. Res.*, **100**, 19023–19040.

Greeley, R., Klemaszewski, J., and Wagner, R. (2000). *Galileo* views of the geology of Callisto, *Planet. Space Sci.*, **48**, 829–853.

Schenk, P.M., McKinnon, W., Gwynn, D., and Moore, J. (2001). Flooding of Ganymede's resurfaced terrains by low-viscosity aqueous lavas, *Nature*, **410**, 57–60.

McCord, T., Hansen, G., and Hibbitts, C. (2001). Hydrated salt minerals on Ganymede's surface: Evidence of an ocean below, *Science*, **292**, 1523–1525.

Hibbitts, C., Klemaszewski, J., McCord, T., Hansen, G., and Greeley, R. (2002). CO_2-rich impact craters on Callisto, *J. Geophys. Res.*, **107**, doi:10.1029/2000JE001412.

Spohn, T. and Schubert, G. (2003). Oceans in the icy Galilean satellites, *Icarus*, **161**, 456–467.

Dombard, A. and McKinnon, W. (2006). Elastoviscoplastic relaxation of impact crater topography with application to Ganymede and Callisto, *J. Geophys. Res.*, **111**, E01001.

Khurana, K., Pappalardo, R., Murphy, N., and Denk, T. (2007). The origin of Ganymede's polar caps, *Icarus*, **191**, 193–202.

Bland, M. and Showman, A. (2007). The formation of Ganymede's grooved terrain: Numerical modeling of extensional necking instabilities, *Icarus*, **189**, 439–456.

Howard, A. and Moore, J. (2008). Sublimation-driven erosion on Callisto: A landform simulation model test, *Geophys. Res. Lett.*, **35**, CiteID L03203.

Bland, M., Showman, A., and Tobie, G. (2009). The orbital thermal evolution and expansion of Ganymede, *Icarus*, **200**, 207–221.

Impact cratering: Satellite ages and interiors

Journal articles

McKinnon, W. and Melosh, H.J. (1980). Evolution of planetary lithospheres: Evidence from multiring basins on Ganymede and Callisto, *Icarus*, **44**, 454–471.

Schenk, P.M. and McKinnon, W.B. (1987). Ring geometry on Ganymede and Callisto, *Icarus*, **72**, 209–234.

Schenk, P.M. (1991). Ganymede and Callisto: Complex crater formation and planetary crusts, *J. Geophys. Res.*, **96**, 15635–15664.

Schenk, P.M. and McKinnon, W.B. (1991). Dark ray and dark floor craters on Ganymede and the provenances of large impactors in the Jovian system, *Icarus*, **89**, 318–346.

Schenk, P.M. (1993). Central pit and dome craters: Exposing the interiors of Ganymede and Callisto, *J. Geophys. Res.*, **98**, 7475–7498.

Zahnle, K., Schenk, P., Sobieszczyk, S., Dones, L., and Levison, H. (2001). Differential cratering of synchronously rotating satellites by ecliptic comets, *Icarus*, **153**, 111–129.

Schenk, P. (2002). Thickness constraints on the icy shells of the Galilean satellites from a comparison of crater shapes, *Nature*, **417**, 419–421.

Schenk, P. and Ridolfi, F. (2002). Morphology and scaling of ejecta deposits on icy satellites, *Geophys. Res. Lett.*, **29**, doi:10.1029/2001GRL013512.

Zahnle, K., Schenk, P., Levison, H., and Dones, L. (2003). Cratering rates in the Outer Solar System, *Icarus*, **163**, 263–289.

Schenk, P., Chapman, C., Zahnle, K., and Moore, J. (2004). Ages and interiors: The cratering record of the Galilean satellites. In *Jupiter*, ed. F. Bagenal, T. Dowling and W. McKinnon. Cambridge University Press, Cambridge/ New York, pp. 427–456.

Appendix 3: Index maps of high-resolution images

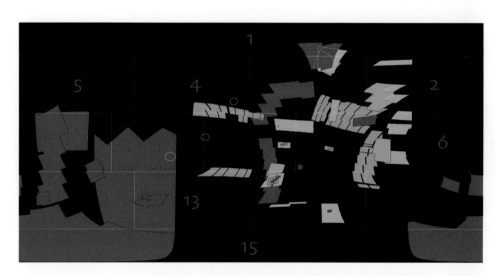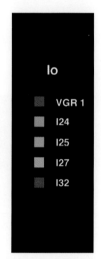

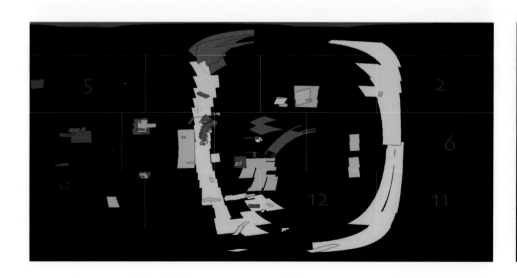

Appendix 3: Index maps of high-resolution images 365

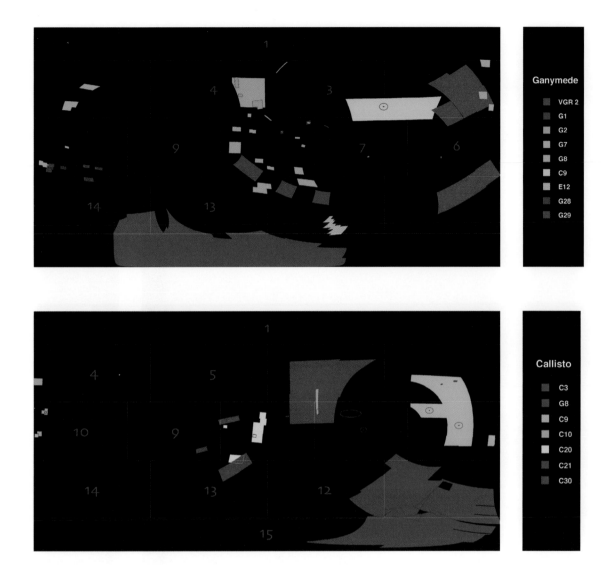

Figure A.1 Footprint maps showing the locations of all high-resolution images of the satellites from both *Voyager* and *Galileo*, with resolutions better than 1 kilometer. It is a very poor sampling compared to Mars, Venus and the Moon, for which we have global mapping coverage at 200–300 meters resolution! Of interest are the north–south swaths on Io and Europa, which were acquired at 300 to 350 meters and at 215 to 250 meters resolution. Of the anticipated near-global medium-resolution mapping, these are virtually the only global mapping mosaics acquired by *Galileo*. The large purple, pink and blue swaths are lower-resolution mapping coverage mosaics at 500 to 1000 meters resolution. Numbered rectangular outlines are quadrangle locations. The color keys identify the encounter during which the images were obtained. Footprint size correlates roughly with image resolution. Very small high-resolution mosaics are highlighted by colored ovals.

Appendix 3: Index maps of high-resolution images

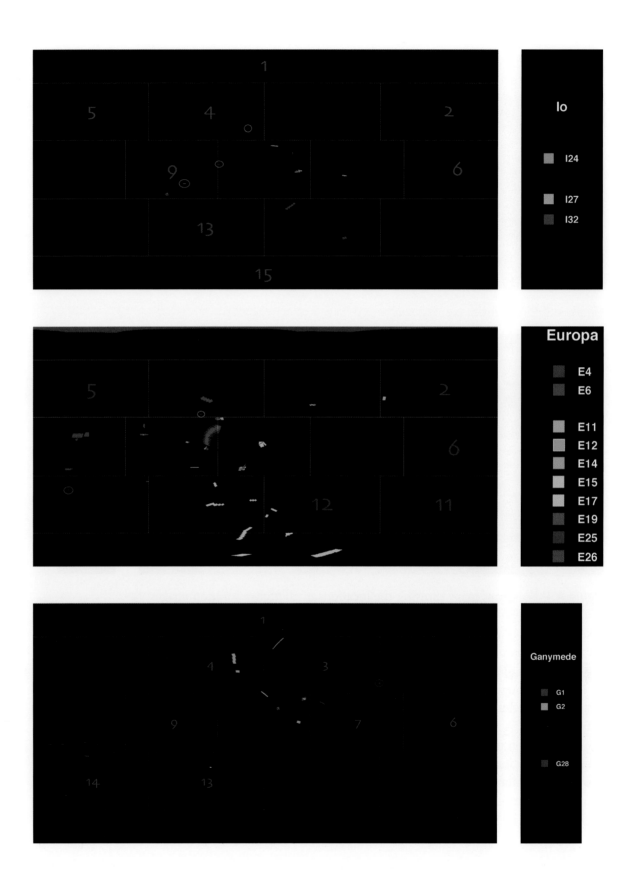

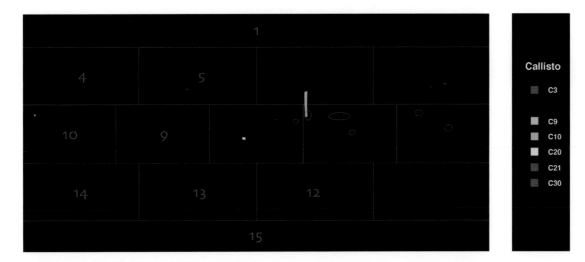

Figure A.2 Footprint maps of locations of high-resolution images better than 100-meter resolution for each satellite. Orbits are indicated by color key at right. Note the overwhelming blackness. A comparable map of Mars would be all color.

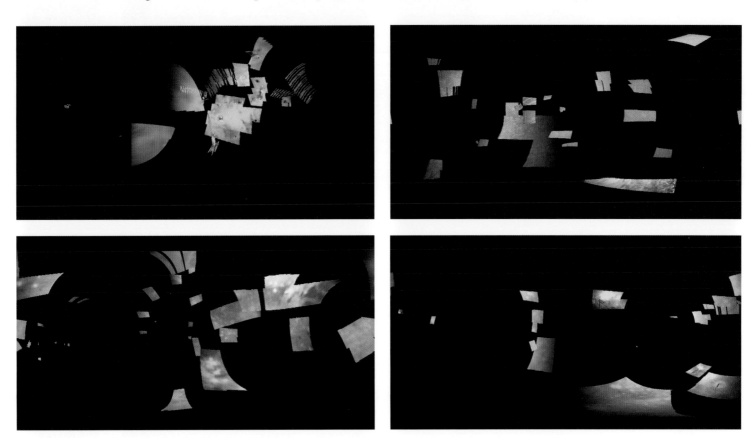

Figure A.3 Footprint maps showing the locations of all high-resolution NIMS images of the Galilean satellites (top row: Io, Europa; bottom row: Ganymede, Callisto). For display, only three of the several hundred NIMS channels are shown. The images have been calibrated but not corrected for photometric or phase angle effects. The blue spots in the Io map are volcanic thermal emission hotspots. Resolutions of these data range from ~1 to ~15 kilometers. Additional low-resolution global mapping data are not shown.

Appendix 4: Data tables

Table A.1 **Physical parameters of the Galilean satellites**

	Moon	Io	Europa	Ganymede	Callisto
Mean radius (km)	1737.1	1821.3[1]	1560.8	2631.2	2410.3
Density (g/cm^3)	3.35	3.53	3.01	1.94	1.83
Distance to planet (km)	384 400	421 000	671 000	1 070 000	1 882 700
Orbital period (days)	27.32	1.77	3.55	7.155	16.69
Mean surface temperature (K)[2]	220	130	103	109	134
Albedo	0.12	0.63	0.67	0.43	0.22

[1] Maximum equatorial, minimum equatorial and polar radii: 1830.0, 1818.7, 1815.0 km
[2] Night temperatures drop to 50 K on Europa and 80 K on Callisto, 100 K on the Moon.

Table A.2 Spacecraft encounters with the Galilean satellites

Encounter	Date	Io	Europa	Ganymede	Callisto
			Distance (in kilometers)		
Voyager 1	1979–3–5	20 600	—	114 700	124 000
Voyager 2	1979–7–9	—	205 700	62 100	215 000
Galileo (J0)	1995–12–07	897[1]	844[1]	—	—
Galileo (G1)	1996–6–27	—	153 500	844	—
Galileo (G2)	1996–9–6	—	—	250	—
Galileo (C3)	1996–11–4	244 000	32 000	—	1104
Galileo (E4)	1996–12–19	—	692	—	—
Galileo (E6)	1997–2–20	—	587	318 000	—
Galileo (G7)	1997–4–5	—	23 245	3059	—
Galileo (G8)	1997–5–7	—	—	1585	33 500
Galileo (C9)	1997–6–25	—	—	79 960	415
Galileo (C10)	1997–9–17	—	—	—	525
Galileo (E11)	1997–11–6	—	2690	—	—
Galileo (E12)	1997–12–16	—	200	14 400	—
Galileo (E14)	1998–3–29	252 000	1650	—	—
Galileo (E15)	1998–5–31	—	2520	—	—
Galileo (E16)	1998–7–21	—	1837[2]	—	—
Galileo (E17)	1998–9–26	—	3600	—	—
Galileo (E18)	1998–11–22	—	2273[2]	—	—
Galileo (E19)	1999–2–1	—	1495[2]	—	—
Galileo (C20)	1999–5–5	—	—	—	1310
Galileo (C21)	1999–6–30	127 000	—	—	1050
Galileo (C22)	1999–8–12	—	—	—	2288
Galileo (I24)	1999–10–11	610[3]	—	—	—
Galileo (I25)	1999–11–26	300[2]	8640	—	—
Galileo (E26)	2000–1–3	—	343	—	—
Galileo (I27)	2000–2–22	198	—	—	—
Galileo (G28)	2000–5–20	—	—	809	—
Galileo (G29)	2000–12–28	—	—	2338	—
Galileo (C30)	2001–5–25	—	—	—	123

Table A.2 (*cont.*)

Encounter	Date	Io	Europa	Ganymede	Callisto
			Distance (in kilometers)		
Galileo (I31)	2001–8–6	194[3]	—	—	—
Galileo (I32)	2001–10–16	184	—	—	—
Galileo (I33)	2002–1–17	102[2]	—	—	—
Galileo (A34)	2002–11–4	45 800[4]	—	—	—
New Horizons	2007–2–28	2 240 000	2 960 000	3 015 000	4 150 000

[1] Tape recorder anomaly resulted in a decision to not record any observations.
[2] Spacecraft safing cancelled half the planned E19 and I25 observations. Safing cancelled E16, E18 and I33 observations entirely.
[3] SSI camera anomaly corrupted most I24 and ruined all I31 SSI observations. Software on the ground partially restored most of the I24 images. NIMS observations were successful.
[4] No remote-sensing observations were allocated during A34, including those of a very close pass of Amalthea.

Appendix 5: Nomenclature gazetteer

Many features can also be found in high-resolution plates associated with the referenced quadrangles.

Name	Latitude	Longitude	Satellite	Quadrangle	Plate
Ababinili Patera	12.79	142.16	Io	Emakong Patera	Ji7, Ji8.2.1
Abydos Facula	33.4	153.4	Ganymede	Galileo Regio	Jg3
Acala Fluctus	8.97	334.59	Io	Loki	Ji10, Ji10.1.2, Ji10.2
Achelous	61.8	11.7	Ganymede	Perrine	Jg2, Jg2.2
Adad	57.5	358.1	Ganymede	Nun Sulci	Jg5
Adal	75.5	79.7	Callisto	Gipul Catena	Jc1
Adapa	73.1	31.3	Ganymede	Etana	Jg1
Adlinda	−48.5	35.6	Callisto	Adlinda	Jc11
Adonis Linea	−61	122.6	Europa	Taliesin	Je12
Aegir	−45.8	103.8	Callisto	Ilma	Jc12
Agave Linea	12.8	273.1	Europa	Castalia Macula	Je9, Je9.3
Agenor Linea	−43.8	213.5	Europa	Agenor Linea	Je13
Agloolik	−47.7	82.4	Callisto	Adlinda	Jc11
Agni Patera	−40.8	333.05	Io	Creidna Patera	Ji14
Agreus	15.9	232.7	Ganymede	Tiamat Sulcus	Jg9
Ägröi	43.2	10.9	Callisto	Vestri	Jc2
Agrotes	60.9	192.5	Ganymede	Philus Sulcus	Jg4
Ah Peku Patera	10.34	107.04	Io	Emakong Patera	Ji7
Ahti	41.4	102.4	Callisto	Asgard	Jc3
Aidne Patera	−1.78	177.09	Io	Prometheus	Ji8
Áine	−43	177.5	Europa	Taliesin	Je12
Ajleke	22.7	101.4	Callisto	Asgard	Jc3
Akhmin Facula	27.7	189.5	Ganymede	Philus Sulcus	Jg4, Jg4.1
Akitu Sulcus	38.9	194.3	Ganymede	Philus Sulcus	Jg4

Name	Latitude	Longitude	Satellite	Quadrangle	Plate
Akycha	72.6	318.7	Callisto	Gipul Catena	Jc1
Aleyin	15.1	134.1	Ganymede	Memphis Facula	Jg7, Jg7.1
Alfr	−9.9	222.7	Callisto	Valfodr	Jc9
Áli	59	55.9	Callisto	Vestri	Jc2
Alphesiboea Linea	−25.1	175.9	Europa	Taliesin	Je12
Altjirra Patera	−34.4	108.97	Io	Tohil Mons	Ji12
Amaethon	13.82	177.47	Europa	Cilix	Je8
Amaterasu Patera	38.13	306.53	Io	Amaterasu Patera	Ji5
Amergin	−14.7	230.6	Europa	Castalia Macula	Je9
Amirani	24.46	114.68	Io	Amirani	Ji3, Ji3.1, Ji7
Ammura	31.7	342.4	Ganymede	Nun Sulci	Jg5
Amon	33.7	220.7	Ganymede	Philus Sulcus	Jg4
Amset	−14.4	178.8	Ganymede	Uruk Sulcus	Jg8, Ji8.6
Ánarr	44	0.5	Callisto	Vestri	Jc2, Jc5
Anat	−4.1	127.9	Ganymede	Memphis Facula	Jg7
Andjeti	−52.8	161.1	Ganymede	Osiris	Jg12
Androgeos Linea	11.7	279.3	Europa	Castalia Macula	Je9, Je9.4
Angpetu Patera	−21.15	8.79	Io	Masubi	Ji11
Angus	−12.6	75.1	Europa	Brigid	Je7
Anhur	32.6	192.3	Ganymede	Philus Sulcus	Jg4, Jg4.1
Annwn Regio	20	320	Europa	Callanish	Je5, Je10
Anshar Sulcus	18	197.9	Ganymede	Uruk Sulcus	Jg4, Jg8, Jg8.5
Antum	5.1	218.9	Ganymede	Tiamat Sulcus	Jg9
Anu	65.2	344.3	Ganymede	Etana	Jg1
Anubis	−84.2	128.9	Ganymede	Hathor	Jg15, Jg15.1
Anzu	63.5	62.5	Ganymede	Perrine	Jg2
Apis Tholus	−10.9	347.88	Io	Loki	Ji10
Apophis	−8	276	Ganymede	Tiamat Sulcus	Jg9
Apsu Sulci	−39.4	234.7	Ganymede	Apsu Sulci	Jg13
Aquarius Sulcus	52.4	3.9	Ganymede	Perrine	Jg2
Aramazd Patera	−73.58	336.81	Io	Haemus Montes	Ji15
Arbela Sulcus	−21.1	349.8	Ganymede	Namtar	Jg10, Jg10.3, Jg14

Name	Latitude	Longitude	Satellite	Quadrangle	Plate
Arcas	−85.6	67.5	Callisto	Keelut	Jc15
Argadnel Regio	−14.6	208.5	Europa	Cilix	Je8
Argiope Linea	−1.7	195.6	Europa	Cilix	Je8
Argos Planum	−47.98	317.81	Io	Creidne Patera	Ji14
Arinna Fluctus	31.24	149.16	Io	Amirani	Ji3
Arran Chaos	13.4	80.5	Europa	Brigid	Je7
Arusha Patera	−39.06	101.48	Io	Tohil Mons	Ji12
Asgard	32.2	139.9	Callisto	Asgard	Jc3
Asha Patera	−8.84	225.69	Io	Pele	Ji9
Ashîma	−39.1	123	Ganymede	Osiris	Jg12
Askr	51.8	324.1	Callisto	Askr	Jc5
Asshur	54.2	333.5	Ganymede	Nun Sulci	Jg5
Asterius Linea	14.9	270.8	Europa	Castalia Macula	Je9, Je9.3
Astypalaea Linea	−75.8	212.1	Europa	Sidon Flexus	Je15
Atar Patera	31.1	278.58	Io	Amaterasu Patera	Ji5
Aten Patera	−48.53	310.02	Io	Creidne Patera	Ji14
Atra-hasis	22.5	254.2	Ganymede	Philus Sulcus	Jg4
Audr	−30.9	80.6	Callisto	Adlinda	Jc11
Austri	−80.9	64.5	Callisto	Keelut	Jc15
Autonoë Linea	18.2	165.1	Europa	Cilix	Je8
Avagddu	1.4	169.5	Europa	Cilix	Je8
Aya	68.3	322.2	Ganymede	Etana	Jg1
Aziren	35.4	178.2	Callisto	Asgard	Jc3
Ba'al	24.9	329.9	Ganymede	Nun Sulci	Jg5
Babbar Patera	−39.79	271.98	Io	Creidne Patera	Ji14
Bactria Regio	−47.6	123.77	Io	Tohil Mons	Ji12
Balder Patera	11.44	156.1	Io	Prometheus	Ji8, Ji8.2.1, Ji8.3.4
Balgatan Regio	−50	30	Europa	Butterdon Linea	Je11
Balkr	28.9	11.7	Callisto	Vestri	Jc2
Balor	−52.8	97.8	Europa	Taliesin	Je12
Barnard Regio	−10.7	19	Ganymede	Dardanus Sulcus	Jg6
Barri	−31.5	70.5	Callisto	Adlinda	Jc11

Name	Latitude	Longitude	Satellite	Quadrangle	Plate
Bau	23	48.7	Ganymede	Perrine	Jg2
Bavörr	49.1	20	Callisto	Vestri	Jc2
Belenus Patera	2.9	157.72	Io	Prometheus	Ji8, Ji8.2.1
Beli	62.6	80.2	Callisto	Vestri	Jc2
Belus Linea	9.3	231.4	Europa	Castalia Macula	Je8.6, Je9, Je9.8
Bes	−25.5	180.9	Ganymede	Apsu Sulci	Jg13
Biflindi	−53.6	74.1	Callisto	Adlinda	Jc11
Bigeh Facula	29	94.3	Ganymede	Galileo Regio	Jg2.4, Jg3
Boösaule Montes	−3.68	269.09	Io	Pele	Ji9
Bochica Patera	−61.5	18.85	Io	Masubi	Ji11
Boeotia Macula	−53.6	166.8	Europa	Taliesin	Je12
Bosphorus Regio	−1.58	120.51	Io	Emakong Patera	Ji7
Bragi	75.5	60.7	Callisto	Gipul Catena	Jc1
Brami	28.8	19	Callisto	Vestri	Jc2
Bran	−24.2	205.6	Callisto	Hoenir	Jc13, Jc13.1
Bress	37.64	98.66	Europa	Tyre	Je3
Brigid	10.8	81.3	Europa	Brigid	Je7
Bubastis Sulci	−72.3	282.9	Ganymede	Hathor	Jg15, Jg15.1
Buga	22.3	323.9	Callisto	Askr	Jc5
Bulicame Regio	34.85	190.82	Io	Lei-Kung Fluctus	Ji4
Buri	−37.5	45.5	Callisto	Adlinda	Jc11
Burr	42.7	134.5	Callisto	Asgard	Jc3
Busiris Facula	15.7	215.4	Ganymede	Uruk Sulcus	Jg8
Buto	13.2	203.5	Ganymede	Uruk Sulcus	Jg8, Jg8.3
Butterdon Linea	−44.7	0.1	Europa	Butterdon Linea	Je11
Byblus Sulcus	37.9	199.9	Ganymede	Philus Sulcus	Jg4
Cadmus Linea	38.7	191.7	Europa	Rhadamanthys Linea	Je4, Jg4.2
Callanish	−16.7	334.5	Europa	Callanish	Je10, Je10.1
Camaxtli Patera	15.25	136.8	Io	Emakong Patera	Ji7
Camulus	−26.5	81.1	Europa	Butterdon Linea	Je11
Capaneus Mensa	−16.46	121.59	Io	Emakong Patera	Ji7
Carancho Patera	1.43	317.2	Io	Loki	Ji10

Name	Latitude	Longitude	Satellite	Quadrangle	Plate
Castalia Macula	−1.6	225.7	Europa	Castalia Macula	Je9, Je9.2
Cataquil Patera	−23.89	16.57	Io	Masubi	Ji11
Catha Patera	−53.76	101.58	Io	Tohil Mons	Ji12
Caucasus Mons	−32.38	238.57	Io	Caucasus Mons	Ji13
Chaac Patera	11.88	157.44	Io	Prometheus	Ji8, Ji8.2, Ji8.3
Chalybes Regio	57.5	86.21	Io	Zal Patera	Ji2
Chors Patera	68.47	249.85	Io	Chors	Ji1
Chrysor	15.3	134.3	Ganymede	Memphis Facula	Jg7, Jg7.1
Chthonius Linea	−1.4	304.2	Europa	Callanish	Je10
Cilicia Flexus	−59.5	171.7	Europa	Taliesin	Je12
Cilix	2.6	181.9	Europa	Cilix	Je8
Cisti	−31.7	64.2	Ganymede	Nabu	Jg11
Cliodhna	−2.5	76.4	Europa	Brigid	Je7
Colchis Regio	1.63	205.01	Io	Prometheus	Ji8
Conamara Chaos	9.7	272.7	Europa	Castalia Macula	Je9
Coptos Facula	9.9	209.2	Ganymede	Uruk Sulcus	Jg8
Corick Linea	17.8	18.3	Europa	Euphemus Linea	Je6
Cormac	−36.9	88.1	Europa	Butterdon Linea	Je11
Creidne Patera	−53.02	342.53	Io	Creidne Patera	Ji14
Crimea Mons	−76.16	241.18	Io	Haemus Montes	Ji15
Cuchi Patera	−0.9	144.57	Io	Prometheus	Ji8
Culann Patera	−20.22	160.17	Io	Prometheus	Ji8, Ji8.5
Cyclades Macula	−62.5	191.3	Europa	Agenor Linea	Je13
Daedalus Patera	19.52	274.35	Io	Pele	Ji9
Dag	58.5	73.3	Callisto	Vestri	Jc2
Dagda	37.35	168.74	Europa	Tyre	Je3
Damkina	−30	5	Ganymede	Nabu	Jg11
Danel	−4.3	21.3	Ganymede	Dardanus Sulcus	Jg6
Danr	62.5	76.9	Callisto	Vestri	Jc2
Danube Planum	−22.73	257.44	Io	Caucasus Mons	Ji13
Dardanus Sulcus	−46.9	17.5	Ganymede	Nabu	Jg6, Jg11
Dazhbog Patera	55.1	301.48	Io	Amaterasu Patera	Ji5

Name	Latitude	Longitude	Satellite	Quadrangle	Plate
Debegey	10.2	166.2	Callisto	Vidarr	Jc8
Deirdre	−65.4	207.3	Europa	Agenor Linea	Je13
Delphi Flexus	−68.2	174.1	Europa	Sidon Flexus	Je15
Dendera	−1.2	255.4	Ganymede	Tiamat Sulcus	Jg9
Dia	73	50.5	Callisto	Gipul Catena	Jc1
Diarmuid	−61.3	102	Europa	Taliesin	Je12
Diment	23.1	351.8	Ganymede	Nun Sulci	Jg5
Dingir Patera	−4.13	341.37	Io	Loki	Ji10, Ji10.2.1
Dodona Planum	−58.8	347.51	Io	Creidne Patera	Ji14
Doh	30.6	141.4	Callisto	Asgard	Jc3, Jc3.3
Donar Fluctus	22.67	187.63	Io	Lei-Kung Fluctus	Ji4
Dorian Montes	−25.89	198.13	Io	Caucasus Mons	Ji13, Ji13.2
Drizzlecomb Linea	7.7	111.7	Europa	Brigid	Je7
Drumskinny Linea	48.3	161	Europa	Tyre	Je3, Je3.1.2
Dryops	80	34.8	Callisto	Gipul Catena	Jc1
Dukug Sulcus	83.5	3.8	Ganymede	Etana	Jg1
Durinn	67	89.1	Callisto	Gipul Catena	Jc1
Dusura Patera	37.47	119.02	Io	Amirani	Ji3
Dyfed Regio	10	250	Europa	Castalia Macula	Je9
Dylan	−55.3	84.4	Europa	Butterdon Linea	Je11
Ea	17.7	148.7	Ganymede	Uruk Sulcus	Jg8, Jg8.1
Echion Linea	−11.6	185.2	Europa	Cilix	Je8
Echo Mensa	−80	355.63	Io	Haemus Montes	Ji15
Edfu Facula	25.7	147.1	Ganymede	Galileo Regio	Jg3
Egdir	33.9	35.9	Callisto	Vestri	Jc2
Egres	42.5	176.6	Callisto	Asgard	Jc3
Egypt Mons	−41.49	257.6	Io	Caucasus Mons	Ji13, Ji13.1
Eikin Catena	−8.9	15.5	Callisto	Valhalla	Jc6
Ekhi Patera	−28.35	88.5	Io	Masubi	Ji11
El	1	151.4	Ganymede	Uruk Sulcus	Jg8
Elam Sulci	58.2	200.3	Ganymede	Philus Sulcus	Jg4
Elathan	−31.9	79.8	Europa	Butterdon Linea	Je11

Name	Latitude	Longitude	Satellite	Quadrangle	Plate
Emakong Patera	−3.33	119.82	Io	Emakong Patera	Ji7, Ji7.1
Enki Catena	38.8	13.6	Ganymede	Perrine	Jg2, Jg2.3
Enkidu	−26.6	325.2	Ganymede	Namtar	Jg14
Enlil	55.3	312.2	Ganymede	Nun Sulci	Jg5
En-zu	11.6	168.4	Ganymede	Uruk Sulcus	Jg8, J8.2
Eochaid	−50.48	233.33	Europa	Agenor Linea	Je13
Epaphus Mensa	−52.97	239.99	Io	Caucasus Mons	Ji13
Epigeus	23.4	180.6	Ganymede	Philus Sulcus	Jg4, Jg4.5
Erech Sulcus	−7.3	179.2	Ganymede	Uruk Sulcus	Jg8, Jg8.6
Erichthonius	−15.3	175.3	Ganymede	Uruk Sulcus	Jg8, Jg8.6
Erlik	66.8	1.3	Callisto	Gipul Catena	Jc1
Eshmun	−17.4	192.1	Ganymede	Uruk Sulcus	Jg8
Estan Patera	21.61	87.71	Io	Zal Patera	Ji2, Ji7.3.1
Etana	74.7	340.5	Ganymede	Etana	Jg1
Ethiopia Planum	−45.3	24.59	Io	Masubi	Ji11
Euboea Fluctus	−45.01	350.96	Io	Creidne Patera	Ji14
Euboea Montes	−48.89	338.77	Io	Creidne Patera	Ji14, Ji14.1
Euphemus Linea	−11.4	45.7	Europa	Euphemus Linea	Je6
Euxine Mons	26.27	126.49	Io	Amirani	Ji3
Fadir	56.6	12.6	Callisto	Vestri	Jc2
Falga Regio	30	210	Europa	Rhadamanthys Linea	Je4
Fili	64.2	349.7	Callisto	Askr	Jc5
Fimbulthul Catena	8.2	64.8	Callisto	Valhalla	Jc6
Finnr	15.5	4.3	Callisto	Valhalla	Jc6
Fjorgynn Fluctus	11.23	358.08	Io	Loki	Ji10
Fo Patera	40.79	192.25	Io	Lei-Kung Fluctus	Ji4
Freki	79.8	351.7	Callisto	Gipul Catena	Jc1
Frodi	68.4	139.9	Callisto	Gipul Catena	Jc1
Fuchi Patera	28.34	327.57	Io	Amaterasu Patera	Ji5
Fulla	74	108.1	Callisto	Gipul Catena	Jc1
Fulnir	60.1	35.3	Callisto	Vestri	Jc2
Göll	57.3	319.7	Callisto	Askr	Jc5

Name	Latitude	Longitude	Satellite	Quadrangle	Plate
Göndul	60	114.1	Callisto	Asgard	Jc3
Gabija Patera	−51.9	202.53	Io	Caucasus Mons	Ji13
Gad	−13.6	137.6	Ganymede	Memphis Facula	Jg7
Galai Patera	−10.87	288.11	Io	Pele	Ji9
Galileo Regio	47	129.6	Ganymede	Galileo Regio	Jg3
Gandalfr	−80.5	63.6	Callisto	Keelut	Jc15
Geb	56.3	182.6	Ganymede	Philus Sulcus	Jg4
Geinos	18.6	219.4	Ganymede	Tiamat Sulcus	Jg9
Geirvimul Catena	48.9	347.2	Callisto	Askr	Jc5
Geri	66.7	353.8	Callisto	Gipul Catena	Jc1
Gibil Patera	−14.99	294.64	Io	Loki	Ji10
Gilgamesh	−62.8	125	Ganymede	Osiris	Jg12
Ginandi	−85.3	52.1	Callisto	Keelut	Jc15
Gipul Catena	68.5	54.2	Callisto	Gipul Catena	Jc1
Gir	34	145.7	Ganymede	Galileo Regio	Jg3
Girru Patera	22.8	239.92	Io	Lei-Kung Fluctus	Ji4
Gish Bar Mons	18.6	87.7	Io	Emakong Patera	Ji7.3
Gish Bar Patera	16.18	90.26	Io	Emakong Patera	Ji7.3
Gisl	57.2	34.6	Callisto	Vestri	Jc2
Glaukos Linea	57.8	230.9	Europa	Rhadamanthys Linea	Je4
Gloi	49	245	Callisto	Gloi	Jc4
Gomul Catena	35.5	47.1	Callisto	Vestri	Jc2, Jc2.1
Gortyna Flexus	−42.1	144.6	Europa	Taliesin	Je12
Govannan	−37.3	302.8	Europa	Pwyll	Je14
Gráinne	−59.7	99.4	Europa	Taliesin	Je12
Grannos Patera	11.17	145.29	Io	Prometheus	Ji8, Ji8.2.1
Grimr	41.5	214.6	Callisto	Gloi	Jc4
Gula	64.1	12.3	Ganymede	Perrine	Jg2, Jg2.2
Gunnr	64.6	104.7	Callisto	Asgard	Jc3
Gunntro Catena	−19.5	343.1	Callisto	Vali	Jc10
Gwern	9.14	344.54	Europa	Callanish	Je10
Gwydion	−60.5	81.6	Europa	Butterdon Linea	Je11

Name	Latitude	Longitude	Satellite	Quadrangle	Plate
Gymir	63.7	48.8	Callisto	Vestri	Jc2
Hábrok	76.2	131.9	Callisto	Gipul Catena	Jc1
Hár	−3.5	358	Callisto	Vali	Jc10, Jc10.2
Haemus Montes	−69.71	46.23	Io	Haemus Montes	Ji15, Ji15.2
Haki	25	315.1	Callisto	Askr	Jc5
Halieus	34.3	167.1	Ganymede	Galileo Regio	Jg2.4, Jg3
Haokah Patera	−20.87	186.65	Io	Prometheus	Ji8
Hapi	−30.6	212.6	Ganymede	Apsu Sulci	Jg13
Harakhtes	35.9	100.2	Ganymede	Galileo Regio	Jg3
Harmonia Linea	28	171.7	Europa	Tyre	Je3
Haroeris	28.5	296.8	Ganymede	Nun Sulci	Jg5
Harpagia Sulcus	−11.7	318.7	Ganymede	Misharu	Jg10, Jg10.5, Jg10.6
Hatchawa Patera	−59.42	31.99	Io	Masubi	Ji11
Hathor	−66.9	268.6	Ganymede	Hathor	Jg15, Jg15.1
Hay-tau	14.5	133.1	Ganymede	Memphis Facula	Jg7, Jg7.1
Hedetet	−33	251.1	Ganymede	Apsu Sulci	Jg13
Heimdall	−63.5	357	Callisto	Lempo	Jc14
Heiseb Patera	29.79	244.85	Io	Lei-Kung Fluctus	Ji4
Heliopolis Facula	18.5	147.2	Ganymede	Uruk Sulcus	Jg8
Heno Patera	−57.14	311.47	Io	Creidne Patera	Ji14
Hephaestus Patera	1.95	289.79	Io	Loki	Ji10
Hepti	64.5	23.4	Callisto	Vestri	Jc2
Hermes Mensa	−43.48	246.3	Io	Caucasus Mons	Ji13
Hermopolis Facula	22.4	195.3	Ganymede	Philus Sulcus	Jg4, Jg4.4
Hershef	47.3	269.5	Ganymede	Philus Sulcus	Jg4, Jg5
Hi'iaka Montes	−5.13	81.8	Io	Emakong Patera	Ji7.4
Hi'iaka Patera	−3.64	79.47	Io	Emakong Patera	Ji7
Hijsi	63.1	171.5	Callisto	Asgard	Jc3
Hiruko Patera	−65.09	328.83	Io	Creidne Patera	Ji14
Hödr	69.1	89.2	Callisto	Gipul Catena	Jc1
Hoenir	−33.7	260.9	Callisto	Hoenir	Jc13
Högni	−11.8	4.8	Callisto	Valhalla	Jc6

Name	Latitude	Longitude	Satellite	Quadrangle	Plate
Höldr	43.9	108.2	Callisto	Asgard	Jc3
Horus Patera	−9.82	338	Io	Loki	Ji10, Ji10.2.1
Humbaba	−55.2	67.3	Ganymede	Nabu	Jg11
Huo Shen Patera	−15.01	328.98	Io	Loki	Ji10, Ji10.2.1
Hursag Sulcus	−9.7	233.1	Ganymede	Tiamat Sulcus	Jg9
Hybristes Planum	−54.41	18.45	Io	Masubi	Ji11
Hyperenor Linea	−12.1	324.4	Europa	Callanish	Je10
Igaluk	5.6	316	Callisto	Vali	Jc10
Ilah	21.9	160.6	Ganymede	Galileo Regio	Jg3
Illyrikon Regio	−69.85	169.04	Io	Haemus Montes	Ji15
Ilma	−29.9	167.2	Callisto	Ilma	Jc12
Ilmarinen Patera	−14.44	1.15	Io	Kanehekili	Ji6
Ilus	−13.4	110.3	Ganymede	Memphis Facula	Jg7
Inachus Tholus	−16.18	347.76	Io	Loki	Ji10
Ino Linea	−1.7	174.6	Europa	Cilix	Je8
Inti Patera	−68.37	347.52	Io	Haemus Montes	Ji15
Ionian Mons	8.61	236.56	Io	Pele	Ji9
Iopolis Planum	−35.31	332.72	Io	Creidne Patera	Ji14
Irkalla	−32.6	114.7	Ganymede	Osiris	Jg12
Ishkur	0.3	8.4	Ganymede	Dardanus Sulcus	Jg6
Isimu	8.4	360	Ganymede	Misharu	Jg10
Isis	−67.3	201.1	Ganymede	Hathor	Jg13, Jg15
Isum Patera	29.82	208.46	Io	Lei-Kung Fluctus	Ji4
Itzamna Patera	−16.13	99.46	Io	Emakong Patera	Ji7
Ivarr	−5.8	321.4	Callisto	Vali	Jc10
Iynx Mensa	−62.25	303.98	Io	Creidne Patera	Ji14
Jalkr	−38.6	82.7	Callisto	Adlinda	Jc11
Janus Patera	−4.56	39.05	Io	Kanehekili	Ji6
Jumal	58.9	118	Callisto	Asgard	Jc3
Jumo	65.7	11.8	Callisto	Gipul Catena	Jc1, Jc2
Kári	48.2	116.3	Callisto	Asgard	Jc3
Kadi	47.7	178.5	Ganymede	Galileo Regio	Jg3

Name	Latitude	Longitude	Satellite	Quadrangle	Plate
Kami-Nari Patera	−8.7	235.08	Io	Pele	Ji9
Kane Patera	−48.39	11.77	Io	Masubi	Ji11
Kanehekili	−18.21	33.61	Io	Kanehekili	Ji6
Kanehekili Fluctus	−17.68	33.56	Io	Kanehekili	Ji6
Karei Patera	0.2	13.08	Io	Kanehekili	Ji6
Karl	56.4	330.6	Callisto	Askr	Jc5
Katreus Linea	−38.8	213.3	Europa	Agenor Linea	Je13
Kava Patera	−16.83	341.33	Io	Loki	Ji10
Keelut	−76.8	90.9	Callisto	Keelut	Jc15
Kennet Linea	−41	312	Europa	Pwyll	Je14
Khalla Patera	5.14	303.56	Io	Loki	Ji10
Khensu	1	152.9	Ganymede	Uruk Sulcus	Jg8, Jg8.7
Khepri	20.4	147.6	Ganymede	Uruk Sulcus	Jg8
Khnum Catena	32.9	349.1	Ganymede	Nun Sulci	Jg5
Khonsu	−37.5	190.8	Ganymede	Apsu Sulci	Jg13
Khumbam	−24.1	335.4	Ganymede	Namtar	Jg14
Kibero Patera	−11.83	305.1	Io	Loki	Ji10
Kingu	−34.8	227.1	Ganymede	Apsu Sulci	Jg13
Kinich Ahau Patera	49.34	310.2	Io	Amaterasu Patera	Ji5
Kishar	72.6	349.7	Ganymede	Etana	Jg1
Kishar Sulcus	−6.4	216.6	Ganymede	Tiamat Sulcus	Jg9
Kittu	0.4	334.6	Ganymede	Misharu	Jg10, Jg10.1
Kol Facula	4.5	282.7	Callisto	Valfodr	Jc9
Kul'	62.9	121.9	Callisto	Asgard	Jc3
Kulla	33.3	113.8	Ganymede	Galileo Regio	Jg2.4, Jg3
Kurdalagon Patera	−49.61	217.99	Io	Caucasus Mons	Ji13
Lagamal	64.4	244.8	Ganymede	Philus Sulcus	Jg1, Jg4
Lagash Sulcus	−10.9	163.2	Ganymede	Uruk Sulcus	Jg8, Jg8.4
Lakhamu Fossa	−11.6	227.7	Ganymede	Tiamat Sulcus	Jg9
Lakhmu Fossae	50.4	128	Ganymede	Galileo Regio	Jg3
Laki-oi Patera	−38.94	61.93	Io	Masubi	Ji11
Larsa Sulcus	3.8	248.7	Ganymede	Tiamat Sulcus	Jg9

Name	Latitude	Longitude	Satellite	Quadrangle	Plate
Latpon	58.8	171.2	Ganymede	Galileo Regio	Jg3, Jg3.2
Lei-Kung Fluctus	40.4	206.47	Io	Lei-Kung Fluctus	Ji4
Lei-zi Fluctus	14.25	46.16	Io	Kanehekili	Ji6
Lempo	−25.2	319.9	Callisto	Lempo	Jc14
Lerna Regio	−62.35	291.86	Io	Creidne Patera	Ji14
Libya Linea	−54	181	Europa	Agenor Linea	Je12.1, Je12.2, Je13
Ljekio	49.1	162.3	Callisto	Asgard	Jc3
Llew Patera	12.16	242.31	Io	Pele	Ji9
Llyr	−1.8	221.8	Europa	Castalia Macula	Je9
Lodurr	−50.8	270.1	Callisto	Lempo	Jc13, Jc14
Lofn	−56.5	22.3	Callisto	Adlinda	Jc11
Loki	18.22	302.56	Io	Loki	Ji10, Ji10.1
Loki Patera	12.97	308.8	Io	Loki	Ji10
Loni	−3.6	214.3	Callisto	Vidarr	Jc8, Jc9
Losy	65.3	323.3	Callisto	Gipul Catena	Jc1, Jc5
Lu Huo Patera	−38.52	353.17	Io	Creidne Patera	Ji14
Luchtar	−40.2	257.57	Europa	Agenor Linea	Je13
Lug	27.99	44.31	Europa	Murias Chaos	Je2
Lugalmeslam	23.8	193.8	Ganymede	Philus Sulcus	Jg4, Jg4.4
Lumha	36	154.3	Ganymede	Galileo Regio	Jg3
Lycaon	−45.4	5.9	Callisto	Adlinda	Jc11
Lyrcea Planum	−41.92	268.83	Io	Caucasus Mons	Ji13
Maa	1.3	203.6	Ganymede	Uruk Sulcus	Jg8, Jg8.8
Maasaw Patera	−40.28	339.09	Io	Creidne Patera	Ji14, Ji14.1
Maderatcha	30.7	95.3	Callisto	Asgard	Jc3
Mael Dúin	−16.8	197.9	Europa	Cilix	Je8, Je8.3, Je8.4
Maeve	58.8	78.9	Europa	Murias Chaos	Je2
Mafuike Patera	−13.52	259.47	Io	Pele	Ji9
Malik Patera	−34.15	129.59	Io	Tohil Mons	Ji12
Mama Patera	−11.28	355.31	Io	Loki	Ji10
Manannán	3.1	239.7	Europa	Castalia Macula	Je9
Manua Patera	35.78	321.73	Io	Amaterasu Patera	Ji5

Name	Latitude	Longitude	Satellite	Quadrangle	Plate
Marduk	−29.29	209.74	Io	Caucasus Mons	Ji13
Marduk Fluctus	−27.59	210.83	Io	Caucasus Mons	Ji13
Marius Regio	6.8	181.2	Ganymede	Uruk Sulcus	Jg8
Masaya Patera	−22.62	344.49	Io	Creidne Patera	Ji10.2.1, Ji14
Mashu Sulcus	29.8	205.7	Ganymede	Philus Sulcus	Jg4, Jg4.4
Masubi	−49.6	56.18	Io	Masubi	Ji11
Masubi Fluctus	−50.49	58.13	Io	Masubi	Ji11
Math	−25.6	183.7	Europa	Agenor Linea	Je8.2, Je13
Maui	19.53	122.31	Io	Emakong Patera	Ji7
Maui Patera	16.61	124.25	Io	Emakong Patera	Ji7
Mazda Paterae	−8.81	313.28	Io	Loki	Ji10
Mbali Patera	−31.5	5.06	Io	Masubi	Ji11, Ji11.1
Media Regio	8.65	59.43	Io	Kanehekili	Ji6
Mehen Linea	56	236.7	Europa	Rhadamanthys Linea	Je4
Mehit	29	164.4	Ganymede	Galileo Regio	Jg3
Melkart	−9.9	186.2	Ganymede	Uruk Sulcus	Jg8, Jg8.10, Jg8.11
Memphis Facula	14.1	131.9	Ganymede	Memphis Facula	Jg7
Menahka Patera	−31.31	344.75	Io	Creidne Patera	Ji14
Menhit	−36.5	140.5	Ganymede	Osiris	Jg12, Jg12.3
Mentu Patera	7	139.36	Io	Emakong Patera	Ji7
Mera	64.1	75.2	Callisto	Vestri	Jc2
Merrivale Linea	−41	299.5	Europa	Pwyll	Je14
Michabo Patera	1.2	167.6	Io	Prometheus	Ji8, Ji8.7.1
Midir	3.65	338.75	Europa	Callanish	Je10
Mihr Patera	−16.51	305.45	Io	Loki	Ji10
Mimir	32.6	53.2	Callisto	Vestri	Jc2
Min	29.2	1.2	Ganymede	Perrine	Jg2
Minos Linea	47.2	195.2	Europa	Rhadamanthys Linea	Je4
Mir	−3.3	230.3	Ganymede	Tiamat Sulcus	Jg9
Misharu	−4.4	335.9	Ganymede	Misharu	Jg10
Mithra Patera	−59	266.46	Io	Caucasus Mons	Ji13
Mitsina	57.5	103.7	Callisto	Asgard	Jc3

Name	Latitude	Longitude	Satellite	Quadrangle	Plate
Modi	66.4	119.3	Callisto	Gipul Catena	Jc1
Monan Mons	15.2	104.5	Io	Emakong Patera	Ji7, Ji7.3.1
Monan Patera	19.72	105.39	Io	Emakong Patera	Ji7, Ji7.3.1
Mongibello Mons	22.75	66.95	Io	Zal Patera	Ji2, Ji6.1
Mont	44.6	311.9	Ganymede	Nun Sulci	Jg5, Jg5.1
Mor	30.5	327.4	Ganymede	Nun Sulci	Jg5
Morvran	−4.9	152.6	Europa	Cilix	Je8
Mot	9.9	165.9	Ganymede	Uruk Sulcus	Jg8
Moytura Regio	−50	294.3	Europa	Pwyll	Je14
Mulungu Patera	17.26	217.87	Io	Pele	Ji9
Murias Chaos	22.4	83.8	Europa	Murias Chaos	Je2, Je2.2
Mush	−15.1	114.8	Ganymede	Memphis Facula	Jg7
Mycenae Regio	−36.07	165.39	Io	Tohil Mons	Ji12
Mysia Sulci	−7	7.9	Ganymede	Dardanus Sulcus	Jg6
Nabu	−45.4	1.2	Ganymede	Nabu	Jg11
Nah-Hunte	−17.8	85.2	Ganymede	Memphis Facula	Jg7
Nakki	−56.4	69.7	Callisto	Adlinda	Jc11
Nama	57	331	Callisto	Askr	Jc5
Namarrkun Patera	10.06	175.52	Io	Prometheus	Ji8
Namtar	−58.3	340.8	Ganymede	Namtar	Jg14
Nanna	−17.6	241.9	Ganymede	Tiamat Sulcus	Jg9
Nanshe Catena	15.4	352.9	Ganymede	Misharu	Jg10, Jg10.2
Nár	−1.5	46	Callisto	Valhalla	Jc6
Narberth Chaos	−26	273	Europa	Pwyll	Je14.1.1
Nefertum	44.3	321.1	Ganymede	Nun Sulci	Jg5, Jg5.1
Neheh	72.1	62.5	Ganymede	Etana	Jg1
Neith	29.4	7	Ganymede	Perrine	Jg2, Jg2.1
Nemea Planum	−72.82	267.72	Io	Haemus Montes	Ji15
Nergal	38.6	200.3	Ganymede	Philus Sulcus	Jg4, Jg4.2
Nerrivik	−16.9	56.4	Callisto	Valhalla	Jc6
Niamh	21.1	216.9	Europa	Rhadamanthys Linea	Je4, Je8.6.1
Nicholson Regio	−33.5	4.8	Ganymede	Nabu	Jg10, Jg11

Name	Latitude	Longitude	Satellite	Quadrangle	Plate
Nidaba	17.7	123.2	Ganymede	Memphis Facula	Jg7
Nidi	66.4	94.9	Callisto	Gipul Catena	Jc1
Nigirsu	−58.2	320.5	Ganymede	Namtar	Jg14
Nile Montes	52.07	250.02	Io	Lei-Kung Fluctus	Ji4
Nina Patera	−38.24	162.6	Io	Tohil Mons	Ji12
Nineveh Sulcus	23.5	53.1	Ganymede	Perrine	Jg2
Ningishzida	14.1	189.8	Ganymede	Uruk Sulcus	Jg8, Jg8.5
Ninkasi	59.2	48.7	Ganymede	Perrine	Jg2
Ninki	−8.2	120.5	Ganymede	Memphis Facula	Jg7
Ninlil	6.2	118.3	Ganymede	Memphis Facula	Jg7
Ninsum	−14.5	140.6	Ganymede	Memphis Facula	Jg7
Ninurta Patera	−16.74	315.25	Io	Loki	Ji10
Nippur Sulcus	36.9	185	Ganymede	Philus Sulcus	Jg4
Nirkes	31.4	164.3	Callisto	Asgard	Jc3
Njord	16.7	132.6	Callisto	Njord	Jc7
Nori	45.2	343.6	Callisto	Askr	Jc5
Norov-Ava	54.6	112.8	Callisto	Asgard	Jc3
Nuada	62.3	272.5	Callisto	Askr	Jc5
Numi-Torum	−50.1	92.9	Callisto	Ilma	Jc12
Nun Sulci	49.5	316.4	Ganymede	Nun Sulci	Jg5
Nusku Patera	−65.01	3.82	Io	Masubi	Ji11
Nut	−54.3	269.3	Ganymede	Apsu Sulci	Jg13, Jg15.1
Nyambe Patera	0.38	343.11	Io	Loki	Ji10, Ji10.1.2
Nyctimus	−62.8	3.9	Callisto	Adlinda	Jc11
Ogma	87.45	287.86	Europa	Ogma	Je1, Je1.1
Oisín	−52.3	213.4	Europa	Agenor Linea	Je13
Oluksak	−47.8	63.5	Callisto	Adlinda	Jc11
Ombos	4.8	236	Ganymede	Tiamat Sulcus	Jg9
Omol'	42.3	116.9	Callisto	Asgard	Jc3
Onga Linea	−38.7	211.3	Europa	Agenor Linea	Je13
Orestheus	−46.7	47.7	Callisto	Adlinda	Jc11
Osiris	−38.1	166.4	Ganymede	Osiris	Jg12, Jg12.1

Name	Latitude	Longitude	Satellite	Quadrangle	Plate
Oski	57.5	269	Callisto	Gloi	Jc4
Ot Mons	4.21	215.64	Io	Prometheus	Ji8, Ji8.6
Ot Patera	−1.1	217.4	Io	Pele	Ji9
Ottar	61.5	103.9	Callisto	Asgard	Jc3
Päive Patera	−45.71	358.54	Io	Creidne Patera	Ji14
Pan Mensa	−50.88	31.77	Io	Masubi	Ji11
Pautiwa Patera	−34.2	345.59	Io	Creidne Patera	Ji14
Pekko	18.3	5.4	Callisto	Valhalla	Jc6
Pelagon Linea	35.5	173.6	Europa	Tyre	Je3
Pele	−18.71	255.28	Io	Pele	Ji9, Ji9.1
Pelorus Linea	−19.8	188.3	Europa	Cilix	Je8
Perrine Regio	33.2	32.5	Ganymede	Perrine	Jg2
Philae Sulcus	65.5	169	Ganymede	Etana	Jg1, Jg3.2
Philus Sulcus	44.1	209.5	Ganymede	Philus Sulcus	Jg4
Phineus Linea	−29.8	319.9	Europa	Pwyll	Je14
Phocis Flexus	−44.5	198.4	Europa	Agenor Linea	Je13
Phoenix Linea	16.6	188.8	Europa	Cilix	Je8
Phrygia Sulcus	12.4	23.5	Ganymede	Dardanus Sulcus	Jg6
Pillan Mons	−8.81	246.72	Io	Pele	Ji9
Pillan Patera	−12.34	243.25	Io	Pele	Ji9, Ji9.2.1
Podja Patera	−18.41	304.74	Io	Loki	Ji10, Ji14.2
Powys Regio	0	145	Europa	Cilix	Je8
Prometheus	−1.52	153.94	Io	Prometheus	Ji8, Ji8.1
Prometheus Mensa	−1.9	151.9	Io	Prometheus	Ji8
Pryderi	−66.1	159.1	Europa	Sidon Flexus	Je15
Ptah	−65.9	217	Ganymede	Hathor	Jg15
Punt	−24.9	239.9	Ganymede	Apsu Sulci	Jg13
Purgine Patera	−2.61	297.3	Io	Loki	Ji10
Pwyll	−25.2	271.4	Europa	Pwyll	Je14, Je14.1
Pyerun Patera	−55.64	251.21	Io	Caucasus Mons	Ji13
Ra Patera	−8.66	324.7	Io	Loki	Ji10, Ji10.2
Radegast Patera	−27.78	159.98	Io	Tohil Mons	Ji8.7.1, Ji12, Ji12.3

Name	Latitude	Longitude	Satellite	Quadrangle	Plate
Randver	−71.9	53.9	Callisto	Keelut	Jc15
Rarog Patera	−41.71	304.41	Io	Creidne Patera	Ji14
Rata Mons	−36.3	201.16	Io	Caucasus Mons	Ji13
Rata Patera	−35.61	199.78	Io	Caucasus Mons	Ji13
Rathmore Chaos	25.4	75	Europa	Murias Chaos	Je2
Reginleif	−66	96.5	Callisto	Keelut	Jc15
Reginn	39.8	90.1	Callisto	Asgard	Jc3
Reiden Patera	−13.4	235.45	Io	Pele	Ji9
Reifnir	−50.8	54.3	Callisto	Adlinda	Jc11
Reshef Patera	27.69	158.06	Io	Amirani	Ji3, Ji3.3
Reshet Patera	0.53	305.48	Io	Loki	Ji10
Rhadamanthys Linea	19.3	200.5	Europa	Cilix	Je4, Je4.1, Je8
Rhiannon	−80.9	194.9	Europa	Sidon Flexus	Je15, Je15.4
Rigr	70.8	244.6	Callisto	Gipul Catena	Jc1
Rongoteus	53.6	106.1	Callisto	Asgard	Jc3
Rota	27.2	108.4	Callisto	Asgard	Jc3
Ruaumoko Patera	14.72	139.74	Io	Emakong Patera	Ji7, Ji8.2.1
Ruti	13.2	308.6	Ganymede	Misharu	Jg10
Ruwa Patera	0.19	1.75	Io	Kanehekili	Ji6
Saga	0.6	325.9	Callisto	Vali	Jc10
Saltu	−14.2	352.7	Ganymede	Misharu	Jg10
Sapas	57.4	33.9	Ganymede	Perrine	Jg2
Sarakka	−3.3	53.5	Callisto	Valhalla	Jc6
Sarpedon Linea	−49.5	92.9	Europa	Taliesin	Je12
Sati	30.9	12.8	Ganymede	Perrine	Jg2
Savitr Patera	48.19	123.36	Io	Amirani	Ji3
Sebek	61.2	356.9	Ganymede	Nun Sulci	Jg5
Sêd Patera	−2.91	303.57	Io	Loki	Ji10
Seima	17.1	215.9	Ganymede	Uruk Sulcus	Jg8
Seker	−39.2	345.5	Ganymede	Namtar	Jg14
Selket	15	105.8	Ganymede	Memphis Facula	Jg7
Sengen Patera	−32.89	303.73	Io	Creidne Patera	Ji14

Name	Latitude	Longitude	Satellite	Quadrangle	Plate
Seqinek	55.5	25.4	Callisto	Vestri	Jc2
Serapis	−12.5	44.1	Ganymede	Dardanus Sulcus	Jg6
Seth Mons	−10.72	134.19	Io	Emakong Patera	Ji7
Seth Patera	−5.35	131.67	Io	Emakong Patera	Ji7
Sethlaus Patera	−52.29	193.83	Io	Caucasus Mons	Ji13
Shakuru Patera	24.12	265.74	Io	Lei-Kung Fluctus	Ji4
Shamash Patera	−34.99	152.59	Io	Tohil Mons	Ji12
Shamshu Mons	−12	71.51	Io	Kanehekili	Ji6, Ji6.1, Ji7.4
Shamshu Patera	−10.07	62.97	Io	Kanehekili	Ji6, Ji6.1
Shango Patera	32.35	100.52	Io	Amirani	Ji3
Sharpitor Linea	65.4	171.7	Europa	Tyre	Je3
Sholmo	53.7	16.2	Callisto	Vestri	Jc2
Shoshu Patera	−20.36	324.67	Io	Loki	Ji10
Shu	43.2	356.8	Ganymede	Nun Sulci	Jg5
Shuruppak Sulcus	−19.3	232.2	Ganymede	Tiamat Sulcus	Jg9, Jg13
Sicyon Sulcus	32.7	18.5	Ganymede	Perrine	Jg2
Sid Catena	49.2	103.9	Callisto	Asgard	Jc3
Sidon Flexus	−66.4	183.4	Europa	Sidon Flexus	Je15
Sigurd Patera	−5.94	97.93	Io	Emakong Patera	Ji7
Sigyn	35.9	29	Callisto	Vestri	Jc2
Silpium Mons	−52.71	272.34	Io	Creidne Patera	Ji14
Sin	52.9	357.5	Ganymede	Nun Sulci	Jg5
Sippar Sulcus	−15.4	189.3	Ganymede	Uruk Sulcus	Jg8, Jg13
Siun Patera	−49.84	0.44	Io	Masubi	Ji11
Siwah Facula	7	143.1	Ganymede	Memphis Facula	Jg7
Sköll	55.6	315.6	Callisto	Askr	Jc5
Skeggold	−49.7	31.9	Callisto	Adlinda	Jc11
Skuld	10	37.9	Callisto	Valhalla	Jc6
Skythia Mons	26.21	99.01	Io	Amirani	Ji3
Sobo Fluctus	14.08	150.59	Io	Prometheus	Ji8, Ji8.2.1
Sparti Linea	59.3	245.5	Europa	Rhadamanthys Linea	Je4
Staldon Linea	−0.8	27.4	Europa	Euphemus Linea	Je6

Name	Latitude	Longitude	Satellite	Quadrangle	Plate
Steropes Patera	15.54	138.85	Io	Emakong Patera	Ji7
Sudri	55.9	135.6	Callisto	Asgard	Jc3
Sui Jen Patera	−19.1	2.65	Io	Kanehekili	Ji6
Sumbur	67.1	325.2	Callisto	Gipul Catena	Jc1
Surt	45.21	336.49	Io	Amaterasu Patera	Ji5
Surya Patera	21.47	151.59	Io	Amirani	Ji3
Susanoo Patera	22.39	219.8	Io	Lei-Kung Fluctus	Ji4
Svarog Patera	−48.66	265.74	Io	Caucasus Mons	Ji13
Svol Catena	10.6	37.2	Callisto	Valhalla	Jc6
Taliesin	−22.8	138	Europa	Taliesin	Je12
Talos Patera	−26.39	354.75	Io	Creidne Patera	Ji14
Tammuz	13.4	230.6	Ganymede	Tiamat Sulcus	Jg9
Tanit	57.5	36.6	Ganymede	Perrine	Jg2
Tapio	30.1	108.6	Callisto	Asgard	Jc3
Tara Regio	−10	75	Europa	Brigid	Je7
Taranis Patera	−71.37	25.53	Io	Haemus Montes	Ji15
Tarsus Regio	−39.69	55.14	Io	Masubi	Ji11
Tashmetum	−39.7	264.5	Ganymede	Apsu Sulci	Jg13
Ta-urt	27.6	304.1	Ganymede	Nun Sulci	Jg5
Taw Patera	−33.65	358.37	Io	Creidne Patera	Ji14
Tawhaki Patera	3.32	76.18	Io	Emakong Patera	Ji7, Ji7.4
Tawhaki Vallis	0.5	72.8	Io	Emakong Patera	Ji7, Ji7.4
Tectamus Linea	26.9	199.2	Europa	Rhadamanthys Linea	Je4
Tegid	0.8	164.4	Europa	Cilix	Je8
Telegonus Mensae	−53.31	115.89	Io	Tohil Mons	Ji12, Ji12.1, Ji12.2
Telephassa Linea	−0.8	177.2	Europa	Cilix	Je8
Terah Catena	7.1	277.5	Ganymede	Tiamat Sulcus	Jg9
Teshub	−68.5	279.6	Ganymede	Hathor	Jg15, Jg15.1
Tettu Facula	37.6	161.2	Ganymede	Galileo Regio	Jg3
Thasus Linea	−66.1	184.0	Europa	Sidon Flexus	Je15
Thebes Facula	7.1	202.4	Ganymede	Uruk Sulcus	Jg8
Thekkr	−80.3	62.0	Callisto	Keelut	Jc15

Name	Latitude	Longitude	Satellite	Quadrangle	Plate
Thera Macula	−46.7	181.2	Europa	Agenor Linea	Je12, Je13, Je13.4
Thomagata Patera	25.67	165.94	Io	Amirani	Ji3, Ji3.3
Thor	39.15	133.14	Io	Amirani	Ji3, Ji3.3
Thorir	−31.9	66.7	Callisto	Adlinda	Jc11
Thoth	−43.3	147.2	Ganymede	Osiris	Jg12
Thrace Macula	−45.9	172.1	Europa	Taliesin	Je12, Je12.2
Thynia Linea	−59.2	154.5	Europa	Taliesin	Je12, Je15.3
Tiamat Sulcus	3.4	208.5	Ganymede	Uruk Sulcus	Jg8, Jg9
Tien Mu Patera	12.31	134.3	Io	Emakong Patera	Ji7, Ji8.2.1
Tiermes Patera	22.38	349.95	Io	Amaterasu Patera	Ji5
Tindr	−2.3	355.5	Callisto	Vali	Jc10, Jc10.2
Tohil Mons	−29.5	160.49	Io	Tohil Mons	Ji12, Ji12.3
Tohil Patera	−25.63	158.66	Io	Tohil Mons	Ji12, Ji12.3
Tol-Ava Patera	1.8	322.14	Io	Loki	Ji10
Tontu	27.6	100.3	Callisto	Asgard	Jc3
Tormsdale Linea	47.7	258	Europa	Rhadamanthys Linea	Je4
Tornarsuk	28.8	127.6	Callisto	Asgard	Jc3
Tros	11.1	27.3	Ganymede	Dardanus Sulcus	Jg6
Tsūi Goab Fluctus	−1.4	163.4	Io	Prometheus	Ji8, Ji8.7.1
Tsūi Goab Tholus	−0.1	163	Io	Prometheus	Ji8
Tuag	59.92	172.36	Europa	Tyre	Je3
Tung Yo Fluctus	−16.6	356.49	Io	Loki	Ji10
Tung Yo Patera	−18.27	0.93	Io	Kanehekili	Ji6
Tupan Patera	−18.73	141.13	Io	Emakong Patera	Ji7, Ji7.2
Tvashtar Mensae	61.6	119.94	Io	Amirani	Ji3, Ji3.2
Tvashtar Paterae	62.76	123.53	Io	Amirani	Ji3, Ji3.2
Tyll	44.8	166.5	Callisto	Asgard	Jc3
Tyn	71.1	232.5	Callisto	Gipul Catena	Jc1
Tyre	33.6	146.6	Europa	Tyre	Je3, Je3.1
Uaithne	−48.5	90.7	Europa	Taliesin	Je12
Udaeus Linea	48.6	239.4	Europa	Rhadamanthys Linea	Je4
Ukko Patera	30.95	18.41	Io	Zal Patera	Ji2

Name	Latitude	Longitude	Satellite	Quadrangle	Plate
Uksakka	−49.5	42.2	Callisto	Adlinda	Jc11
Ülgen Patera	−40.85	287.31	Io	Creidne Patera	Ji14
Umma Sulcus	4.1	250	Ganymede	Tiamat Sulcus	Jg9
Upuant	46.4	319.5	Ganymede	Nun Sulci	Jg5, Jg5.1
Ur Sulcus	49.8	177.5	Ganymede	Galileo Regio	Jg3
Uruk Sulcus	0.8	160.3	Ganymede	Uruk Sulcus	Jg8
Uta Fluctus	−33.05	16.65	Io	Masubi	Ji11
Uta Patera	−35.86	22.4	Io	Masubi	Ji11, Ji11.2
Utgard	45	134	Callisto	Asgard	Jc3
Vahagn Patera	−24.11	350.75	Io	Creidne Patera	Ji14
Valfödr	−1.3	247	Callisto	Valfodr	Jc9
Valhalla	14.7	56	Callisto	Valhalla	Jc6
Vali	9.7	325.3	Callisto	Vali	Jc10
Vanapagan	39.5	158.5	Callisto	Asgard	Jc3
Veralden	33.3	95.5	Callisto	Asgard	Jc3
Vestri	45.3	52.5	Callisto	Vestri	Jc2
Vidarr	12.1	193.4	Callisto	Vidarr	Jc8
Viracocha Patera	−61.77	280.09	Io	Creidne Patera	Ji14
Vitr	−22.1	349.4	Callisto	Lempo	Jc14
Vivasvant Patera	75.14	293.98	Io	Chors	Ji1
Volund	28.62	172.5	Io	Amirani	Ji3
Vu-Murt	21.5	170.3	Callisto	Asgard	Jc3
Vutash	31.6	102.3	Callisto	Asgard	Jc3
Wabasso Patera	−22.9	166.7	Io	Tohil Mons	Ji8.7.1, Ji12, Ji12.3
Wayland Patera	−32.8	225.23	Io	Caucasus Mons	Ji13
We-ila	−12.4	290.3	Ganymede	Misharu	Jg10
Wepwawet	−69.9	59.8	Ganymede	Hathor	Jg15
Xibalba Sulcus	43.8	71.1	Ganymede	Perrine	Jg2
Yaw Patera	9.9	132.2	Io	Emakong Patera	Ji7
Yelland Linea	−16.7	196	Europa	Cilix	Je8, Je8.3, Je8.4
Ymir	51.5	99.7	Callisto	Asgard	Jc3
Yuryung	−54.7	85.7	Callisto	Adlinda	Jc11

Name	Latitude	Longitude	Satellite	Quadrangle	Plate
Zakar	31.2	333.7	Ganymede	Nun Sulci	Jg5
Zal Montes	38.08	77	Io	Zal Patera	Ji2, Ji2.1
Zal Patera	40.25	74.5	Io	Zal Patera	Ji2
Zamama	18.43	172.59	Io	Prometheus	Ji3.3, Ji8, Ji8.4
Zaqar	58.2	37.3	Ganymede	Perrine	Jg2
Zu Fossae	38.5	150.5	Ganymede	Galileo Regio	Jg3

Latest updates to the catalog of names can be found at http://planetarynames.wr.usgs.gov.

Index

See also Appendix 5 for specific place names.

ages, surface,
 Ganymede, 27, 35
 Callisto, 27, 29
 Europa, 27, 37
asteroids, 27
 aurora, Ganymede, 34

bands (*see also* dark bands; dilational bands; ridged bands), 257, 258, 259
basin, impact (*see also* multi-ring basins), 18, 99, 109, 124, 126, 129, 142, 165, 168
biology, Europa, 49, 50, 51
bright band, 180, 182, 203, 245, 247
bright-ray craters, 27, 58, 67, 68, 78–79, 109, 110, 132, 135, 147, 148, 150, 154, 165, 166, 172, 174, 184, 214, 217, 224, 225, 236, 241
bright terrain, 22, 34, 36, 108, 118, 120, 124, 126, 132, 133, 140, 143, 144, 148, 157, 174, 177, 178, 211

calderas (*see also* patera), 42, 275, 279, 288, 289, 290, 291, 292, 293, 294, 295, 297, 301, 306, 308, 314, 320, 322, 325, 326, 327, 332, 334, 342, 343, 348, 349
 Ganymede, 19, 20, 48, 51, 133, 144, 162, 163, 172, 173, 175, 176, 178
 Triton, 51
central dome crater, 28, 70, 88, 89, 98, 101, 104, 110, 115, 128, 135, 147, 148, 149, 150, 164, 165, 166, 168
central peak crater, 98, 116, 162, 182, 189, 209, 216, 245, 251, 261
central pit crater, 98, 128, 154, 158, 170
chaos, 181, 182, 183, 189, 191, 195, 197, 201, 202, 204, 206, 217, 224, 227, 228, 230, 236, 240, 242, 246, 247, 249, 261

color,
 Callisto, 17
 Europa, 16
 Ganymede, 17
 Io, 16
comets, 25, 26, 31, 65, 75, 126, 138, 145, 150, 151, 152, 154
composition,
 Callisto, 31, 101
 Europa, 40, 49, 50, 203
 Ganymede, 32, 34, 35, 131, 147, 156
 Io, 41, 42, 265, 308, 321
 surface (*see also* NIMS), 12, 14
compression, data, 11, 12
control networks, 12
convection, 22, 43
Copernicus, 2
crater chains, 25, 31, 32, 61, 65, 73, 90, 93, 117, 126, 144, 150, 154
craters, Europa, 40
craters (*see also* secondary craters; ejecta, crater; basin, impact; multi-ring basins; specific crater typesfractures related to), 63
 global asymmetry of, 25, 62, 112
 impact, 26, 28
cycloidal ridges (*see also* double ridges), 189, 209, 233, 242, 255, 257, 259

dark bands, 180, 182, 184, 192, 198, 199, 211, 212, 217, 224, 233, 241, 247, 250
dark-floor craters, 126, 140, 145, 164, 175
dark-ray craters, 17, 112, 132, 138, 150, 151, 152, 153
dark spot, 217
dark terrain, Ganymede, 24, 30, 33, 35, 90, 108, 110, 111, 114, 124, 139, 147, 154, 157, 159, 164, 174, 211

degradation (*see also* sublimation; erosion; segregation; ice-rock), 85, 96, 139, 154
depressions,
 Europa (*see also* troughs), 199, 201, 206, 211, 221, 233, 236, 242, 247, 248
 Io, 313
diapirs, 23, 24, 40, 48, 191, 198, 199, 203, 241, 249
differentiation, 32, 33
dilational bands, 22, 154, 180, 181, 184, 202, 207, 211, 212, 213, 217, 223, 234–235, 242, 243, 244, 245, 255, 257, 259
double ridges (*see also* cycloidal ridges), 186, 188, 191, 195, 198, 213, 219, 221, 229, 234, 238

ejecta, crater, 71, 73, 75, 98, 100, 105, 116, 117, 124, 131, 134, 138, 141, 145, 152, 162, 164, 177, 182, 189, 209, 218, 219, 238, 245, 250, 251, 252, 261
Enceladus, 37, 51, 52, 232
erosion (*see also* sublimation; degradation), 71, 76, 84, 89, 92, 97, 142, 171, 228, 287, 313, 314, 336, 337, 343, 351
eruption, 265, 266, 269, 274, 279, 280, 282, 283, 344
Europa, new missions, 54

faulting, 20, 21, 36, 38, 45, 66, 156, 212, 215, 221, 229, 230, 246–247, 258, 345
Flandro, Gary, 4
flows, icy, 246
folding, 21, 243, 259
fractures (fissures), 90, 93, 125, 126, 127, 130, 133, 141, 143, 146, 147, 149, 154, 157, 158, 159, 160, 161, 191, 195, 197, 202, 219, 228, 229, 243, 247, 248, 249, 253, 259, 279, 290, 332, 334, 338, 342
furrows (*see also* multi-ring basins), 35, 108, 109, 113, 114, 119, 126, 138, 139, 141, 142, 143

Galilei, Galileo, 1–2
graben, 67–68, 71, 78–79, 84, 99, 109, 119, 139, 165, 195, 238
Grand Tour, ix
grooves, Ganymede, *see* bright terrain

heat flow, 115, 127, 135, 138
High-Gain Antenna (HGA), ix, 4, 5
hotspots, thermal, 43, 44, 271, 272, 277, 279, 285, 293, 294, 295, 297, 300, 304, 305, 316, 320, 325, 331, 333, 342

ice shell, Europa, 28, 37, 40, 49, 195, 201, 203, 207, 211, 218, 221, 236, 242, 251, 257
impact melt, 130, 149
interiors,
    Callisto, 32
    Europa, 37
    Ganymede, 33, 34, 35
    Io, 43, 44

landslides, 44, 70, 76, 87, 89, 116, 338, 343, 347, 348
LaPlace, 2
lava, icy, 143, 147, 154
lava channel, 293, 296, 324
lava flows, 19, 39, 42, 269, 273, 275, 277, 285, 286, 287, 291, 292, 293, 295, 298, 300, 301, 303, 304, 305, 306, 308, 309, 310, 312, 314, 316, 322, 323, 325, 327, 328, 331, 332, 333, 335, 347, 348, 349
lava fountain, 279, 320
lava lake, 267, 269, 286, 288, 292, 294, 316, 320, 326, 327
lenticulae, 188, 197, 198, 199, 224

magnetic fields (magnetospheres), 25, 33, 43, 108, 112, 123, 150, 203
Marius, Simon, 2
Mars, 49, 53, 116
mass wasting, *see* erosion; degradation; sublimation; landslides
matrix (*see also* chaos), 186, 197, 224, 227, 228, 239, 253, 258
mountains, 43, 45, 46, 264, 268, 270, 271, 273, 274, 279, 286, 288, 289, 290, 291, 295, 296, 297, 313, 316, 322, 326, 330, 331, 332, 335, 342, 343, 344, 345, 346, 347, 350, 351

multi-ring basins (*see also* furrows; graben), 30, 58, 60, 61, 64, 67, 74, 75, 93, 99, 100, 103, 114, 165, 171, 182, 184, 192, 193, 195, 216, 236, 238

New Horizons, x, 43, 51, 52, 274, 282
NIMS, 6, 12, 35, 40, 43, 53, 68, 100, 106, 131, 147, 156, 272, 277, 281, 285, 292, 293, 294, 295, 300, 304, 305, 307, 308, 310, 314, 320, 321, 324, 342
nonsynchronous rotation, 23, 25
    Europa, 38, 192
    Ganymede, 53

ocean,
    Europa, 37, 38, 40, 49, 50, 52, 54, 195, 221
    Ganymede, 33, 34
outgassing, 265

palimpsest (crater), 59, 66, 90, 93, 96, 102, 113, 119, 136, 137, 138, 143, 151, 154
patera, 273, 298, 300, 305, 328, 342, 350
penedome crater, 72, 86, 88, 93, 98, 99, 101, 104, 114, 115, 135, 158, 170, 173
penepalimpsest (crater), 132, 136, 141, 178
Pioneer, 3
plasma bombardment, 25, 112, 123
plate tectonics, 38, 44
plateau, 189, 217, 219, 221, 223, 298, 301, 332, 336
plumes, volcanic,
    Europa, 232
    Io, 42, 43, 44, 265, 266, 267, 270, 275, 279, 282, 283, 284, 285, 288, 297, 298, 303, 304, 310, 312, 316, 320, 322, 325, 326, 327, 331, 332, 344
polar caps,
    Ganymede, 14, 26, 60
    Io, 41, 266, 268, 270, 274
polar frosts, 26, 59, 64, 106, 108, 111, 113, 119, 123, 174, 176
polar wander, 23, 25
    Europa, 23, 39, 180, 186, 189, 192, 201, 211, 236, 242, 245, 258
    Ganymede, 53
Prometheus, mythology, 41
pull-apart faulting, *see* dilational bands

radiation effects on *Galileo*, 6, 10, 11
regolith, 25

relaxation, viscous, 24, 28, 117, 127, 136, 139, 146, 147, 154, 171
resonance, *see also* LaPlace), 3, 23, 32, 35, 41, 42, 48
reticulate terrain, 147, 148
ridged bands, 215, 224, 227, 234, 240, 261
ridges (*see also* cycloidal ridges; double bands), Io, 303, 305, 308

safing, *Galileo*, 11
saturation, image, 10
scrambling, image, 11
secondary craters, 76, 89, 90, 93, 98, 100, 116, 121, 123, 124, 125, 128, 131, 134, 140, 148, 152, 159, 165, 166, 168, 170, 171, 173, 175, 176, 177, 192, 193, 195, 197, 218, 228, 238, 243, 252, 254, 257, 259
segregation, ice-rock, 26, 65, 81, 139, 140, 142
Shoemaker-Levy 9, 31, 32, 117 Comet, 47
silicates, 18
sputtering, 108
sublimation (*see also* segregation; ice-rock), 26, 31
sulfur, 184, 203, 292, 308, 326, 330, 332, 351, 349
sulfur dioxide, 265, 286, 288, 298, 303, 307, 312, 326, 332, 335, 345, 348, 350, 351, 352

talus, 228, 230
tectonism (*see also* folding; fractures; faulting; multi-ring basins), 20, 63, 81, 105, 143, 163, 174
tidal heating, Europa, 38, 49
tides, 23, 49, 189, 209, 229, 255
Titan, 33, 52
triple bands, *see* dark bands; double ridges
Triton, 51
troughs (*see also* depressions), 189, 191

viscous relaxation, *see* relaxation, viscous
volcanism, 19, 42, 92, 95, 96, 106, 127, 133, 144, 157, 161, 162, 163, 173, 186, 240, 271, 273, 297
volcanoes, 19, 42, 43, 265, 267, 268, 284, 285, 298, 309, 326, 327, 329, 330, 344, 347

wedge-shaped bands, *see* dilational bands